Mathematical Engineering

Today, the development of high-tech systems is unthinkable without mathematical modeling and analysis of system behavior. As such, many fields in the modern engineering sciences (e.g. control engineering, communications engineering, mechanical engineering, and robotics) call for sophisticated mathematical methods in order to solve the tasks at hand.

The series Mathematical Engineering presents new or heretofore little-known methods to support engineers in finding suitable answers to their questions, presenting those methods in such manner as to make them ideally comprehensible and applicable in practice.

Therefore, the primary focus is—without neglecting mathematical accuracy—on comprehensibility and real-world applicability.

To submit a proposal or request further information, please use the PDF Proposal Form or contact directly: Dr. Thomas Ditzinger (thomas.ditzinger@springer.com)

Indexed by SCOPUS, zbMATH, SCImago.

Yuri N. Skiba · David Parra Guevara

Mathematical Methods for the Assessment and Control of Industrial Emissions

Springer

Yuri N. Skiba
Instituto de Ciencias de la Atmósfera y
Cambio Climático
National Autonomous University of Mexico
Mexico, Distrito Federal, Mexico

David Parra Guevara
Instituto de Ciencias de la Atmósfera y
Cambio Climático
National Autonomous University of Mexico
Mexico, Distrito Federal, Mexico

ISSN 2192-4732 ISSN 2192-4740 (electronic)
Mathematical Engineering
ISBN 978-3-031-58111-3 ISBN 978-3-031-58109-0 (eBook)
https://doi.org/10.1007/978-3-031-58109-0

This Springer imprint is published by the registered company Springer Nature Switzerland AG
The registered company address is: Gewerbestrasse 11, 6330 Cham, Switzerland

Preface

Industrial pollution is a serious problem worldwide, both in developed and developing countries. The rapid growth of the population and significant development of production and transport have led to the release of toxic pollutants that have a negative impact on the environment and human health. The problem of environmental pollution is exacerbated by large industrial complexes whose emissions exceed sanitary standards, as well as accidental or hidden discharges of pollutants. This applies equally to the processes taking place on the continent and in the ocean. Since many gases and aerosols affect the environment in one way or another, the immediate task is to assess the strength and localization of emissions in order to reduce their negative impact.

In recent years, the problem of protecting and restoring the environment has become one of the most important tasks of many sciences, the development of which is stimulated by the increasing pace of technological progress in all countries. There are two approaches to reduce emissions and protect the environment in the major industrial regions. The former, so-called "technological way" deals with "green" technologies that clean air or remove pollutants. The second approach including analytical methods of prevention is of great mathematical interest and is more cost-effective than the application of end-of-pipe treatment technologies. The methods use mathematical models to describe the dispersion of pollutants, and the concentrations of harmful pollutants are regulated by a control of the intensity of emissions from the pollution sources. The advantage of this approach is that it makes it possible to simulate on a computer the concentrations of various primary and secondary pollutants, to determine the sources of excessive pollution of particularly vulnerable areas (residential areas, forests, etc.), and to develop emission control strategies for operating industrial enterprises. Also, this approach makes it possible to plan the placement of a new industrial zone and the installation of safety devices in high-risk areas to prevent an accident or unauthorized discharge.

This book is intended to introduce the reader to modern mathematical methods for preventing dangerous environmental pollution. It discusses various dispersion models and emission control strategies. The overall goal of all strategies is to limit emissions from pollution sources in order to comply with sanitary standards and prevent risks to human health. Since sanitary standards are usually associated with temporal and spatial averages, the proposed control strategies aim to reduce average concentrations of pollutants to acceptable levels in a particular zone and time interval. An effective control strategy should be applied whenever a short-term forecast model predicts a violation of sanitary standards. The focus is on various mathematical aspects related to industrial pollution, including the dispersion and assessment of pollutants, the control of their emissions, and the identification of the basic parameters of unknown sources. Because these problems are interdisciplinary, the reader needs a basic knowledge of differential equations, mathematical modeling, optimization theory, and numerical methods. The attached bibliography contains extensive literature on these environmental issues.

To illustrate the basic ideas of mathematical control, we will often use a simple two-dimensional (vertically integrated) model that describes the dispersion of quasi-passive pollutants. Despite this, any control strategy can also be applied to a three-dimensional dispersion model. Various examples of air and marine pollution control are presented in Chaps. 5–7. In particular, the application of dual assessments to oil spill problems is given in Sects. 4.1 and 4.2, while the assessment of vehicular emissions is described in Sect. 4.3. A method of detection of industrial plants that violate established emission rates is proposed in Sect. 4.4, while a method of optimal location of a new industrial plant is described in Sect. 4.5. Chapter 5 is devoted to methods of control of industrial air pollution, and Chap. 6 considers various mathematical problems of bioremediation of marine oiled shorelines. Chapter 7 is devoted to solving inverse problems related to the identification of the location and intensity of unknown point sources, and Chap. 8 describes numerical methods used to simulate the dispersion of pollutants in 2D and 3D problems, which are based on a coordinate-wise splitting of the model operator.

Given its breadth of coverage, the book will benefit a wide variety of readers, including researchers, professionals, aspirants, and students.

Mexico City, Mexico

Yuri N. Skiba
David Parra Guevara

Acknowledgements This work was partly supported by grants 14539 and 25170 of the National System of Researchers (Sistema Nacional de Investigadores, CONAHCYT, Mexico). We are sincerely grateful to Dr. Valentina Davydova-Belitskaya for her kind assistance in carrying out numerical experiments to study vehicle emissions on the main roads in the Guadalajara metropolitan area and Dr. Tonatiuh Sánchez-Vizuet for providing Figs. 8.5.1–8.5.5. We would like to thank the staff at Springer, in particular Senior Editors Dr. Mayra Castro and Dr. Robinson dos Santos, and also Project Coordinator Mr. Ashok Arumairaj, for their help and support.

Contents

Chapter 1
Introduction

Air pollution is a general term that covers a broad range of contaminants (gases and particles) in the atmosphere. Pollution can occur from natural causes or from human activities. The most characteristic sources of air pollution have always been combustion proceses that produce such primary pollutants as sulfur dioxide or smoke [1]. Although the environment has its own mechanisms to assimilate wastes and reintegrate nature, these mechanisms become ineffective in the conditions of growing anthropogenic activity [2]. Today, many of the key pollutants in the urban atmospheres are secondary pollutants produced by processes initiated through photochemical reactions. These pollutants form mixtures that cause various damages in humans and ecosystems [2–4].

Each air pollutant, depending on its concentration and toxicity, entails various public health problems, ranging from discomfort of airways to an increase in deaths among the most vulnerable population (cardiac patients, children and elderly, etc.). In either event, contamination is a factor that worsens the quality of life of humans. Unfortunately, the impact of a mixture of pollutants on ecosystems is not only local, as in the case of urban photochemical smog [5]. It can also be regional (acid deposition) [6, 7], or global (destruction of the ozone layer and global climate change) [8]. Today's society cannot return to pre-industrial times to solve the problems caused by air pollution, and therefore it is important to control emissions in order to reduce the concentration of harmful substances in the atmosphere to recommended maximum permissible levels.

Water from seas and rivers has traditionally been used as a means of removing human waste, and the biological cycle of water ensures the reabsorption of this recyclable organic wastes. But today, in addition to these organic wastes, a large number of harmful chemicals are released into rivers and seas, which destroy aquatic flora and fauna, cancel the action of bacteria and algae in the process of biodegradation of organic and chemical water pollutants. In particular, petroleum products are often from ships or coastal sites and are absorbed by marine life. Being carcinogenic and

Y. N. Skiba and D. Parra Guevara, *Mathematical Methods for the Assessment and Control of Industrial Emissions*, Mathematical Engineering,
https://doi.org/10.1007/978-3-031-58109-0_1

mutagenic, polycyclic aromatic hydrocarbons contained in crude oil, pose a danger to aquatic organisms and, ultimately, to humans (when consuming fish and shellfish).

One way of regulation of concentrations of harmful pollutants in the atmosphere is based on using mathematical models of dispersion of polluting substances [9, 10]. The advantage of this approach is that it permits a computer simulation of concentrations of various primary and secondary pollutants and identification of the sources that cause the excessive pollution of particularly sensitive areas (residential areas, forests, etc.). Besides, the use of adjoint models and estimates enables the development of strategies to control emissions of operating industrial plants [11–14]. Also, this approach makes it possible to plan the location of a new industrial zone [9, 15] and installation of safety devices in high-risk areas to prevent an accident or illicit dumping.

In this book, various dispersion models and emission control strategies are considered. The overall objective of all strategies is to limit emissions from industrial sources in order to meet air quality standards and reduce the risks to human health. As the health standards are usually related with temporal and spatial averages, the proposed control strategies aim at reducing mean concentrations of pollutants to acceptable levels in a particular region and time interval. A control strategy should be applied whenever prognostic model predicts a violation of sanitary standards.

Generally, there are two approaches to reduce emissions and protect the environment in the major industrial regions. The former, so-called "technological path", deals with "green" technologies (based on various filtering methods [16–18]) which allow maintaining the lowest levels of emissions of harmful pollutants. The second approach provides control of emission rates from the main pollution sources and is of considerable mathematical interest. To illustrate the basic ideas of mathematical control, we will often use a simple two-dimensional (vertically integrated) model that describes the dispersion of quasi-passive pollutants (in this model, the chemical reactions are ignored or described by a linear law). Of course, every control strategy can also be applied to any three-dimensional dispersion model. Besides, experience with these strategies has pointed to new areas of application. Various examples of pollution control in the atmosphere and the marine environment are presented in Chaps. 5 and 6.

This chapter provides general information on the most important air and water pollutants (Sects. 1.1–1.3) and the main processes of their dispersion (Sect. 1.4). The main features of the long-term and short-term programs of environmental pollution control are explained in Sect. 1.5.

1.1 Pollution and Adverse Health Outcomes

In a historical sense, air pollution has become a serious problem since the smoke generated by the incomplete combustion of wood began to enter the atmosphere. For civilizations in Western Europe, the introduction of bituminous coal as an energy source in the fourteenth century, added a new aspect to the problem of air pollution.

Coal remains an important energy source, due to its low cost and abundance when compared to other fuels, particularly for electricity generation. Air pollution from coal-fired power plants is large and varied and contributes to a significant number of negative environmental and health impacts. A typical coal plant generates 3.5 million tons of CO_2 per year. Coal combustion releases nitrogen oxides, sulfur dioxide, particulate matter, mercury and dozens of other substances known to be hazardous to human health. Burning coal is also a leading cause of smog and acid rain.

In general, the atmospheric pollutant is a substance that natural or anthropogenic sources emit into the atmosphere in quantities that in the short, medium or long term have adverse effects on living organisms, ecosystems and materials. Natural events, such as volcanic eruptions, forest fires and anaerobic fermentation processes, can produce large quantities of pollutants such as sulfur dioxide (SO_2), carbon dioxide (CO_2), nitrogen oxides (NO_x), methane (CH_4), hydrochloric acid (HCl), ammonia (NH_3), hydrosulfuric acid (H_2S), solid particles, etc. [1]; however, these events are sporadic. Due to the diversity and intensity of emitted substances, human activities such as agriculture, power generation, transportation, mining, metallurgy, chemical industry, electronics, and waste treatment are the main generators of the following pollutants: sulfur dioxide (SO_2), carbon monoxide (CO), carbon dioxide (CO_2), nitrogen oxides (NO_x), methane (CH_4), hydrochloric acid (HCl), ammonia (NH_3), ozone (O_3), sulfuric acid (H_2SO_4), nitric acid (HNO_3), heavy metals (Pb, Cd, Zn), hydrocarbons, etc. [4].

Analysis of the phenomenon of air pollution at the regional level and in the medium term shows that the five of the total number of pollutants emitted into the atmosphere are the most important, since they account for 95% of pollution [2, 3]. These pollutants are carbon monoxide (CO), sulfur dioxide (SO_2), nitrogen oxides (NO_x), hydrocarbons and particulate matter. Of these, the first three are chemicals that have a defined composition; hydrocarbons are a family of substances; and finally, particulate matter involves liquid or solid aerosols which are suspended in the air with dimensions ranging from 0.001 to 10,000 microns. Carbon monoxide, nitrogen oxides and hydrocarbons originate mainly from the combustion of gasoline in motor vehicles, while sulfur dioxide and particulate matter are emitted by industrial sources such as power plants, incinerators and cement works. The concentrations of these five contaminants, together with the concentration of ozone, whose formation and accumulation depends on the presence of nitrogen oxides, hydrocarbons and solar radiation with a wavelength of 430 nm [5] are used as an index of air quality [3].

There is no doubt that excessive levels of pollution are causing a lot of damage to human and animal health, tropical rainforests, as well as the wider environment. Pollution effects depend on the toxicity, reactivity and amount of pollutants, and on the environmental conditions (atmospheric stability, moisture, etc.). The reactivity is important, as common air pollutants can react with one another to form highly reactive and toxic substances [19]. Typical examples of secondary pollutants are chlorine gases, ozone and sulfuric and nitric acids. In particular, the sulfuric and nitric acids are formed when sulfur dioxide and nitrogen dioxide react with the radicals OH and water [6]. These acidic substances may impact far from the source, since the reactions which form them are slow [7]. Thus, emissions occurring in

Table 1.1 Health standards for the principal pollutants

Compound	Health standards ($\mu g/m^3$)	Time interval
CO	10000	Average value for 8 h
	40000	Average value for 1 h
NO_x	100	Average value for 1 year
	200	Average value for 1 h
Hydrocarbons	160	Average value for 3 h
SO_2	80	Average value for 1 year
	365	Average daily
Particulate matter	75	Average value for 1 year
	260	Average daily
Ozone	240	Average value for 1 h

Germany can cause acid rain in Sweden. In contrast, the reactions involved in the formation of ozone are fast, and therefore their impact is local. For example, when the sun shines on air polluted with automobile exhaust, photochemical reactions of nitrogen dioxide, non-methane hydrocarbons and carbon monoxide produce "Los Angeles-type smog" containing high concentrations of ozone and other oxidants [5].

Mathematical modelling of the dispersion and control of pollutants is a very difficult problem, because the equations describing the reactions of substances are usually nonlinear [11]. For the sake of simplicity, in the dispersion models considered in this text, the reactivity is parametrized by a linear term. In order to control air pollution, the maximum allowable average concentrations (sanitary standards) are set for priority pollutants [3, 4]. Table 1.1 shows the health standards for the six principle pollutants.

Moreover, it is noteworthy that the impact of various contaminants on the health of humans is a subject that is still only partially known, because the effects of substances in large quantities within a few hours are understood, but in general the impact that the medium or low concentrations produce over years or decades is unknown.

For example, experiments carried out on animals have shown that concentrations of 2 ppm (parts per million) of ozone for 3 h are lethal [3]. Moreover, the concentrations of 0.1 ppm during episodes of photochemical smog in cities for periods of 4–8 h in a day, cause eye irritation and asthma attacks. However, the health effects when humans are exposed to this pollutant for years are unjustly ignored. The polynuclear hydrocarbons such as benzopyrene are another example. These substances are formed during combustion of coal, fuel oil and gasoline, and at high concentrations are identified as potent cancerogens. However, their impact on humans is not known when they are continuously present at concentrations of a few nanograms per cubic meter, which is typical for large cities.

One factor which complicates the impact of pollutants on living beings is the fact that hundreds of harmful substances coexist in the environment at the same time, and their combined effect (known as the synergistic action) can be greater than the sum of their individual effects. The most striking example is the air pollution episode in December 1952 in London, where nearly 4000 people died as a result of the synergistic action of the two substances formed during the combustion of coal in homes and factories: sulfur dioxide and solid particulate matter [4]. Finally, it should be noted that contaminants from natural sources may represent serious short-term air quality problems when they are generated in significant amounts near human settlements.

1.2 Main Atmospheric Pollutants

We now consider various types of air pollutants in more detail.

Smog. The term *smog* was first coined during the 1950s when it was used to describe a mixture of smoke and fog experienced in London. Smog occurs when emissions from industry, motor vehicles, incinerators, open burning and other sources accumulate under certain climatic conditions. The term smog, as used now, is widely applied in the pollution conditions, characterized by a significant reduction in visibility.

Haze. The term *haze* is traditionally an atmospheric phenomenon where dust, smoke and other dry particles obscure the clarity of the sky. Actually, haze and smog are closely related. Both conditions represent a reduction in visibility. They differ in terms of the intensity and geography. Haze generally refers to the reduction of visibility that is not as intense as smog. The term smog is often used to describe the marked reduction of visibility in cities or large metropolitan areas. On the other hand, haze refers to a wide scale of unfavorable pollution levels, which cause reduced visibility during the summer months.

Gases and particles. Due to important impact on the atmosphere, vegetation, human health and materials, the traditional air pollutants, gases and particles have received considerable research and attention. Although the terms gases and particles are used, they actually represent the three phases of matter. The point is that the particles represent two phases: solid and liquid. When fine solid particles and liquid droplets are dispersed in the atmosphere, the resulting suspension is characterized as an aerosol. Aerosols may reduce visibility, stain materials and affect health. Aerosols can be natural or not. Examples of natural aerosols are fog, forest exudates and geyser steam. Examples of artificial aerosols are haze, dust, particulate air pollutants and smoke. Aerosols can be generated in different ways. For example, gaseous aerosols are produced by the combustion and subsequent condensation of metal vapors. And photochemical aerosol is produced by the condensation of gases which results from atmospheric photochemical reactions. Most acid aerosols which contribute to the formation of the slight haze and acidification of precipitation are apparently also produced by photochemical reactions.

Although the particles are the most visible manifestation of air pollution, they constitute only a relatively small part of the emissions, since 90% of the pollution substances released into the atmosphere are gases. These gaseous pollutants can be produced by burning fuels and other materials, smelting an ore, or evaporation of volatile liquids or solids.

Sulfur dioxide. Sulfur dioxide (SO_2) is historically the most prominent of pollutant gases, having achieved this reputation as a respiratory irritant. Globally, most of SO_2 is produced by volcanoes and by the oxidation of sulfur gases, caused by the decomposition of plants. Since the "natural" sulfur dioxide is emitted to the atmosphere mainly at high altitudes or away from the densely populated centers, the background concentration of gas in the clean air is quite small (about 1 ppm). However, at present a considerable amount of sulfur dioxide is emitted at ground level, particularly, in inland areas of the Northern Hemisphere. The main anthropogenic source of SO_2 is the combustion of coal, a solid that, depending on the geographical area where it is extracted, contains from 1 to 9% of sulfur dioxide. In many countries, the use of coal is important for the generation of electricity. Typically, half or more of the sulfur is retained in the coal; if coal is pulverized prior to combustion, this type of sulfur can be removed mechanically. Many point sources of SO_2 are associated with the smelting industry of nonferrous metals.

Nitrogen dioxide. Nitrogen dioxide is the chemical compound with the formula NO_2. It is considered harmful to plants and is known to be toxic to humans, but not at concentrations that normally occur in the atmosphere. The only physical property of nitrogen oxides which is of great interest in the air pollution is its color, NO_2 is the yellow/brown colored gas [20]. This typical color becomes visible when the NO_2 emissions in the chimney exceed the 75 ppm. Exposure to 20 ppm is immediately dangerous to life and health. People with asthma, as well as children and the elderly are generally at greater risk for the health effects of NO_2. In addition, its negative effect is the enormous corrosion it can cause due to acidification of the environment.

Carbon monoxide. Carbon monoxide (CO) is the specific chemical compound, the simplest among the six principal pollutants. Carbon monoxide is a colorless, odorless and tasteless gas that is slightly less dense than air. The most important property of CO is its ability to reduce the oxygen carrying capacity of our blood. Carbon monoxide is the result of inefficient combustion of carbonaceous fuels. It is not produced in large quantities by most stationary sources of fuel, such as boilers and heaters because they are designed and tuned to operate as efficiently as possible. The largest CO emissions result from unplanned combustions such as wildfires and solid waste combustion, which are inefficient because the fuel is not uniform and has high moisture content. A part of CO also comes from various industrial processes, particularly smelters, refineries and paper mills. The main source of carbon monoxide is the combustion of gasoline in motor vehicles. Combustion in a motor is very inefficient, in part by the intrinsic properties of the motor, and partly because the cars are designed to produce velocity and power instead of economy. The cars produce more than half of the nearly 100 million tons of carbon monoxide emitted in a year [20].

Ozone. Ozone (O_3) is colorless or slightly bluish gas. It is a normal constituent of the atmosphere with peak concentrations in the middle stratosphere. As a consequence of anthropogenic influences, O_3 represents two different but very important roles. In the stratosphere, the quasi-stationary concentration of ozone acts as a filter for ultraviolet radiation. But in the troposphere, at slightly elevated levels, ozone is a pollutant. In both atmospheres, urban and non-urban, the concentrations of O_3 are often higher than those produced by photolysis of NO_2. The key to high levels of tropospheric ozone is the chemical reactions that convert NO into NO_2 without consuming O_3. In relatively clean atmosphere, during the hottest months of the year, the concentrations of background level of O_3 are estimated in the range of 20–50 ppb (parts per billion). The main natural sources of O_3 are the photochemical processes and movement of stratospheric ozone in regions named *tropopause folds* in the lower atmosphere, particularly during the months of spring.

Pollution sources. Pollutants are classified as primary or secondary. Primary pollutants are emitted directly from a source (ash from a volcanic eruption, carbon monoxide gas from motor vehicle exhaust, sulfur dioxide from a factory, etc.). Secondary pollutants are not emitted directly. Rather, they are formed in the air when primary pollutants react or interact.

The sources of primary pollutants can be classified as mobile or stationary sources, combustion or non-combustion sources, point or area sources, direct or indirect sources. Major point sources include power or industrial plants, refineries, wastewater treatment plants, chemical or nuclear explosions, oil spill from a damaged oil tanker, etc. Emissions from motor vehicles, aircrafts, trains and ships along their routes are examples of mobile line sources. While point sources of pollution were studied since the late nineteenth century, linear sources did not receive much attention from scientists until the late 1960s, when environmental regulations for highways and airports began to emerge. An area source is a collection of individually small emission sources within a single geographical area that produce similar air pollutants. Locomotives operating on certain linear tracks are examples of a line source, whereas locomotives operating within a railyard are an example of an area source of pollution. Other area sources of air pollution are open burning, forest fires, evaporation losses from large spills of volatile liquids. The sources that do not emit themselves, but indirectly contribute to high pollution levels, are indirect sources. As examples we can call malls or stadiums, which contribute to a high level of vehicular traffic in the zones of their location.

1.3 Water Pollutants

Water is essential for life as we know it. Without water is hard to imagine any life. The water cycle in our planet, also called hydrological cycle, in essence consists in the fact that the water enters the atmosphere by evaporation or breathability and returns to the soil by condensation and precipitation. Fortunately, the Earth is almost flooded: a total volume of about 1400 million cubic kilometers covers 71% of the land surface. Yet in many parts is not easy to obtain the necessary quantities of sufficiently pure water. The fact that terrestrial ecosystems, humans and agriculture depend more on accessible fresh water, which constitutes only 0.4% of the total volume, implies that the problems of scarcity and water quality are multiplied. An estimated 580 people in India die of water pollution related illness every day.

The highest concentration of the world's human population occurs on the banks of rivers and coastal areas, with the consequent impact of human activity on water. Practically all the wastewater (85%) are discharged into the coastal strip, not only in developing countries but also in developed countries such as US [21], particularly in estuaries and bays where heavy contamination is common near large cities because of nutrients, toxic substances and microorganisms. As occurs in the atmosphere, diffuse sources of marine pollution are difficult to register. One must distinguish between what is denoted by point source and diffuse source. Point source is any single identifiable source that discharges substances harmful to aquatic life, recreation and human consumption. Examples of a point source in a bay are a ship, a municipal treatment plant, the river mouth or a submarine diffuser. A diffuse source of water pollution is commonly more difficult to identify and to control than a point source because it arises in a catchment from many different sources that do not have an obvious discharge point. Diffuse source water pollution is caused when pollutants from a range of dispersed urban and rural land use activities contaminate our waterways. Atmospheric deposition of pollutants is an important diffuse source of acid, nutrients, metals and other substances. In the coastal strip, the berths areas, drains of untreated water, as well as rain runoffs that drag oil, grease, dust, dirt, debris, animals, microorganisms, and chemicals such as metals, pesticides, and fertilizers, can be considered as diffuse sources. The contributions of highly toxic contaminants constitute a serious threat, contribute to the deterioration of coastal ecosystems and represent a huge challenge for health authorities, as it is very difficult to quantify and monitor their effects. When the ground cover has not been disturbed by human activity, rainwater percolates through the soil, which absorbs and detoxifies many pollutants. But when the soil is not sufficiently covered with natural vegetation, water drains freely, washing on its way to coastal waters solids and liquids.

According to the definition of a pollutant, it is considered that pollution is generated in water by the addition of any substance in sufficient quantity to cause measurable harmful effects on flora and fauna (including human). The presence in the environment of one or more contaminants, or any combination thereof, impair or disrupt the life, health and human welfare, flora and fauna, or degrade the quality of air, water, land, assets or resources of the nation. One can speak of four basic types of contamination: physical contamination (noise, infrasound, thermal pollution and radioisotopes), chemical contamination (hydrocarbons, detergents, plastics, pesticides, heavy metals, sulfur derivatives and nitrogen), biological contamination (bacteria, fungi, viruses, parasites, introduction of animals and plants from other areas) and elements that damage the aesthetics (landscape degradation and the introduction of industries).

Often the taste, smell and appearance indicate that water is contaminated, but the presence of hazardous pollutants can only be detected by special and precise chemical and biological tests. Among the factors that generate pollution and characterize industrial civilization are: the growth of production and excessive energy consumption, the growth in the metallurgical industry; the growth of road, air and water circulation, and the growth of the amount of waste and scrap being thrown and/or incinerated.

The water of the seas and rivers has traditionally been used as a means of disposal of human waste, and the biological cycles of water ensure the reabsorption of this recyclable organic waste. But today, apart from these organic waste, large quantities of harmful chemicals are thrown into rivers and seas, which destroy aquatic plant and animal life, and cancel the action of bacteria and algae in the process of biodegradation of organic and chemical pollutants of water.

The most serious is that petroleum derivatives are often ejected from ships or coastal industries and absorbed by marine fauna and flora. While being carcinogenic and mutagenic (Table 1.2), polycyclic aromatic hydrocarbons, the components of crude oil, represent a risk to aquatic organisms and ultimately to humans (through fish and shellfish consumption).

Physical contaminants. These contaminants affect the appearance of the water, and when they float or sink they interfere with aquatic flora and fauna. They are insoluble liquids or solids of natural origin and synthetic products that are thrown into the water as a result of human activities, as well as foams, oil residues and heat (thermal pollution).

Chemical contaminants. Chemical pollutants include organic and inorganic compounds dissolved or dispersed in the water. Inorganic contaminants are various products that come from domestic, agricultural and industrial discharges or soil erosion. The main ones are chlorides, sulfates, nitrates and carbonates, and also acidic waste, alkaline waste and toxic gases dissolved in the water, such as sulfur oxides, nitrogen, ammonia, chlorine and hydrogen sulfide. Many of these pollutants are released directly into the atmosphere and fall washed away by rain. This acid rain has harmful effects which can be observed both in vegetation and buildings and monuments in the industrialized cities.

Organic contaminants. Organic pollutants also dissolve or disperse in water coming from domestic, agricultural, industrial waste and soil erosion. They are human and animal waste, waste from slaughterhouse, waste processing of food for humans and animals, various industrial chemicals of natural origin, such as oils, greases, tars and dyes, and various synthetic chemicals such as paints, herbicides, insecticides, and so on. Organic contaminants consume oxygen dissolved in the water and affect aquatic life (eutrophication). Abnormal levels of nitrogen compounds in the water, such as ammonia or chloride, are used as an index of the presence of these contaminating impurities in the water.

Biological contaminants. Biological pollutants include fungi, bacteria and viruses that cause diseases, algae and other aquatic plants. Some bacteria are harmless and others participate in the degradation of organic matter in the water. There are certain bacteria that decompose inorganic substances. The elimination of the viruses that are transported in water is a very difficult and expensive job.

Other contaminants. It should also be noted such important contaminants as heavy metals (lead, cadmium, mercury), certain pesticides, cyanides, hydrocarbons, arsenic and phenol, which provoke the destruction of aquatic ecosystems and serious damage to people who drink water or its products contaminated with this class of chemicals.

Table 1.2 Toxic ingredients in everyday products that contaminate water

Product	Ingredients	Effect
Household cleaners	Powder abrasive cleaners and sodium phosphate, ammonia, ethanol	Corrosive, toxic and irritating
Cleaners with ammonia	Ammonia, ethanol	Corrosive, toxic and irritating
Bleachers	Sodium hydroxide, potassium hydroxide, hydrogen peroxide, sodium hypochlorite or calcium	Toxic and corrosive
Disinfectants	Ethylene and methylene glycol, sodium hypochlorite	Toxic and corrosive
Openers	Sodium hydroxide, potassium hydroxide, sodium hypochlorite, hydrochloric acid, petroleum distillates	Extremely corrosive and toxic
Polishers of floors and furniture	Ammonia, diethylene glycol, petroleum distillates, nitrobenzene, naphtha and phenols	Flammable and toxic
Cleaners and polishers of metals	Sulfuric acid	Corrosive and toxic
Oven cleaners	Potassium hydroxide, sodium hydroxide, ammonia	Corrosive and toxic
Toilet bowl cleaners	Oxalic acid, muriatic acid, paradichlorobenzene and sodium hypochlorite	Corrosive, toxic and irritating
Carpet cleaners	Naphthalene, perchlorethylene, oxalic acid and diethylene	Corrosive, toxic and irritating
Aerosol products	Hydrocarbons, flammable	Toxic and irritating
Pesticides and insect repellents	Organophosphates, carbonates and pyrethrins	Toxic and poisonous
Adhesives	Hydrocarbons	Flammable and irritant
Antifreeze	Ethylene glycol	Toxic
Gasoline	Lead tetraethyl	Toxic and flammable
Engine Oil	Hydrocarbons, heavy metals	Toxic and flammable
Transmission fluid	Hydrocarbons, heavy metals	Toxic and flammable
Wiper fluid	Detergents, methanol	Toxic
Batteries	Sulfuric acid, lead	Toxic
Brake fluid	Glycols, glycol ethers	Flammable
Car body wax	Naphtha	Flammable and irritant

According to the origin, it is considered that pollution is of two types:

(a) the pollution caused by natural or geochemical reasons, and
(b) the pollution caused by human activities and called anthropogenic pollution.

Among the harmful effects on microorganisms, populations and ecosystems, are the following:

(a) damage to human health (poisoning, infectious and chronic diseases, death),
(b) damage to flora and fauna (eutrophication, disease and death),
(c) alterations of ecosystems (erosion, eutrophication, persistent accumulation of harmful compounds, destruction), and
(d) aesthetic discomfort (bad smells, tastes and unpleasant appearance)

Ocean pollution. The ocean is currently the "dustbin of the world", which will bring adverse effects in the future. Most of the world's coastal areas are polluted mainly due to discharges of sewage, chemicals, waste, radioactive waste, oil and sediment. The countries whose seas are the most polluted are Bangladesh, India, Pakistan, Indonesia, Malaysia, Thailand and the Philippines. In particular, dolphins, sea lions and sea turtles die when they eat waste or are trapped in bags, ropes and other plastic garbage thrown into the sea.

Biofilms. You might not be familiar with the word "biofilm", however every day you are in contact with these substances. The plaque that forms on your teeth and cause dental caries and periodontal disease is a type of biofilm. Also, the biofilm causes blockage in the pipes and covers stones in rivers or streams. The biofilm is formed when bacteria adhere to the surfaces of some kind of water environment and begin to secrete a slimy, sticky substance that can stick all kinds of materials—metal, plastic, soil particles, medical implant materials, biological tissues. Biofilms may be formed of a single species of bacteria but generally consist of many species of bacteria, fungi, algae, protozoa, debris and corrosion products. Essentially, biofilms may form on any surface exposed to bacteria and water.

Associations of microorganisms participate in many biotechnological processes, including wastewater treatment, where they form a structured community, enclosed in an extracellular polymer matrix (biofilm). The introduction of microsatellites into biofilms (bioaugmentation) stimulates the activity of microorganisms-destroyers and protects them from the extremal environmental factors. Under natural conditions, the majority of microorganisms exist as structured associations called biofilms. Similar associations are created in many biotechnological processes: soil bioremediation, microbial leaching of metals from ores, in the process of wastewater treatment, and in many other cases.

For example, in the biofilters that represent primary systems of wastewater treatment, water is filtered through a solid support covered with a dense microbial biofilm, whose microorganisms remove various contaminants, oxidize ammonia, absorb phosphates and heavy metals, and restore nitrates.

Oil spills. Oil spills represent a serious problem for aquatic systems and can eliminate or adversely affect fish, other aquatic organisms, birds and mammals. Spills can kill or reduce populations of organisms living in coastal sands and rocks, and can kill the worms and insects that serve as food for birds and other animals. When oil is introduced into the coastal marshes, it damages or kills fish, shrimp, birds and other animals. An oil spill may also contaminate beaches used for swimming and recreation, and even when sophisticated techniques of collection of hydrocarbons are applied, they generate waste contaminated areas (beaches, seabed). In the study of oil degradation with microorganisms that oxidize hydrocarbons, it has been shown that in the reconstructed biofilms (containing besides microorganisms their natural satellites which are able to oxidize oil), the rate of oxidation of hydrocarbon is many times higher than the rate of oxidation in the absence of satellites [22]. This is due to the fact that the satellites emit in the environment substances that activate the vital functions of the oil oxidants. The role of satellites in a biofilm can also be in the protection of microorganisms-destructors from environmental stressors. Therefore, the controlled application of biofilms can improve the performance of sewage systems and industrial water systems without significant additional costs. Using this approach, it is equally important both the formation of biofilm, and the regulation of the composition of its community (bioaugmentation).

Therefore, one way to remove oil residues is through biostimulation, consisting of the addition of nutrients (nitrogen, phosphorus) to facilitate the increase of native oil-degrading microorganisms in the environment (bacteria, fungi). Such microorganisms consume hydrocarbons and produce water, carbon dioxide and biomass, and thereby the contaminants are reincorporated to the environment. An important factor to achieve maximum growth rates of microorganisms is reaching critical concentrations of nutrients and keeping them for a suitable time. This is a difficult goal, since the concentration in contaminated areas depends on the place where nutrients are supplied, the rate of discharge and the physical influences of the aquatic system (dispersion conditions). For example, the distribution and movement of dissolved nutrients, and hence their concentration and residence time in the field affected by oil also depend on the water waves [23]. In particular, the authors note that in the presence of water waves, the residence time of nutrients in the tidal zone is reduced by about 25%.

Bioaugmentation is the addition of bacterial cultures required to speed up the rate of degradation of a contaminant. In case of petroleum, microorganisms capable of degrading hydrocarbons are used if some contaminants cannot be degraded by the existing microorganisms. There are many different types of hydrocarbon degrading microorganisms, with over 200 known species of bacteria, fungi, and yeast that are capable of degrading different compounds, from a simple such as methane up to compounds with more than 40 carbon atoms [24, 25].

Since most natural environments contain microorganisms, bioaugmentation is not as effective for cleaning oil polluted regions, since the addition of non-native organisms into the environment often leads to their competition with the existing beneficial microorganisms. That is why bioaugmentation has never given a positive effect in cleaning oil contaminated waters [26]. Also, bioaugmentation has been criticized, since the introduction of non-native species into the environment is not safe. Unlike bioaugmentation, the primary advantage of biostimulation is that the environment is modified to stimulate already present native microorganisms to achieve maximum degradation rate of toxic oil compounds.

1.4 Main Processes of Pollution Dispersion

Since humans discovered the usefulness of fire, the atmosphere has been used as a repository of waste products of their activities. For thousands of years, the atmosphere, with its huge volume and constant movement, has readily accepted and dispersed relatively small amounts of polluting substances. The dispersion process depends on the local weather conditions (speed and direction of winds, turbulence, atmospheric stability, etc.) and the topography of the region [27]. But in the modern era of big cities and intense industrialization, the ability of the atmosphere to disperse pollutants has been exceeded. Adverse weather conditions lead to a deterioration of the dispersion of pollutants and relatively high pollution levels. This causes significant human health and environmental damages.

Wind-induced advection. Advection is the horizontal movement of the atmosphere, and prevailing winds are known as advective winds. The horizontal winds play a significant role in the transport of pollutants. With the wind speed increases, the volume of air passing per unit time through a unit area also increases. If the emission rate from a source is relatively constant, but the wind speed becomes twice as much, then the concentration of contaminants is reduced by half, i.e., the pollution concentration is inversely proportional to the wind speed.

In the surface (boundary) layer, horizontal wind velocity depends on the friction. Therefore, boundary layer is also known as the friction layer, where friction is an important force. The friction force is proportional to the surface roughness, which in turn is determined by terrain topography (presence of mountains, valleys, rivers, lakes, forests, crop areas, buildings, etc.). On a clear, calm night, the boundary layer over a lake may extend only a few hundred meters upward, but may extend a few thousand meters upward over a city on a hot, windy afternoon. On average, wind speed over smooth surfaces such as lakes and crop areas is generally higher than over the mountains and buildings. Dispersion of pollutants is also affected by wind direction variability. If the wind direction varies significantly, the pollutants will be distributed over a larger area than in the case when the wind direction is relatively constant. In this case, the concentration of pollutants in a given area in the first case will be less. Moreover, the concentration of pollutants over a given area will be smaller in the first case. It should be noted that substantial changes in the wind direction can occur in a short period of time. For example, a change in the wind direction of 30° in one hour is typical, and during a period of 24 h the wind direction can be changed by 180°. Similarly, seasonal changes may result in wind direction changes by as much as 360° [27].

Atmospheric turbulence. A plume emitted from a source is carried by mean winds. But this process is also affected by atmospheric turbulence. Atmospheric turbulence is small-scale, irregular air movements characterized by winds that vary in speed and direction. Turbulence is important because it mixes and churns the atmosphere and causes water vapor, smoke, pollutants and other substances, as well as energy, to become distributed both vertically and horizontally. Thus, under the influence of atmospheric turbulence, a plume spreads over a large area with time. There are two main mechanisms for generating atmospheric turbulence. These are mechanical and convective turbulence.

The mechanical turbulence is generated as the air flows over obstacles on the earth surface, such as trees, buildings, and hills. Its intensity increases with increasing wind speed and with increasing surface roughness and decreases with height above the ground. When vertical heat flux is small, so that most of the atmospheric turbulence is mechanically generated, the atmosphere is said to be neutral or in a state of neutral stability.

The convective turbulence is related with atmospheric instability. As solar radiation heats the earth surface, the lower layers of the atmosphere increase in temperature and convection begins, driven by buoyancy forces. The convective turbulence is usually associated with large eddies, the effects of which are often visible on plumes from stacks. At night, when the surface of the earth cools, temperature increases with height, and turbulence tends to be suppressed.

Atmospheric precipitation. With the exception of CO_2, clorofluorocarbons *CFCs*, and some other chemically stable compounds, no appreciable accumulation of pollutants in the atmosphere occurs, even if the pollutants are continuously emitted into the atmosphere. This is due to the processes of dry and wet deposition. Dry deposition is the removal of gaseous or particulate material from the pollution plume by gravitational sedimentation of particles of relatively large size and absorption of particles of gas or liquid in liquid or solid material. Wet deposition is the removal of pollution plume components by the action of rain. For example, the wet deposition of radionuclides in a pollution plume by a burst of rain often forms so called hot spots of radioactivity on the underlying surface. Meteorological phenomena that increase the frequency and intensity of rainfall in a region significantly reduce pollutant levels and improve air quality. Precipitations associated with migratory cyclones are particularly effective in cleaning up the atmosphere. On the other hand, pollution problems are aggravated in the regions that get little precipitation.

Sedimentation. Sedimentation is the tendency for particles in suspension to settle out of the fluid in which they are entrained, and come to rest against a barrier. For example, the atmospheric particles fall to the ground from the air. Washout of particles by snowflakes, rain, hail, sleet, mist or fog is a common form of agglomeration and sedimentation. Particles with diameters less than 20 microns are treated as a gas, and their effects, despite their velocity, are generally ignored, while the larger particles already have perceptible sedimentation rates.

Inversion layer. Atmospheric inversion layer is a layer, within which an atmospheric property is inverted, i.e., its change with height is deviated from the normal pattern. Usually, within the lower atmosphere the air near the Earth is warmer than the air above it, largely because the atmosphere is heated from below as solar radiation warms the earth surface, which in turn then warms the layer of the atmosphere directly above it. Thus, the atmospheric temperature normally decreases with increasing altitude. The inversion almost always refers to a temperature inversion, i.e., an increase in temperature with height. Therefore, the air cannot rise in an inversion layer, since it is cooler and thus denser at the bottom. In other words, temperature inversion stops atmospheric convection (which is normally present) from happening in the affected area. This can become a problem in big cities (for example, Greate Smog of 1952 in London).

1.5 Short-Term and Long-Term Pollution Control Problems

Control of air pollution control in cities is a complex issue that deserves attention due to the negative impact of high concentrations of pollutants on human health. One approach to solving this problem is to implement regional air pollution control programs. Pollution control is regulated by various environmental agencies, which set limits on the release of pollutants into the air, water, and land. It should be noted that each country has different environmental agencies and a completely different set of laws to limit emissions of industrial pollution. In addition, any efforts to measure and study health and environment impacts may prove important in the future, and every individual should take responsibility for preserving the environment.

In general, the goal of air pollution control programs is usually to meet a set of air quality standards, or at least reduce the number of days per year when these standards are violated. Depending upon programs length, they can be divided into two categories [28]: short-term control and long-term control. These control programs supplement each other, however, the air quality objectives of long- and short-term control strategies may be quite different.

The duration of a short-term control problem can vary from several hours to several days, usually under atmospheric inversions that promote the accumulation of contaminants. The actions that characterize this type of control include immediate reduction of pollutant emissions and even suspension of production. A goal of short-term control is ordinarily to keep the maximum concentration of a certain pollutant below a given value (existing sanitary standard) on that particular day or during the whole period of atmospheric stability. Weather forecasts and pollutant concentrations measured continuously at a number of stations are used to regulate the intensity of pollution sources and prevent the occurrence of dangerously high concentrations. An ideal control system should be able to predict the weather and pollution levels from few hours to a couple of days in advance, and use the information from this prediction in deciding what action to take.

Examples of short-term control strategies are:

(a) introducing restrictions on the use of vehicles,
(b) prohibiting the use of certain fuels (replacement of coal or fuel oil to natural gas),

(c) prohibiting certain activities, such as incineration of refuse, and
(d) restricting emission rates from industrial sources (see Chap. 5)

Item d) is an interesting mathematical problem. It is important to note that a mathematical control strategy is only useful if it is able to determine the necessary reduction in emissions from pollution sources to prevent excessive concentrations of harmful pollutants. At present, such preventive methods are not adequately incorporated into the existing short-term control programs to combat air pollution, as most control strategies are applied only after the concentrations of pollutants have already reached dangerous levels [29].

On the other hand, the duration of a long-term control program can vary from several months to several years, and such a program usually focuses on the strategies that reduce the total mass of pollutants emitted during the whole period. Some examples of long-term air pollution control policies are:

(a) enforcing standards that restrict the pollutant content of combustion exhaust,
(b) introducing new technologies of production and cleaning (filters or catalytic converters),
(c) requiring motor vehicles to meet certain emissions standards,
(d) prohibiting or encouraging the use of certain fuels in power plants, and
(e) closing or relocating some industrial plants in a region

For long-term control, a typical objective might be to reduce to a specified value the expected number of days per year that the maximum hourly average concentration of a certain pollutant exceeds a given value (air quality standard). Thus, the use of long-term strategies does not necessarily prevent dangerous levels of contamination, which may occur on certain days. This is because the set of actions are generally designed with the aim not only to maintain air quality standards, but also to meet cost–benefit criteria. In other words, the set of long-term policies are designed to minimize

$$C = C_e + C_s$$

where C_e is the cost of implementation of control activities, and C_s is the cost of damage to human health and ecosystems. Furthermore, this minimization process is realized under the constraint that the number of days per year when air quality standards J_α^0 are violated must not exceed a maximum allowable number $N_{\max}$, i.e.

$$N(J_\alpha > J_\alpha^0) \leq N_{\max}$$

When J_α^0 decreases, cost C_e increases, while cost C_s decreases, and therefore an optimal set of actions corresponds to a minimal value of C. Obviously, in this long-term control problem, the requirement to reduce emission rates was included only indirectly.

1.6 Mean and Point Pollution Control

As has already been noted, the goal of air pollution control programs is to meet a set of air quality standards, or at least reduce the number of days per year when these standards are violated. We also noted that all control programs can be divided into two categories: long-term and short-term control strategies [28]. These programs supplement each other, but differ in the methods and duration of their implementation. For example, the use of long-term strategies does not necessarily prevent dangerous levels of contamination, which may occur on certain days. For such short events (hours to days), short-term local (city) control should be applied to keep pollutant concentrations below air quality standards. Such control actions must include immediate reduction of pollutant emissions and even suspension of certain activities. In the event of an unfavorable forecast from a numerical air pollution model, the first step in a short-term control strategy is to determine the amount of reduction in the intensity of each pollution source so that a reforecast with reduced sources intensities would no longer violate air quality standards. The second step is the technical implementation of such reduction, which may consist of different procedures depending on the type of emission source. Typically, emission reduction parameters (damping parameters) are calculated using mathematical models that combine an air quality model with an optimization model.

The idea of using mathematical programming models for environmental quality control arose in the late 60 s [30–33]. In particular, the linear programming models proposed by Teller [34] and Kohn [35] for air pollution control are the first optimal strategies that take into account economic and environmental variables. Since then, many air pollution control models have been developed for different purposes and polluted regions. The following elements characterize the optimization model: (i) the air pollutants considered, i.e. primary or secondary substances, which determine the complexity of chemical reactions that should be considered in the dispersion model; (ii) an objective or cost function that estimates the cost of changes in emission rates; (iii) environmental restrictions that must be observed in a specific limited area (control zone); and (iv) technological constraints that define the upper and lower limits of scarce resources or fuel materials. Of course, different formulations of these elements create linear or nonlinear programming models with many variables and constraints. However, the common goals of all air pollution control models are to meet environmental goals, technological constraints, and minimize the costs associated with reducing the emission intensity of sources [36–41].

It is important to note that while many studies have analyzed long-term air quality plans, fewer studies have addressed the issue of air pollution episodes [42]. Short-term emission control is a complex problem because its optimal solution requires a series of simulations using a regional air quality model to determine pollutant concentrations for different emission control scenarios [42–44]. In practice, such a procedure can increase response time to short-term pollution events to the extent that mitigation actions can be determined after a pollution crisis. In such cases, rapid response is critical.

To formulate a control strategy, it is assumed that the short-term forecast obtained using the dispersion model is unfavorable, that is, the concentration of a pollutant emitted from several sources exceeds the corresponding air quality standard in the control zone. An optimization problem is then formulated to calculate the damping coefficients and determine the optimum emission reduction at the lowest cost.

Control strategies can be implemented to achieve average pollutant concentrations or point concentrations in an area Ω. In the first case, consider the following functional to estimate the average concentration

$$J(\phi) = \frac{1}{\tau|\Omega|} \int_{T-\tau}^{T} \int_{\Omega} \phi(r, t) dr dt \tag{1.6.1}$$

Here $\phi(r, t)$ is the pollutant concentration at point r and moment t, and $(T - \tau, T)$ is the time interval in which the forecast of the average concentration is unfavorable, i.e., exceeding the air quality standard: $J(\phi) > J_0$. Chapter 5 presents some optimal and non-optimal short-term control strategies. Each of these strategies describes a procedure for reducing the intensity of emissions from pollution sources in order to comply with air quality standard $J(\phi) \leq J_0$ in the area Ω and time interval $(T - \tau, T)$.

On the other hand, control of point concentrations is a more rigorous strategy that requires more computational effort and time to obtain an answer. This strategy is now briefly described. The corresponding analysis is shown in detail in [40].

We assume that in a region D there are N sources with emission rates $f_i(r, t)$ $(i = 1, \ldots, N)$. It is also assumed that meteorological conditions and air quality are predicted in D over a time interval $(0, T)$ using an appropriate dispersion model in combination with a weather forecast model. Let us assume that the atmospheric air quality forecast obtained with the emission rates $f_i(r, t)$ $(i = 1, \ldots, N)$ is unfavourable, that is, $\phi(r, t) > J_0$ at some points of a zone $\Omega \subset D$ and at some points in time interval $(0, T)$, where ϕ is the concentration of a pollutant in the region D, and J_0 is the corresponding air quality standard. Then, in order to prevent excessive concentration of the pollutant in the zone Ω, it is necessary to apply a short-term control to establish the appropriate intensity of all pollution sources within $(0, T)$. In other words, we should determine in the time interval $(0, T)$ reduced emission rates, optimal in the sense that the air quality standard will be satisfied:

$$\phi(r, t) \leq J_0 \text{, for any } r \in \Omega \text{ and } t \in (0, T) \tag{1.6.2}$$

To simplify the discussion, we write the linear dispersion model in the form

$$\frac{\partial \phi}{\partial t} + A\phi = \sum_{i=1}^{N} f_i(r, t), \; \phi(r, 0) = \phi^0(r) \text{ in } D \tag{1.6.3}$$

The boundary conditions of the problem are omitted because they do not depend on the forcing and initial distribution of the pollutant $\phi^0(r)$ in$D$, which are the main parameters in the control strategies. It is assumed that model (1.6.3) is well-posed, i.e., its solution is unique and continuously depends on the forcing and the initial condition. We will assume that the solutions of the dispersion model for various forcing and initial conditions are continuous functions. Moreover, the turbulent diffusion, included in the definition of the linear operator$A$, guarantees not only the continuity of the solution of model (1.6.3), but also the continuity of its derivatives.

Short-term control of emission rates of pollutant sources consists of finding nonnegative damping factors $\lambda_i \le 1$ that define new emission rates as

$$\lambda_i f_i(r,t), \quad i = 1, \ldots, N \tag{1.6.4}$$

where $(1 - \lambda_i) \times 100$ represents the percentage reduction in emissions of the ith pollution source. The purpose of introducing the coefficients λ_i is to fulfil the environmental condition (1.6.2) for the solution φ of the following dispersion problem:

$$\frac{\partial \varphi}{\partial t} + A\varphi = \sum\nolimits_{i=1}^{N} \lambda_i f_i(r,t), \quad \varphi(r,0) = \phi^0(r) \text{ in } D \tag{1.6.5}$$

Since this goal can be achieved with different sets of such parameters, the optimal values will be obtained with a suitable optimization process.

Let us now introduce concentration functions $C_i = C_i(r,t)$ $i = 0, 1, \ldots, N$, as solutions to the following dispersion problems:

$$\frac{\partial C_0}{\partial t} + AC_0 = 0, \quad C_0(r,0) = \phi^0(r) \text{ in } D \tag{1.6.6}$$

and

$$\frac{\partial C_i}{\partial t} + AC_i = f_i(r,t), \quad C_i(r,0) = 0 \text{ in } D, \quad i = 1, \ldots, N \tag{1.6.7}$$

Note that if $C_j(r,t) = 0$ for any $r \in \Omega$ and any instant $t \in (0,T)$ then the jth source can be excluded from the control problem. In the future, we believe that such sources have already been identified and eliminated, i.e. all N sources in the dispersion model are responsible for the contamination of the zone Ω. Since the dispersion model (1.6.3) is linear and has a unique solution, the solution to problem (1.6.4) can be written as a linear combination of the concentration functions C_i:

$$\varphi(r,t) = C_0(r,t) + \sum_{i=1}^{N} \lambda_i C_i(r,t), \quad r \in D \text{ and } t \in (0,T) \tag{1.6.8}$$

for any set of nonnegative damping coefficients, $\lambda_i \le 1$, $i = 1, \ldots, N$. By virtue of Eq. (1.6.8), the environmental condition (1.6.2) is equivalent to the following relation:

$$\sum_{i=1}^{N} \lambda_i C_i(r,t) \le J_0 - C_0(r,t), \;\; r \in \Omega \;\text{ and }\; t \in (0,T) \tag{1.6.9}$$

It should be noted that the continuous function

$$\alpha(r,t) = J_0 - C_0(r,t), \;\; r \in \Omega \;\text{ and }\; t \in (0,T) \tag{1.6.10}$$

defined by the right side of inequality (1.6.9) indicates various cases of the control problem. To do this, we define the parameter α_0 as follows:

$$\alpha_0 = \min\{\alpha(r,t);\;\; (r,t) \in \overline{\Omega \times (0,T)}\} \tag{1.6.11}$$

Control Cases:

1. If $\alpha_0 \le 0$ then the damping coefficients should be taken as $\lambda_i = 0$ for any index i (a trivial solution), namely, all emissions should be stopped to prevent dangerous concentrations of the pollutant. This case appears when the initial pollutant concentration $\phi^0(r)$ is high enough.
2. If $\alpha_0 > 0$ then the control problem has non-trivial solutions. Therefore, by setting the optimal criterion, it is possible to select the optimal damping parameters λ_i.

Now, let's define the cost function as

$$F(\lambda_1, ..., \lambda_N) = \sum_{i=1}^{N} c_i^2 \frac{\int_0^T \int_D [f_i(r,t) - \lambda_i f_i(r,t)]^2 drdt}{\int_0^T \int_D [f_i(r,t)]^2 drdt} = \sum_{i=1}^{N} c_i^2 (1-\lambda_i)^2 \tag{1.6.12}$$

where each coefficient $c_i > 0, \;\; i = 1, ..., N$, represents the price that the ith source must pay in accordance with the control strategy, i.e., by reducing its emissions by $(1 - \lambda_i) \times 100$ percent. Considering the cost function (1.6.12), the optimal control problem is to determine the damping coefficients λ_i by solving the optimization problem

$$\text{Minimize}\, F(\lambda_1, \ldots, \lambda_N) = \sum_{i=1}^{N} c_i^2 (1-\lambda_i)^2 \tag{1.6.13}$$

$$\text{subject to: } \left\{ \begin{matrix} \varphi(r,t) \le J_0, \;\; r \in \Omega \text{ and } t \in (0,T) \\ 0 \le \lambda_i \le 1, \;\; i = 1, \ldots, N \end{matrix} \right\} \tag{1.6.14}$$

where φ is the solution of the dispersion model (1.6.5).

Taking into account (1.6.8) and (1.6.10), the optimization problem (1.6.13)–(1.6.14) can be formulated as a quadratic programming problem:

$$\text{Minimize}\, F(\lambda_1, \ldots, \lambda_N) = \sum_{i=1}^{N} c_i^2 (1 - \lambda_i)^2 \tag{1.6.15}$$

$$\text{subject to: } \begin{array}{l} \sum_{i=1}^{N} \lambda_i C_i(r_j, t_k) \le \alpha(r_j, t_k), \;\; r_j \in \Omega \text{ and } t_k \in (0, T) \\ 0 \le \lambda_i \le 1, \;\; i = 1, \ldots, N \end{array} \tag{1.6.16}$$

where $x_{jk} = (r_j, t_k)$ are the grid nodes in the domain $\Omega \times (0, T)$. In the formulation of the problem, the quadratic objective function estimates the cost of control, and linear constraints ensure compliance with the air quality standard at each point of the control zone and at any moment in the control time interval.

Note that the dispersion model solutions C_i calculated independently for each source play an important role in determining this control strategy. Thus, with an increase in the number of pollution sources, the computational complexity and response time to the control problem increase, so a special procedure for setting the optimization problem is necessary. Indeed, the solution $\{\lambda_i^*\}$ of the quadratic programming problem (1.6.15)–(1.6.16) can be calculated using the *qpsolve* optimization function, which belongs to the *Scilab* environment. This solver is implemented in Fortran 77 and is based on the stable primal–dual algorithm developed by Goldfarb and Idnani [45]. The routine is capable of solving strictly convex quadratic programming problems for thousands of variables and constraints in just a few minutes. Therefore, the task that requires more computational time to obtain a solution $\{\lambda_i^*\}$ is the iterative resolution of the dispersion model. According to (1.6.16), it is necessary to calculate the functions C_i, $i = 1, ..., N$, by solving the dispersion problem (1.6.7) for each emission source f_i. Thus, the calculation time increases with the number of sources. Since the short-term control problem requires a fast response, we propose to solve this problem using a parallel procedure. If we use a computing system with M processors under the *SPMD* (simple program multiple data) protocol, the required computational time to solve this problem is $T_p = (N/M + 1) * \Delta\tau$, where $\Delta\tau$ is the computational time required to solve the dispersion problem using one processor. The computation time T_p is acceptable when the N/M ratio is small. When the N/M ratio is large, the number of sources can be reduced by creating n source clusters and assuming that a damping parameter is designed for each cluster. Note that the sources composing a cluster do not necessarily have to be in the same localized area of region$D$, and this separation of sources occurs regardless of the control problem. The purpose of the separation is to get a small n/M ratio.

References

1. Doménech, X.: Química de la Contaminación. Miraguano Ediciones, Madrid (1999)
2. Seinfeld, J.H.: Atmospheric Chemistry and Physics of Air Pollution. Wiley, New York (1992)
3. Caselli, M.: La contaminación atmosférica. Siglo Veintiuno, México (1996)
4. Wark K., Warner, C.F.: Air Pollution, its Origin and Control. Harper & Row (1999)

5. Bravo, H.A., Sosa, R.E., Torres, R.J.: Ozono y lluvia ácida en la Ciudad de México. Ciencias **22**, 33–40 (1991)
6. Beilke, S., Elshout, A.J.: Acid deposition. In: Proceedings of the CEC workshop. D. Reidel Publishing Company (1983)
7. Rodhe, H., Crutzen, P., Vanderpol, A.: Formation of sulfuric and nitric acid in the atmosphere during long-range transport. Tellus **33**, 132–141 (1981)
8. Rivera, A.M.: El Cambio Climático. Tercer Milenio (CONACULTA), México (1999).
9. Marchuk, G.I.: Mathematical Models in Environmental Problems. Elsevier, New York (1986)
10. Skiba, Y.N.: Air pollution estimates. World Resour. Rev. **9**(4), 542–556 (1997)
11. Jhih-Shyang, S.: An optimization model for photochemical air pollution control. Eur. J. Oper. Res. **106**, 1–14 (1998)
12. Parra-Guevara, D., Skiba, Y.N.: Industrial pollution transport. Part II: control of industrial emissions. Environ. Model. Assess. **5**(3), 177–184 (2000)
13. Parra-Guevara, D., Skiba, Yu.N.: Optimización de emisiones industriales para la protección de zonas ecológicas. Atmósfera **13**(1), 27–38 (2000)
14. Penenko, V.V., Raputa, V.F.: Some models for optimizing the operation of the atmospheric-pollution sources. Sov. Meteorol. Hidrology **2**, 46–54 (1983)
15. Marchuk, G.I.: Mathematical issues of industrial effluent optimization. J. Meter. Soc. Jpn. **60**, 481–485 (1982)
16. Coulon, F., McKew, B.A., Osborn, A.M., McGenity, T.J., Timmis, K.N.: Effects of temperature and biostimulation on oil-degrading microbial communities in temperate estuarine waters. Env. Microbiol. **9**(1), 177–186 (2006)
17. Panos, G., Seinfeld, J.H.: Mathematical modeling of turbulent reacting plumes I: general theory and model formulation. Atmos. Environ. **20**, 1791–1807 (1986)
18. Swannell, R.P.J., Mitchell, D., Lethbridge, G., Jones, D., Heath, D., Hagley, M., Jones, M., Petch, S., Milne, R., Croxford, R., Lee, K.: A field demonstration of the efficacy of bioremediation to treat oiled shorelines following the Sea Empress incident. Environ. Technol. **20**(8), 863–873 (1999)
19. Kawada, J.: Health effects of air pollutants and their management. Atmos. Environ. **18**, 613–620 (1984)
20. Lynn, D.: Air Pollution, Threat and Response. Addison-Wesley, Reading (1976)
21. Hill, M.K.: Understanding of Environmental Pollution. Cambridge University Press, Cambridge (1997)
22. Nikolaev, Y.A., Plakunov, V.K.: Microbial biofilms: potential of using in waste water treatment. Water: Chem. Ecol. **2**, 11–13 (2008)
23. Boufadel, M.C., Suidan, M.T., Venosa, A.D.: Tracer studies in a laboratory beach subjected to waves. J. Environ. Eng. **133**(7), 722–732 (2007)
24. Zhu, X., Venosa, A.D., Suidan, M.T., Le, K.: Guidelines for the bioremediation of marine shorelines and freshwater wetlands. U.S. Environmental Protection Agency, Cincinnati, OH (2001)
25. ZoBell, C.E.: Microbial degradation of oil: present status, problems, and perspectives. In: Ahearn D.G., Meyers, S.P. (Eds.), The Microbial Degradation of Oil Pollutants, Publication No. LSU-SG-73-01, Proceedings Center for Wetlands Resource, Louisiana State University, Baton Rouge, LA, 1973, pp. 3–16.
26. Venosa, A.D.: Oil spill bioremediation on coastal shoreline: a critique. In: Sikdar, S.K., Irvine, R.I. (Eds.), Bioremediation: Principles and Practice. Volume III. Bioremediation Technologies. Technomic, Lancaster, PA, 1998, pp. 259–301
27. Godish, T.: Air Quality. Lewis Publishers Inc., Chelsea, Michigan (1991)
28. Flagan, R.C., Seinfeld, J.H.: Fundamentals of Air Pollution Engineering. Prentice-Hall, Englewood Cliffs (1988)
29. Rossin, A.D., Roberts, J.J.: Episode control criteria and strategy for carbone monoxide. J. Air Pollut. Control Assoc. **22**, 254–259 (1972)
30. Cooper, W.W., Hemphill, H., Huang, Z., Li, S., Lelas, V., Sullivan, D.W.: Survey of mathematical programming models in air pollution management. Eur. J. Oper. Res. **96**(1), 1–35 (1997)

31. Freitas, F., Assis, J.J., Almeida, T., Oliveira, I., Reis, E.: Simulation optimization for analysis of sustainable logistics systems. Pesqui. Oper. **37**(1), 145–171 (2017)
32. Greenberg, H.J.: Mathematical programming models for environmental quality control. Oper. Res. **43**(4), 578–623 (1995)
33. Rodrigues, S., Kaviski, E.: Daily scheduling of small hydro power plants dispatch with modified particles swarm optimization. Pesqui. Oper. **35**(1), 25–37 (2015)
34. Teller, A.: The use of linear programming to estimate the cost of some alternative air pollution abatement policies. In: Number 20 in Proceedings of IBM Scientific Computing Symposium on Water and Air Research Management, IBM, New York, pp. 345–353 (1968)
35. Kohn, R.E.: A Linear Programming Model for Air Pollution Control. MIT Press Inc., Cambridge, Mass (1978)
36. Álvarez-Vázquez, L.J., García-Chan, N., Martínez, A., Vázquez-Méndez, M.E.: Optimal control of urban air pollution related to traffic flow in road networks. Math. Control. Relat. Fields **8**(1), 177–193 (2018)
37. Cheng, S.Y., Li, J.B., Feng, B., Jin, Y.Q., Hao, R.X.: A Gaussian-box modeling approach for urban air quality management in a Northern Chinese City—II. Pollutant emission abatement. Water Air Soil Poll. **178**, 15–36 (2007)
38. Feng, L., Fei, H., Jiang, Z.: Adjoint method for the optimum planning of industrial pollutant sources. Science in China, Ser. D: Earth Sciences **48**(8), 1270–1279 (2005)
39. Parra-Guevara, D., Skiba, Yu.N.: On optimal solution of an inverse air pollution problem: theory and numerical approach. Math. Comput. Modelling **43**(7–8), 766–778 (2006)
40. Parra-Guevara, D., Skiba, Y.N.: Quadratic programming formulation for controlling the emissions of air pollution point sources, 207–247. In: Clary, T.S. (ed.) Horizons in Computer Science Research, vol. 14. Nova Science Publ., New York (2017)
41. Parra-Guevara, D., Skiba, Yu.N., Pérez-Sesma, A.: A linear programming model for controlling air pollution. Int. J. Appl. Math. **23**(3), 549–569 (2010)
42. Thunis, P., Degraeuwe, B., Pisoni, E., Meleux, F., Clappier, A.: Analyzing the efficiency of short-term air quality plans in European cities, using the CHIMERA air quality model. Air Qual. Atmos. Health **10**(2), 235–248 (2017)
43. Liao, K.J., Hou, X.: Optimization of multipollutant air quality management strategies: a case study for five cities in the United States. J. Air & Waste Manag. Assoc. Taylor Fr. **65**(6), 732–742 (2015)
44. Yu, S., Li, P., Wang, L., Wu, Y., Wang, S., Liu, K., Zhu, T., Zhang, Y., Hu, M., Zeng, L., Zhang, X., Cao, J., Alapaty, K., Wong, D.C., Pleim, J., Mathur, R., Rosenfeld, D., Seinfeld, J.H.: Mitigation of severe urban haze pollution by a precision air pollution control approach. Sci. Rep. Nature **8**, 8151 (2018)
45. Goldfarb, D., Idnani, A.: A numerically stable dual method for solving strictly convex quadratic programs. Math. Program. **27**, 1–33 (1983)

Chapter 2
Dispersion Problems of Different Degrees of Complexity

As we will see in the following chapters, mathematical methods of control of emissions rates from pollution sources are based on the use of dispersion and adjoint models. They also require solving inverse problems.

Most problems on the dispersion of pollutants can be reduced to the problems of transport (diffusion and convection) of impurities. This chapter introduces the reader to dispersion models of various levels of complexity. Simple one-dimensional dispersion problems and some analytical methods for their solution are given in Sects. 2.1–2.3. In general, the concentration of harmful substances can be created by point sources (industries, power plants, refineries, oil spills, etc.), line sources (roads, highways, railways) or areal sources (forest fires, large railway stations, etc.). The two-dimensional (2D) dispersion problem in a limited area is described in Sect. 2.4. A mathematical problem is called well-posed (in the sense of Hadamard) if it has a unique solution that continuously depends on the forcing and the initial and boundary conditions. The correctness of the two-dimensional dispersion problem is studied in Sects. 2.5–2.9. Particular attention is paid to conditions at the open boundary. The positive definiteness of the problem operator is shown in Sect. 2.5. Section 2.6 proves the uniqueness and stability of the classical solution of the 2D problem, and the existence, uniqueness and stability of a weak solution are studied in Sects. 2.7–2.9. The dispersion problem in a three-dimensional limited area is considered in Sect. 2.10. Its well-posedness can be proved in exactly the same way as the well-posedness of the two-dimensional problem.

2.1 One-Dimensional Linear Advection Problem

Before proceeding to the study of pollutant dispersion processes, it is useful to examine simple one-dimensional problems, and then gradually complicate the mathematical formulation of the pollution dispersion problem.

Y. N. Skiba and D. Parra Guevara, *Mathematical Methods for the Assessment and Control of Industrial Emissions*, Mathematical Engineering,
https://doi.org/10.1007/978-3-031-58109-0_2

Let $D = \{0 < x < 1;\ 0 < t < 1\}$ be the space–time domain. Let us now consider a one-dimensional transport equation

$$\frac{\partial \phi}{\partial t} + u\frac{\partial \phi}{\partial x} = 0 \tag{2.1.1}$$

in the domain D, where u is the constant transport velocity. As the initial condition at $t = 0$ we take

$$\phi(x, 0) = f(x), \quad x \in [0, 1] \tag{2.1.2}$$

where $f(x)$ is a given function. It is easy to verify that

$$\phi(x, t) = f(x - ut) \tag{2.1.3}$$

is the exact solution of problem (2.1.1)–(2.1.2) [1, 2]. So, the solution is invariable along each characteristic line

$$x - ut = x_0 \tag{2.1.4}$$

i.e.,

$$\phi(x, t) = f(x_0) = Const$$

Due to (2.1.3), the solution does not change its initial geometric form, defined by function $f(x)$. The form moves with time in the x-axis direction if u is positive, and in the opposite direction if u is negative. In particular, if $f(x)$ is a wave then (2.1.3) describes the propagation of this wave with velocity u (Fig. 2.1).

It should be noted that the boundary conditions for Eq. (2.1.1) cannot be set arbitrarily. These conditions depend on the slope of the characteristic lines (2.1.4). Suppose, for example, that $u > 0$. Then the characteristic line $x - ut = 0$ divides the domain D into two zones (Fig. 2.2) depending on the value of x_0 that defines the characteristic line $x - ut = x_0$: zone A ($0 < x_0 \leq 1$) and the zone B ($-1 \leq x_0 < 0$). In the zone A, solution is determined by the initial condition (2.1.2), and in the zone B, solution is determined by the characteristic lines which pass through the points $t \in (0, 1]$ of the t-axis. Therefore, problem (2.1.1)–(2.1.2) requires the boundary condition

$$\phi(0, t) = g(t), \quad t \in [0, 1] \tag{2.1.5}$$

where $g(t)$ is a given function. Thus, the initial condition (2.1.2) and boundary condition (2.1.5) determine the solution in the whole domain [3]. If velocity $u = u(x, t)$ is not constant, then the solution to the problem can usually be found only numerically.

Fig. 2.1 Initial value $f(x_0)$ is transported unchanged along the characteristic line (2.1.4) of problem (2.1.1)

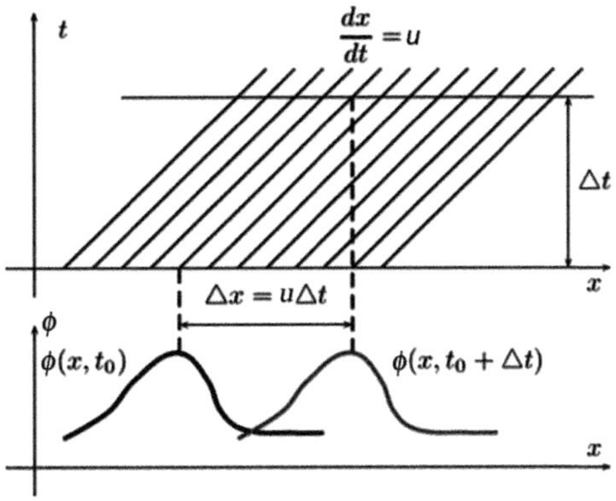

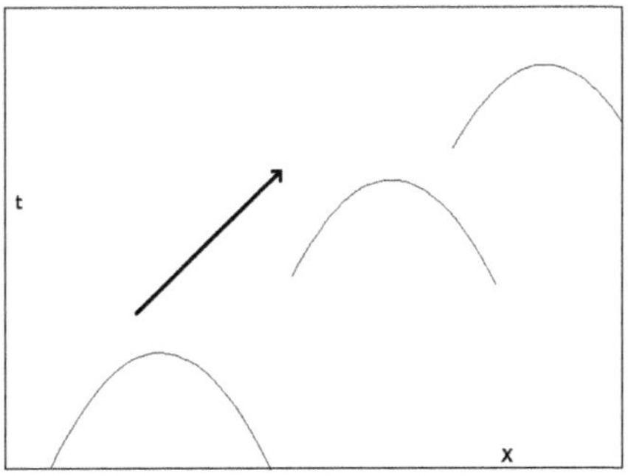

Fig. 2.2 Determination of a solution to Eq. (2.1.1) by the method of characteristic lines (2.1.4)

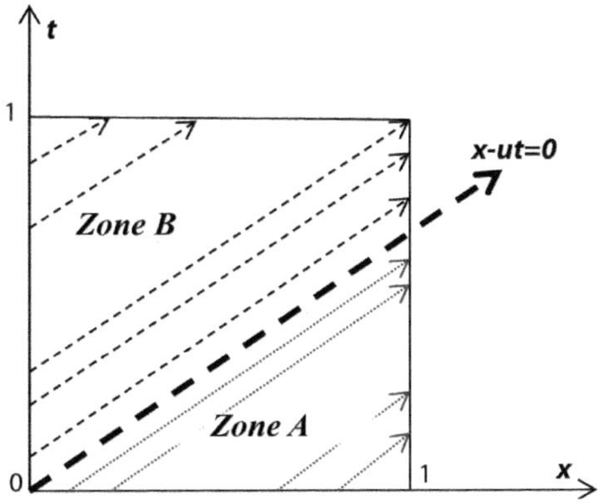

2.2 Simplest 1D Advection–Diffusion Problems

This section considers one-dimensional stationary and evolutionary models of different complexity, which describe processes of dissipation and diffusion.

1. Model of exponential dissipation. The simplest model describing the unsteady process of dissipation of a substance ϕ can be written as

$$\frac{\partial \phi}{\partial t} + \sigma \phi = f, \quad \phi = \phi_0 \quad \text{at } t = 0 \tag{2.2.1}$$

where f is the source. In the case of time-independent source f, the solution is

$$\phi = \phi_0 \exp(-\sigma t) + \frac{f}{\sigma}\big[1 - \exp(-\sigma t)\big] \tag{2.2.2}$$

Thus, solution $\phi(t)$ decays exponentially with time (the decay rate depends on the damping coefficient σ), and eventually reduces to solution $\phi = f/\sigma$ of the stationary problem

$$\sigma \phi = f$$

Despite its simplicity, Eq. (2.2.1) is quite reasonable to describe the process of sedimentation of particles in the atmospheric models.

2. Stationary diffusion model. We now consider a very simple stationary air pollution problem. Suppose that the wind velocity is zero and the diffusion coefficient μ is constant [4]:

$$\sigma \phi = \mu \frac{d^2 \phi}{dx^2} + f, \quad f = Q\,\delta(x - x_0) \tag{2.2.3}$$

The problem describes diffusion of a substance in an infinite medium from a source. Suppose that the source is localized at a point x_0, and Q is its intensity. As the boundary condition, we use the hypothesis of the boundedness of the solution in the whole region. In order to eliminate the delta function, it is convenient to use the integral formulation of the problem. For this purpose, we integrate Eq. (2.2.3) in the interval $(x_0 - \varepsilon, x_0 + \varepsilon)$:

$$\sigma \int\limits_{x_0-\varepsilon}^{x_0+\varepsilon} \phi\, dx = \mu \frac{d\phi}{dx}\Big|_{x_0+\varepsilon} - \mu \frac{d\phi}{dx}\Big|_{x_0-\varepsilon} + Q$$

Passing to the limit as $\varepsilon \to 0$, we obtain

$$\mu \frac{d\phi}{dx}\Big|_{x_0+0} - \mu \frac{d\phi}{dx}\Big|_{x_0-0} + Q = 0 \tag{2.2.4}$$

Denote as ϕ_- and ϕ_+ the solution in the domains $-\infty < x \leq x_0$ and $x_0 \geq x < \infty$, respectively. Then we obtain two problems:

$$\mu \frac{d^2\phi_+}{dx^2} - \sigma\phi_+ = 0\,, \quad \phi_+ = 0 \text{ in the limit } x = \infty \tag{2.2.5}$$

and

$$\mu \frac{d^2\phi_-}{dx^2} - \sigma\phi_- = 0\,, \quad \phi_- = 0 \text{ in the limit } x = -\infty \tag{2.2.6}$$

which are linked by the relation

$$\mu \frac{d\phi_+}{dx} - \mu \frac{d\phi_-}{dx} + Q = 0 \ \text{ at } x = x_0 \tag{2.2.7}$$

If we assume that solution is continuous in the whole domain $-\infty < x < \infty$ including the point $x = x_0$, we obtain one more condition:

$$\phi_+ = \phi_- \ \text{ at } x = x_0 \tag{2.2.8}$$

It is easy to show that the solutions of problems (2.2.5) and (2.2.6) are

$$\phi_+ = C_+ \exp\{-\sqrt{\sigma/\mu}(x - x_0) \quad \text{and} \quad \phi_- = C_- \exp\{-\sqrt{\sigma/\mu}(x_0 - x) \tag{2.2.9}$$

Substituting (2.2.9) into (2.2.7) and (2.2.8), and solving linear equations with respect to C_+ and C_- we obtain $C_+ = C_- = Q/2\sqrt{\sigma\mu}$. Therefore, the solution to problem (2.2.3) is

$$\phi(x) = \frac{Q}{2\sqrt{\sigma\mu}} \begin{cases} exp\{-\sqrt{\sigma/\mu}(x - x_0)\,, & \text{if } x \geq x_0 \\ exp\{-\sqrt{\sigma/\mu}(x_0 - x)\,, & \text{if } x \leq x_0 \end{cases} \tag{2.2.10}$$

Thus, as a result of the diffusion process, the concentration of substance decreases exponentially and symmetrically in both directions from the source point $x = x_0$ (Fig. 2.3).

It is easy to see that

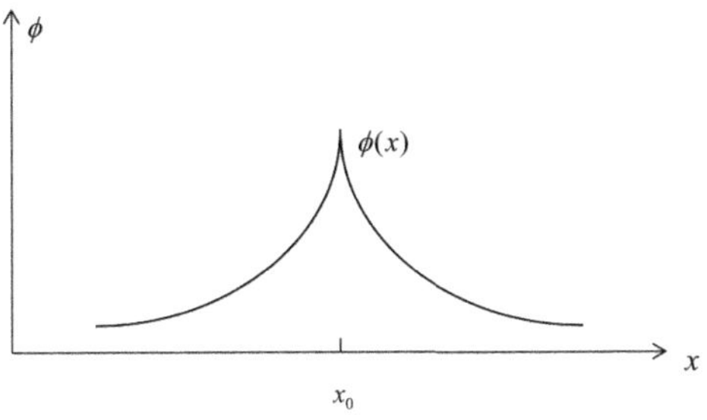

Fig. 2.3 Distribution of $\phi(x)$ in the process of diffusion from the source localized at x_0

$$\int_{-\infty}^{\infty} \phi \, dx = \frac{Q}{\sigma}$$

3. Stationary advection–diffusion model. We now consider a more interesting case when air velocity u is nonzero. Suppose that $u = const > 0$:

$$u\frac{d\phi}{dx} + \sigma\phi = \mu\frac{d^2\phi}{dx^2} + Q\,\delta(x - x_0) \tag{2.2.11}$$

By analogy with the previous case, the Eq. (2.2.11) reduces to two problems:

$$\mu\frac{d^2\phi_+}{dx^2} - u\frac{d\phi_+}{dx} - \sigma\phi_+ = 0\,, \quad \phi_+ = 0 \text{ in the limit } x \to \infty \tag{2.2.12}$$

$$\mu\frac{d^2\phi_-}{dx^2} - u\frac{d\phi_-}{dx} - \sigma\phi_- = 0\,, \quad \phi_- = 0 \text{ in the limit } x \to -\infty \tag{2.2.13}$$

which are linked by two relationships:

$$\mu\frac{d\phi_+}{dx} - \mu\frac{d\phi_-}{dx} + Q = 0 \quad \text{and} \quad \phi_+ = \phi_- \quad \text{if } x = x_0 \tag{2.2.14}$$

The solutions of problems (2.2.12) and (2.2.13) are sought of the form

$$\phi_+ = C_+ \exp\left\{-\left(\sqrt{\frac{\sigma}{\mu} + \frac{u^2}{4\mu^2}} - \frac{u}{2\mu}\right)(x - x_0)\right\}, \quad x \ge x_0 \tag{2.2.15}$$

$$\phi_- = C_- \exp\left\{-\left(\sqrt{\frac{\sigma}{\mu} + \frac{u^2}{4\mu^2}} + \frac{u}{2\mu}\right)(x_0 - x)\right\}, \quad x \le x_0 \tag{2.2.16}$$

Substituting (2.2.15) and (2.2.16) into (2.2.14) we obtain

$$C_+ = C_- = Q/\sqrt{4\sigma\mu + u^2}$$

As a result, the solution to problem (2.2.11) is

$$\phi(x) = \frac{Q}{\sqrt{4\sigma\mu + u^2}} \begin{cases} exp\left\{-\left(\sqrt{\frac{\sigma}{\mu} + \frac{u^2}{4\mu^2}} - \frac{u}{2\mu}\right)(x - x_0)\right\}, & \text{if } x \ge x_0 \\ exp\left\{-\left(\sqrt{\frac{\sigma}{\mu} + \frac{u^2}{4\mu^2}} + \frac{u}{2\mu}\right)(x_0 - x)\right\}, & \text{if } x \le x_0 \end{cases} \tag{2.2.17}$$

Marchuk [4]. The graphics of function $\phi(x)$ for the cases $u = const > 0$ and $u = const < 0$ are presented in Fig. 2.4a and b, respectively. One can see that the

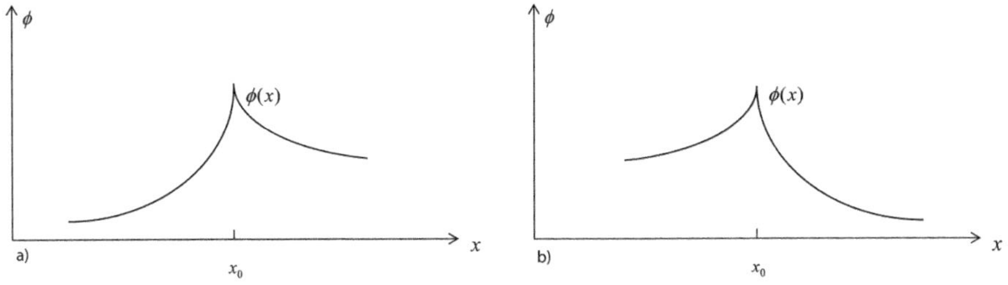

Fig. 2.4 Distribution of solution $\phi(x)$ during transport and diffusion from a source located at point x_0; **a** $u = const > 0$; **b** $u = const < 0$.

non-zero velocity results in the spread of the substance and its slower decrease in the direction of velocity, which destroys the symmetry of Fig. 2.1.

Finally, let us consider a more complicated situation, when $u = u_1 > 0$ for a long period, and then the wind direction is changed: $u = u_2 < 0$. In such a case we have two solutions:

$$\phi_1(x) = \frac{Q}{\sqrt{4\sigma\mu + u_1^2}} \begin{cases} exp\left\{-\left(\sqrt{\frac{\sigma}{\mu} + \frac{u_1^2}{4\mu^2}} - \frac{|u_1|}{2\mu}\right)(x - x_0)\right\}, & \text{if } x \geq x_0 \\ exp\left\{-\left(\sqrt{\frac{\sigma}{\mu} + \frac{u_1^2}{4\mu^2}} + \frac{|u_1|}{2\mu}\right)(x_0 - x)\right\}, & \text{if } x \leq x_0 \end{cases} \tag{2.2.18}$$

and

$$\phi_2(x) = \frac{Q}{\sqrt{4\sigma\mu + u_2^2}} \begin{cases} exp\left\{-\left(\sqrt{\frac{\sigma}{\mu} + \frac{u_2^2}{4\mu^2}} + \frac{|u_2|}{2\mu}\right)(x - x_0)\right\}, & \text{if } x \geq x_0 \\ exp\left\{-\left(\sqrt{\frac{\sigma}{\mu} + \frac{u_2^2}{4\mu^2}} - \frac{|u_2|}{2\mu}\right)(x_0 - x)\right\}, & \text{if } x \leq x_0 \end{cases} \tag{2.2.19}$$

If the wind velocity was positive ($u_1 > 0$) during τ_1 days and negative ($u_2 < 0$) during τ_2 days, then the solution $\phi(x)$ can be found with the linear interpolation:

$$\phi(x) = \frac{\tau_1}{\tau_1 + \tau_2}\phi_1(x) + \frac{\tau_2}{\tau_1 + \tau_2}\phi_2(x) \tag{2.2.20}$$

2.3 Analytical Solutions of 1D Diffusion Problems

The unsteady diffusion equation is a parabolic differential equation that describes the dispersion of substances by diffusion (for example, the redistribution of the temperature in a medium or material object by heat conduction). In the one-dimensional

(1D) case, the equation has the form

$$\frac{\partial \phi}{\partial t} = \frac{\partial}{\partial x}\mu(x, t)\frac{\partial \phi}{\partial x} + f(x, t)$$

A numerical method should be applied in this case (see Chap. 8 and [5, 6]). If the diffusion coefficient is constant ($\mu = const$) then the above equation reduces to

$$\frac{\partial \phi}{\partial t} = \mu\frac{\partial^2 \phi}{\partial x^2} + f(x, t)$$

1. **Homogeneous Diffusion Equation with Homogeneous Boundary Conditions**. We now consider the following problem

$$\begin{aligned} &\phi_t = \mu\phi_{xx}, \quad 0 < x < l,\ 0 < t \le \mathrm{T} \\ &\phi(x, 0) = \varphi(x); \quad 0 \le x \le l \\ &\left.\begin{aligned}\phi(0, t) = 0 \\ \phi(l, t) = 0\end{aligned}\right\} \quad 0 \le t \le T \end{aligned} \tag{2.3.1}$$

The objective is to find the solution $\phi(x, t)$ in the domain $0 < x < l,\ 0 < t \le T$. Applying the method of separation of variables, we will search l solution as the product of two functions:

$$\phi(x, t) = X(x)R(t) \tag{2.3.2}$$

Substitution (2.3.2) into the original equation gives

$$X(x)R'(t) = \mu X''(x)R(t)$$

Dividing the expression by $\mu X(x)R(t)$ we get

$$\frac{1}{\mu}\frac{R'(t)}{R(t)} = \frac{X''(x)}{X(x)} = -\lambda\,, \quad \lambda = const \tag{2.3.3}$$

The number λ is a constant because the left part of (2.3.3) is a function of t, while the right part is a function of x. Thus, we obtain two linear ordinary differential equations:

$$\begin{aligned} X''(x) + \lambda X(x) = 0, \\ R'(t) + \mu\lambda R(t) = 0 \end{aligned} \tag{2.3.4}$$

Taking into account the boundary conditions $\phi(0, t) = X(0)R(t) = 0$ and $\phi(l, t) = X(l)R(t) = 0$ for any t $(0 < t < T)$, we obtain $X(0) = 0$ and $X(l) = 0$ ($R(t) \ne 0$, since we seek only nontrivial solutions). Thus, one must solve the Sturm–Liouville problem

$$\begin{aligned} X''(x) + \lambda X(x) &= 0, \\ X(0) = 0\,,\ X(l) &= 0 \end{aligned} \tag{2.3.5}$$

In solving this problem, three values of λ are possible: $\lambda < 0\,,\ \lambda = 0$ and $\lambda > 0$. In the first two cases, the solution is zero. If $\lambda > 0$ then the general solution is

$$X(x) = C_1 \cos\left(\sqrt{\lambda} x\right) + C_2 \sin\left(\sqrt{\lambda} x\right) \tag{2.3.6}$$

The condition $X(0) = 0$ leads to $C_1 = 0$, while condition $X(l) = 0$ gives

$$X(l) = C_2 \sin\left(\sqrt{\lambda} l\right) = 0$$

Since only nonzero solutions are sought, $C_2 \neq 0$ and hence,

$$\begin{aligned} &\sin\left(\sqrt{\lambda} l\right) = 0, \\ &\sqrt{\lambda} l = \pi n, \quad n = 1,\ 2,\ \ldots \\ &\lambda_n = \left(\tfrac{\pi n}{l}\right)^2, \quad n = 1,\ 2,\ \ldots \end{aligned} \tag{2.3.7}$$

Therefore, the solutions corresponding to these eigenvalues are

$$X_n(x) = C_n \sin\left(\frac{\pi n}{l} x\right), \quad n = 1,\ 2,\ \ldots$$

Solving the differential equation for $R(t)$

$$R'(t) + \mu \left(\frac{\pi n}{l}\right)^2 R(t) = 0$$

we obtain

$$R_n(t) = D_n \exp\left(-\mu \left(\frac{\pi n}{l}\right)^2 t\right), \quad D_n = const$$

The products of the functions $X_n(x)$ and $R_n(t)$ constitute an infinite family of solutions of the homogeneous diffusion Eq. (2.3.1) with homogeneous boundary conditions:

$$\begin{aligned} \phi_n(x,\ t) = X_n(x) R_n(t) = C_n \sin\left(\tfrac{\pi n}{l} x\right) exp\left(-\mu \left(\tfrac{\pi n}{l}\right)^2 t\right), \\ (n = 1,\ 2,\ \ldots) \end{aligned} \tag{2.3.8}$$

These solutions are called fundamental solutions. Since Eq. (2.3.1) is a linear equation and the boundary conditions are homogeneous, any finite linear combination of these solutions is also a solution. Moreover, the infinite series

$$\phi(x, t) = \sum_{n=1}^{\infty} \phi_n(x) = \sum_{n=1}^{\infty} C_n \sin\Big(\frac{\pi n}{l}x\Big)\exp\Big(-\mu\Big(\frac{\pi n}{l}\Big)^2 t\Big) \tag{2.3.9}$$

is a solution as well. The constants C_n are determined from the initial condition

$$\phi(x, 0) = \varphi(x) = \sum_{n=1}^{\infty} C_n \sin\Big(\frac{\pi n}{l}x\Big)$$

where

$$C_n = A_n = \frac{2}{l}\int_0^l \varphi(\xi)\sin\Big(\frac{\pi n}{l}\xi\Big)d\xi\,, \quad (n = 1,\, 2,\, \ldots) \tag{2.3.10}$$

Therefore, the solution of the initial boundary value problem (2.3.1) with homogeneous boundary conditions is given by series (2.3.9) with the coefficients determined by the formulas (2.3.10). In principle, the solution obtained is a formal series, but it can be shown that in the interval $0 < x < l$ and for any moment $t > 0$, the series converges uniformly to a continuous function that solves the problem (2.3.1), besides,

$$\lim_{t\to\infty} \phi(x, t) = 0 \tag{2.3.11}$$

Note that $0 < x < l$ is a limited domain, and according to the principle of maximum, solution $\phi(x, t)$ of the homogeneous problem (2.3.1) can accept extreme values only at the initial moment or at the boundary points $x = 0$ and $x = l$.

2. **Non-homogeneous Diffusion Equation with Homogeneous Conditions.** We now consider a non-homogeneous diffusion equation with homogeneous boundary and initial conditions:

$$\begin{aligned} \phi_t = \mu\phi_{xx} + f(x, t), \quad & 0 < x < l,\ 0 < t \le T \\ \phi(x, 0) = 0, \quad & 0 \le x \le l \\ \left.\begin{aligned}\phi(0, t) = 0 \\ \phi(l, t) = 0\end{aligned}\right\} \quad & 0 \le t \le T \end{aligned} \tag{2.3.12}$$

Applying the method of separation of variables, we suppose that

$$\begin{aligned} \phi_n(x, t) &= X_n(x)R_n(t), \\ f_n(x, t) &= X_n(x)F_n(t), \\ X_n(x) &= \sin\big(\tfrac{\pi n}{l}x\big) \end{aligned} \tag{2.3.13}$$

Then by using the obvious relation

$$X_n''(x) = -\left(\frac{\pi n}{l}\right)^2 X_n(x)$$

we can rewrite the original equation as

$$X_n(x)R_n'(t) = -\mu\left(\frac{\pi n}{l}\right)^2 X_n(x)R_n(t) + X_n(x)F_n(t),$$

and finally, as

$$R_n'(t) = -\mu\left(\frac{\pi n}{l}\right)^2 R_n(t) + F_n(t)$$

We solve the last inhomogeneous linear equation by the Lagrange method of variation of constants. First, we find the general solution of the homogeneous linear equation

$$R_n'(t) = -\mu\left(\tfrac{\pi n}{l}\right)^2 R_n(t),$$
$$R_n(t) = Dexp\left(-\mu\left(\tfrac{\pi n}{l}\right)^2 t\right)$$

In the general case, we replace the constant D by the variable $D(t)$ in the original equation:

$$R_n(t) = D(t)\exp\left(-\mu\left(\frac{\pi n}{l}\right)^2 t\right),$$

$$D_n'(t)exp\left(-\mu\left(\tfrac{\pi n}{l}\right)^2 t\right) - \mu\left(\tfrac{\pi n}{l}\right)^2 exp\left(-\mu\left(\tfrac{\pi n}{l}\right)^2 t\right)D_n(t)$$
$$= -\mu\left(\tfrac{\pi n}{l}\right)^2 exp\left(-\mu\left(\tfrac{\pi n}{l}\right)^2 t\right)D_n(t) + F_n(t),$$

$$D_n'(t)\exp\left(-\mu\left(\frac{\pi n}{l}\right)^2 t\right) = F_n(t),$$

$$D_n(t) = \int_0^t F_n(t)\exp\left(\mu\left(\frac{\pi n}{l}\right)^2 t\right)dt + A_n,$$

$$R_n(t) = A_n\exp\left(-\mu\left(\frac{\pi n}{l}\right)^2 t\right) + \exp\left(-\mu\left(\frac{\pi n}{l}\right)^2 t\right)\int_0^t F_n(t)\exp\left(\mu\left(\frac{\pi n}{l}\right)^2 t\right)dt$$

Due to the initial condition, we have

$$\phi_n(x,\ 0) = X_n(x)R_n(0) = 0,$$
$$R_n(0) = 0$$

Taking into account the conditions for R, we obtain

$$R_n(t) = \int_0^t \exp\left(-\mu\left(\frac{\pi n}{l}\right)^2 (t-\tau)\right) F_n(\tau) d\tau$$

Further,

$$f_n(x,\ t) = X_n(x) F_n(t) = \sin\left(\frac{\pi n}{l} x\right) F_n(t)$$

where

$$F_n(t) = \frac{2}{l} \int_0^l f(\xi, t) \sin\left(\frac{\pi n}{l}\xi\right) d\xi$$

As a result, the general formula is:

$$\phi(x,\ t) = \sum_{n=1}^{\infty} X_n(x) R_n(t)$$

$$= \sum_{n=1}^{\infty}\left[\int_0^t \exp\left(-\mu\left(\frac{\pi n}{l}\right)^2 (t-\vartheta)\right)\left\{\frac{2}{l}\int_0^t f(\xi,\ \vartheta)\sin\left(\frac{\pi n}{l}\xi\right) d\xi\right\} d\vartheta\right] \sin\left(\frac{\pi n}{l} x\right) \tag{2.3.14}$$

3. **Non-Homogeneous Diffusion Equation with Non-Homogeneous Conditions**. In many cases it is possible to solve the non-homogeneous diffusion equation

$$\begin{aligned} \phi_t &= \mu\phi_{xx} + f(x,\ t), \\ \phi(x,\ 0) &= \varphi(x), \\ \phi(0,\ t) &= k_1(t), \\ \phi(l,\ t) &= k_2(t) \end{aligned} \tag{2.3.15}$$

with non-homogeneous conditions using the methods described above and the following procedure. The desired solution $\phi(x,\ t)$ is represented as the sum of two functions:

$$\begin{aligned} \phi(x,\ t) &= \tilde{\phi}(x,\ t) + \Phi(x,\ t), \\ \tilde{\phi}(x,\ 0) &= \phi(x,\ 0) - \Phi(x,\ 0) = \varphi(x) - \Phi(x,\ 0), \\ \tilde{\phi}(0,\ t) &= 0, \\ \tilde{\phi}(l,\ t) &= 0 \end{aligned} \tag{2.3.16}$$

where $\Phi(X, t)$ is searched as

$$\begin{gathered}\Phi(x,\, t) = Ax + b, \\ \Phi(0,\, t) = b = k_1(t), \\ \Phi(l,\, t) = Al + k_1 = k_2 \Rightarrow A = \tfrac{k_2(t) - k_1(t)}{l}, \\ \Phi(x,\, t) = \tfrac{k_2(t) - k_1(t)}{l} x + k_1(t)\end{gathered} \tag{2.3.17}$$

Thus, the original problem is reduced to the following:

$$\begin{gathered}\tilde{\phi}_t = \mu\tilde{\phi}_{xx} + f(x,\, t) - \tfrac{k_2'(t) - k_1'(t)}{l} x - k_1'(t), \\ \tilde{\phi}(x,\, 0) = \varphi(x) - \tfrac{k_2(0) - k_1(0)}{l} x - k_1(0), \\ \tilde{\phi}(0,\, t) = 0, \\ \tilde{\phi}(l,\, t) = 0\end{gathered} \tag{2.3.18}$$

This problem was solved in the previous subsections (see (2.3.1) and (2.3.12)). Thus, solution $\phi\,(x,\, t)$ is the sum of the function $\Phi(x,\, t)$ obtained in (2.3.17) and $\tilde{\phi}(x,\, t)$:

$$\phi(x,\, t) = \tilde{\phi}(x,\, t) + \frac{k_2(t) - k_1(t)}{l} x + k_1(t) \tag{2.3.19}$$

2.4 A 2D Dispersion Problem in a Limited Area

We now consider a two-dimensional dispersion problem supposing that the atmospheric wind velocity is known (for example, the wind can be climatic or calculated with an atmospheric circulation model).

1. **Formulation of the problem.** Let D be a limited two-dimensional domain with an open boundary S, which contains N industrial plants located at points $r_i = (x_i, y_i)$, $(i = 1, 2, ..., N)$. Denote by $\phi(r, t)$ the concentration of a pollutant in point $r = (x, y)$ and time moment t. To study the dispersion of pollutant, we consider in D and time interval $(0, T)$ the two-dimensional (vertically averaged) advection–diffusion equation

$$\frac{\partial}{\partial t}\phi + U \cdot \nabla\phi + \sigma\phi \; - \nabla \cdot \mu\; \nabla\phi = f(r, t) \tag{2.4.1}$$

where

$$f(r, t) \equiv \sum_{i=1}^{N} Q_i(t)\, \delta(r - r_i) \tag{2.4.2}$$

is the forcing representing the sum of emission rates $Q_i(t)$ of N industrial sources $(i = 1, 2, ..., N)$; $U(r, t) = \{u(r, t), v(r, t)\}$ is the nondivergent wind velocity (assumed to be known from observations or some dynamic model), i.e.,

$$\nabla \cdot U = 0 \tag{2.4.3}$$

$\sigma(r, t) > 0$ is the coefficient that characterizes the damping rate of $\phi(r, t)$ due to various physical and chemical processes, and $\mu(r, t) > 0$ is the coefficient of turbulent diffusion, ∇ is the 2D gradient operator and $\delta(r - r_i)$ is the Dirac function. Equation (2.4.1) is solved with the initial condition

$$\phi(r, 0) = \phi^0(r) \tag{2.4.4}$$

Typically, the flow of pollution across the open boundary S of the limited region D is unknown. To obtain a well-posed problem in the sense of Hadamard [7], care must be taken in the use of conditions at the boundary S [4, 8]. In order to set physically and mathematically adequate boundary conditions, we denote as $U_n = U \cdot n$ the projection of the velocity vector U on the unit outward normal n to the boundary S, and divide S into the part S^- (where $U_n < 0$, and the flow of pollutant is directed into the region D) and the part S^+ (where $U_n \geq 0$, and the flow of pollutant is directed out of the region D) (Fig. 2.5). Assume that domain D is sufficiently large to include all the pollution sources, and there are no pollution sources outside of the region D. Then we set [9]

$$\mu \frac{\partial \phi}{\partial n} - U_n \phi = 0 \quad \text{at } S^- \tag{2.4.5}$$

$$\mu \frac{\partial \phi}{\partial n} = 0 \quad \text{at } S^+ \tag{2.4.6}$$

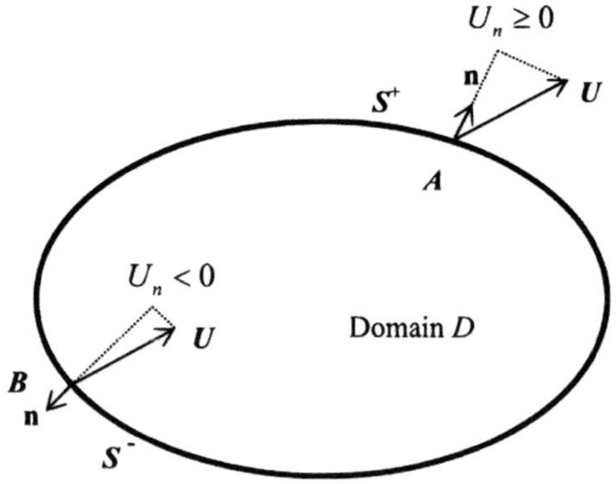

Fig. 2.5 A limited area D with open boundary $S = S^+ \cup S^-$

From a physical point of view, condition (2.4.5) means that the combined pollution flow (the diffusion flow plus advective flow) is zero at the part S^- (i.e., there are no pollution sources outside the domain D) and condition (2.4.6) means that the diffusion flow $\mu \partial\phi/\partial n$ at the part S^+ is negligible compared with the advective flow $U_n\,\phi$. In the limit when there is no diffusion ($\mu = 0$), condition (2.4.5) reduces to $\phi = 0$ (there is no pollution at S^-), and condition (2.4.6) disappears, as it should be. Indeed, the problem of pure advection ($\mu = \sigma = 0$) does not require a condition at the output part S^+ where solution ϕ is determined by the method of characteristic lines [1]. In the following sections, we will see that problem (2.4.1)–(2.4.6) is also well posed from a mathematical point of view, since it has a unique solution that continuously depends on the initial distribution $\phi^0(r)$, as well as on the number N, intensity $Q_i(t)$ and location r_i of the sources.

2. **Integral balance equations**. By integrating Eq. (2.4.1) over region D and taking into account (2.4.3), (2.4.5) and (2.4.6) we obtain the balance equation for the total mass of the pollutant

$$\frac{d}{dt}\int_D \phi dr = \sum_{i=1}^{N} Q_i(t) - \int_D \sigma\phi dr - \int_{S^+} U_n\phi dS \tag{2.4.7}$$

And if we multiply Eq. (2.4.1) by $\phi(r, t)$, then integrate the result over region D and again use conditions (2.4.3), (2.4.5) and (2.4.6) we obtain

$$\begin{aligned}\frac{d}{dt}\int_D \phi^2 dr &= 2\sum_{i=1}^{N} Q_i(t)\phi(r_i, t) - 2\int_D (\sigma\phi^2 + \mu|\nabla\phi|^2)dr - \int_{S^+} U_n\phi^2 dS \\ &+ \int_{S^-} U_n\phi^2 dS\end{aligned} \tag{2.4.8}$$

This equation governs the evolution of the norm $\|\phi\| = \left(\int_D \phi^2 dr\right)^{1/2}$ of solution $\phi(r, t)$.

Since $U_n < 0$ at S^- and $U_n \geq 0$ at S^+, Eq. (2.4.8) can be written as

$$\frac{d}{dt}\int_D \phi^2 dr = 2\sum_{i=1}^{N} Q_i(t)\phi(r_i, t) - 2\int_D (\sigma\phi^2 + \mu|\nabla\phi|^2)dr - \int_S |U_n|\phi^2 dS \tag{2.4.9}$$

According to (2.4.7) and (2.4.8), the total pollutant concentration $\int_D \phi\, dr$ in the domain D and the norm $\|\phi\|$ of solution increase if $Q_i(t) > 0$ and decrease due to dissipative processes ($\sigma > 0$, $\mu > 0$), and the advective flow of pollutant through the boundary of region D.

In the absence of the pollution sources in D ($Q_i(t) = 0$ for all i), both $\int_D \phi\, dr$ and $\|\phi\|$ decrease with time. And if in addition to this, there is no dissipative processes ($\sigma = 0,\ \mu = 0$) and $U_n = 0$ at the boundary S (of course, this is an ideal case) then both integrals are conserved with time:

$$\frac{d}{dt}\int_D \phi dr = 0, \quad \frac{d}{dt}\int_D \phi^2 dr = 0 \tag{2.4.10}$$

Despite the fact that relations (2.4.10) are fulfilled only under artificial conditions, they are very useful for testing the numerical algorithms and computer programs [10].

3. **Pollution sources**. The form of function $f(r, t)$ depends on the type of sources. For example, the pollution sources of N industrial plants are considered as the point sources (Fig. 2.6), and

$$f(r, t) = \sum_{i=1}^{N} Q_i(t)\, \delta(r - r_i) \tag{2.4.11}$$

where $r_i = (x_i, y_i)$ and $Q_i(t)$ characterize the location and the emission rate of the i-th source, respectively ($i = 1, 2, ..., N$). And if vehicle emissions are considered along N roads R_i (Fig. 2.6), then the line sources are defined as

$$f(r, t) = \begin{cases} Q_i(r, t)\,, & \text{if } r \in R_i \\ 0\,, & \text{if } r \notin R_i \end{cases} \quad (i = 1, 2, ..., N) \tag{2.4.12}$$

where $Q_i(r, t)$ is a continuous function that describes the emission rates of a pollutant along the road R_i ($i = 1, 2, ..., N$).

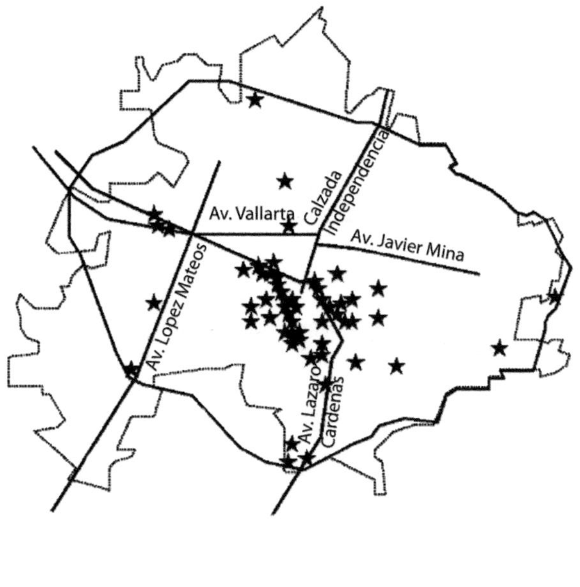

Fig. 2.6 Main avenues of the Guadalajara City and locations of 52 main industrial plants that burn fossil fuels and emit sulfur dioxide (SO_2)

The form (2.4.12) can also be used to describe area sources. For example, in case of fire, R_i is a two-dimensional zone of fire and $Q_i(r, t)$ is a function that describes the intensity of fire in the zone R_i.

However, it should be noted that in the process of discretization, the line and area sources can also be described in the same way as the point sources [11, 12]. In fact, one can divide each line (or area) R_i into sufficiently small non-overlapping parts and then discretize the function $Q_i(r, t)$. Therefore, (2.4.11) will be considered as the general form of forcing $f(r, t)$ in the pollution problems.

2.5 Positive Definiteness of the Problem Operator

We now prove that the linear differential operator

$$A\phi = U \cdot \nabla\phi + \sigma\phi - \nabla \cdot \mu\nabla\phi \tag{2.5.1}$$

with domain

$$\Phi = \{\phi \in C^2(\overline{D})|\ \mu\frac{\partial\phi}{\partial n} - U_n\phi = 0 \text{ at } S^- \text{ and } \mu\frac{\partial\phi}{\partial n} = 0 \text{ at } S^+\} \subset L_2(\overline{D}) \tag{2.5.2}$$

is positive semidefinite ($\overline{D} = D \cup S$, and $L_2(\overline{D})$ is the Hilbert space of square integrable functions on $\overline{D}$). This result is important for proving the uniqueness and stability of the solution of problem (2.4.1)–(2.4.6).

Defining the inner product in $L_2(\overline{D})$ by

$$(u, v) = \int\limits_D u(r)v(r)dr$$

we obtain

$$(A\phi, \phi) = \int\limits_D \phi U \cdot \nabla\phi\, dr + \int\limits_D \sigma\phi^2 dr - \int\limits_D \phi\nabla \cdot \mu\nabla\phi\, dr \tag{2.5.3}$$

Since $\nabla \cdot U = 0$, we have $\nabla \cdot (\phi^2 U) = 2\phi U \cdot \nabla\phi$. If this result is integrated over D, then applying the divergence theorem leads to

$$\int\limits_D \phi U \cdot \nabla\phi\, dr = \frac{1}{2}\int\limits_S \phi^2 U \cdot n\, dS = \frac{1}{2}\left[\int\limits_{S^+} \phi^2 U_n dS + \int\limits_{S^-} \phi^2 U_n dS\right] \tag{2.5.4}$$

Further, integrating the relation $\nabla \cdot (\phi\mu\nabla\phi) = \phi\nabla \cdot (\mu\nabla\phi) + \mu|\nabla\phi|^2$ over D and using the divergence theorem, we obtain

$$-\int_D \phi\nabla \cdot (\mu\nabla\phi)\, dr = \int_D \mu|\nabla\phi|^2 dr - \int_S \phi\mu\nabla\phi \cdot n\, dS \tag{2.5.5}$$

Representing the integral over S as the sum of two integrals over S^+ and S^-, and applying the boundary conditions (2.4.5) and (2.4.6) we obtain

$$-\int_D \phi\nabla \cdot (\mu\nabla\phi) dr = \int_D \mu|\nabla\phi|^2 dr - \int_{S^-} \phi^2 U_n dS \tag{2.5.6}$$

Using formulas (2.5.4) and (2.5.6) in (2.5.3) leads to

$$(A\phi,\, \phi) = \int_D [\sigma\phi^2 + \mu\, |\nabla\phi\, |^2]\, dr + \frac{1}{2}\int_S \phi^2 |U_n| dS > 0 \tag{2.5.7}$$

Thus, the operator A of problem (2.4.1)–(2.4.6) is positive definite. It is positive semidefinite if $\sigma = 0$ and $U_n = 0$ at the boundary S, and it is skew-symmetric if $\sigma = 0,\ \mu = 0$ and $U_n = 0$ at the boundary S.

2.6 Uniqueness and Stability of the Classical Solution

We now demonstrate the uniqueness and stability of the classical solution of the diffusion problem (2.4.1)–(2.4.6). In operator form, the problem can be written as

$$\frac{\partial\phi}{\partial t} + A\phi = f \tag{2.6.1}$$

$$\phi(0) = \phi^0 \tag{2.6.2}$$

where $A\ :\ \Phi \subset H \to H$ is a linear positive semidefinite differential operator with a domain Φ belonging to a Hilbert space H, i.e., the inner product $(A\phi, \phi)$ is nonnegative for any nonzero function ϕ. Indeed, taking the inner product of Eq. (2.6.1) with ϕ we get

$$\left(\frac{\partial\phi}{\partial t}, \phi\right) = (f, \phi) - (A\phi, \phi) \tag{2.6.3}$$

Using the Schwarz inequality [13] and the fact that $(A\phi, \phi) \geq 0$, we obtain

$$\left(\frac{\partial \phi}{\partial t}, \phi\right) \leq \|f\| \|\phi\| \quad \text{where} \quad \|\phi\| = \sqrt{(\phi, \phi)} \tag{2.6.4}$$

Since

$$\left(\frac{\partial \phi}{\partial t}, \phi\right) = \frac{1}{2} \frac{\partial}{\partial t} \|\phi\|^2 = \|\phi\| \frac{\partial}{\partial t} \|\phi\| \tag{2.6.5}$$

(2.6.4) leads to

$$\frac{\partial}{\partial t} \|\phi\| \leq \|f\| \tag{2.6.6}$$

Integrating (2.6.6) with respect to time from $t = 0$ to a moment t, and using the initial condition we obtain

$$\|\phi\| \leq \int_0^t \|f\| \, d\tau + \left\|\phi^0\right\| \tag{2.6.7}$$

Assuming that the norm of f is limited in the time interval $(0, T)$, one can write

$$\|\phi\| \leq T \cdot \max_{0 \leq t \leq T} \|f(r, t)\| + \left\|\phi^0\right\| \tag{2.6.8}$$

Since problem (2.6.1) is linear, the last inequality means the stability of solution with respect to perturbations in the initial distribution ϕ^0 and forcing f.

The same reasoning guarantees the uniqueness of solution to problem (2.4.1)–(2.4.6). Really, if there exist two solutions of problem (2.6.1)–(2.6.2) with the same functions ϕ^0 and f, then the difference $\delta\phi$ is a new solution that satisfies the conditions $\delta\phi^0 = 0$ and $\delta f = 0$. Therefore, $\delta\phi \equiv 0$ due to (2.6.8), and the two solutions coincide to each other.

It should be noted that the inequality $(A\phi, \phi) \geq 0$ is a sufficient condition for the stability of the solution. Also, this inequality is a necessary condition for stability. Indeed, if this condition is violated, then the operator A has at least one negative eigenvalue, and any perturbation that has a nonzero projection on the corresponding eigenfunction will grow with time.

It is possible to discretize the problem so that the matrix of the discrete problem is also positive semidefinite. Then the discrete problem also has a unique stable solution. In addition, this property makes it possible to apply the component-wise splitting methods to develop an efficient numerical algorithm for solving two-dimensional (and also three-dimensional) dispersion problems.

2.7 Weak Formulation of the Dispersion Problem

Effective strategies for monitoring and controlling pollutant concentrations cannot be developed without using a suitable dispersion model. In particular, the model must be mathematically well-posed according to Hadamard. Our goal now is to prove that a weak solution to problem (2.4.1)–(2.4.6) exists, is unique, and depends continuously on the forcing and initial conditions. Note that even in case of a simple dispersion equation, it is not trivial to formulate a well-posed problem in a limited area. In fact, the flow of pollution through the open boundary is typically unknown, and therefore any errors in the boundary conditions will spread inside the region by means of advection and diffusion. This process may essentially modify or even destroy the solution. Errors in the emissions rates and initial distribution of pollutant in the region may also distort the solution. Therefore, it is very important to set such boundary conditions that are appropriate, both physically and mathematically. Conditions (2.4.5) and (2.4.6) correspond to these requirements (see Theorem 2.8.1).

1. **Functional spaces.** We now introduce some functional spaces, and an embedding theorem to be used in the following sections. The solutions are defined in a three-dimensional space–time domain $Q = D \times (0, T)$. Denote by $L_2(Q)$ the Hilbert space of real functions in Q with the inner product

$$\langle f, g\rangle = \int_Q f(r,t)\, g(r,t)\, dr\, dt \tag{2.7.1}$$

and norm

$$\|g\| = \langle g, g\rangle^{1/2} \tag{2.7.2}$$

Let M be a set of twice continuously differentiable functions $\phi(r,t)$ in a closed domain $\overline{Q} = \overline{D} \times [0, T]$ which satisfy the zero initial condition (2.4.4): $\phi(r, 0) = 0$, and have a finite norm

$$\|\phi\|_H = \left(\|\phi_t\|^2 + \|\phi_x\|^2 + \left\|\phi_y\right\|^2 + \|\phi\|_S^2\right)^{1/2} \tag{2.7.3}$$

Hereinafter ϕ_t, ϕ_x and ϕ_y denote the partial derivatives of function ϕ with respect to t, x and y, respectively, and

$$\|\phi\|_S = \left\{\int_0^T \int_{S^+} |\phi|^2 dS dt\right\}^{1/2} \tag{2.7.4}$$

is the norm in the space $L_2\left(S^+\right)$ of the trace of function $\phi(r,t)$ at the boundary S^+. We introduce in M the inner product

$$\langle \varphi, \phi \rangle_H = \langle \varphi_t, \phi_t \rangle + \langle \varphi_x, \phi_x \rangle + \langle \varphi_y, \phi_y \rangle + \langle \varphi, \phi \rangle_S \tag{2.7.5}$$

where

$$\langle \varphi, \phi \rangle_S = \int_0^T \int_{S^+} \varphi \phi dS dt \tag{2.7.6}$$

and denote by $H(Q)$ the Hilbert space obtained by closing the set M in the norm (2.7.3)

Lemma 2.7.1 ([14]) *In the space $H(Q)$, the norm $\|g\|_H$ is equivalent to the norm $\left(\|g\|^2 + \|g\|_H^2\right)^{1/2}$.*

Proof To prove the equivalence of two norms [15], it is sufficient to demonstrate that

$$\left(\|g\|^2 + \|g\|_H^2\right)^{1/2} \le C \, \|g\|_H \tag{2.7.7}$$

Let $g \in M$. Since $g(r, 0) = 0$, we have

$$g(r, t) = \int_0^t \frac{\partial g}{\partial \tau}(r, \tau) d\tau$$

and by (2.7.2),

$$\|g\|^2 = \int_Q \left(\int_0^t \frac{\partial g}{\partial \tau} d\tau \right)^2 dr \, dt \le T^2 \|g_t\|^2 \le T^2 \|g\|_H^2$$

Therefore, (2.7.7) is satisfied for $C = \sqrt{1 + T^2}$. The lemma is proved.

Let us define a set V of all functions

$$g(r, t) = \int_0^t \psi(r, \tau) \, d\tau \tag{2.7.8}$$

where $\psi(r, t)$ belongs to $H(Q)$. The set V is dense in $H(Q)$, since M is dense in $H(Q)$, and the closure $\overline{V}$ of set V in the norm (2.7.5) contains M.

2. **Weak formulation of the pollution transport problem**. A week formulation of the problem expands the domain in which the solution of the problem is sought. This expansion normally simplifies the proof of the existence of a solution, but, to some extent, complicates the proof of its uniqueness. Note that if a weak solution is sufficiently smooth, then it is a classical solution of problem. Now we give

a weak statement of problem (2.4.1)–(2.4.6), and in the next two sections we prove the existence and uniqueness of its solution and its continuous dependence on the initial condition and forcing (see Theorem 2.8.1). The conditions of the theorem impose restrictions on the magnitude and smoothness of the coefficients of the model. These limitations must be considered when parameterizing various physicochemical processes in the model.

Definition 2.7.1 A function $\phi(r, t)$ of space $H(Q)$ is said to be a week solution of dispersion problem (2.4.1)–(2.4.6) if it satisfies the identity.

$$\begin{aligned}&\left\langle\phi_t, g_t e^{-\delta t}\right\rangle+\left\langle\mu \phi_x, g_{tx} e^{-\delta t}\right\rangle+\left\langle\mu \phi_y, g_{ty} e^{-\delta t}\right\rangle+\left\langle\sigma \phi, g_t e^{-\delta t}\right\rangle \\ &+\int_0^T \int_{S+} U_n \phi g_t e^{-\delta t} dS dt+\left\langle u \phi_x, g_t e^{-\delta t}\right\rangle+\left\langle v \phi_y, g_t e^{-\delta t}\right\rangle=\left\langle f, g_t e^{-\delta t}\right\rangle\end{aligned} \tag{2.7.9}$$

for any function $g(r, t)$ of the set V and some $\delta > 0$.

Since $\phi(r, t)$ belongs to $H(Q)$, and $g(r, t)$ belongs to V, all terms of the identity (2.7.9) make sense. Formally, this identity can be derived if we multiply Eq. (2.4.1) by $g_t\, e^{-\delta t}$ and integrate the result by parts with boundary conditions (2.4.5) and (2.4.6). The solution defined in such a manner is a generalization of the classical solution of problem (2.4.1)–(2.4.6).

Note that if $g_t e^{-\delta t} \equiv 1$ (or $g_t e^{-\delta t} = 2\phi$) then identity (2.7.9) coincides with the balance Eq. (2.4.7) (Eq. (2.4.8)) pre-integrated over $(0, T)$. Also note that in (2.7.9), the function $e^{-\delta t}$ is nonzero in the whole interval $(0, T)$, and the product $g_t\, e^{-\delta t}$ can be replaced by a function $z_t(r, t)$. However, the form $g_t\, e^{-\delta t}$ is a common practice for various estimates [14].

2.8 Existence of a Weak Solution

Theorem 2.8.1 given below states the well-posedness of dispersion problem (2.4.1)–(2.4.6). We now prove the existence of a weak solution, and its uniqueness and continuous dependence on the forcing and initial condition will be demonstrated in Sect. 2.9. For simplicity, let us assume that the velocity $U(r)$ and diffusion coefficient $\mu(r)$ are time-independent. First, we prove the theorem for the case when the initial condition (2.4.4) is equal to zero, and then generalize the result to a nonzero initial condition.

Theorem 2.8.1 *Suppose that velocity $U(r)$ and diffusion coefficient $\mu(r)$ are time-independent, $U(r)$ is non-divergent, emission rates $Q_i(t) \in L_2(0, T)$ for all i $(i = 1, 2, \ldots, N)$, $\phi^0(r) \equiv 0$ and*

$$\max_{\overline{Q}}\{\mu,\ \sigma,\ |\sigma_t|,\ |U|\} = \beta < \infty, \quad \min_{\overline{Q}}\{\mu,\ \sigma\} = \alpha > 0$$

Then problem (2.4.1)–(2.4.6) has a unique week solution $\phi(r, t)$ in space $H(Q)$, which depends continuously on the forcing and initial condition.

Proof We now use the functional method to prove the existence of a weak solution. Denote by $a_i(\phi, g)$ the i-th term in the left-hand side of identity (2.7.9), $i = 1, 2, \ldots, 7$, and by $a_8(f, g)$ the term in its right-hand side. For a given $g(r, t)$ of set V, each left-hand term $a_i(\phi, g)$ is a linear bounded functional (of ϕ) on $H(Q)$, $i = 1, 2, \ldots, 7$. For example,

$$\left|a_1(\phi, g)\right| \equiv \left|\left\langle \phi_t, g_t e^{-\delta t}\right\rangle\right| \le \left\|g_t e^{-\delta t}\right\| \cdot \|\phi_t\| \le C_1(g) \cdot \|\phi\|_H \tag{2.8.1}$$

$$\begin{aligned}\left|a_2(\phi, g)\right| &\equiv \left|\left\langle \mu\phi_x, g_{tx} e^{-\delta t}\right\rangle\right| \le \beta\left\|g_{tx} e^{-\delta t}\right\| \cdot \|\phi_x\| = \beta\left\|\psi_x e^{-\delta t}\right\| \cdot \|\phi_x\| \\ &\le \beta\|\psi_x\| \cdot \|\phi_x\| \le \beta\|\psi\|_H \cdot \|\phi\|_H \le C_2(g) \cdot \|\phi\|_H\end{aligned}$$

and

$$|a_5(\phi, g)| \equiv \left|\int_0^T \int_{S^+} U_n \phi g_t e^{-\delta t} dS dt\right| \le \beta\|\phi\|_S\ \|\psi\|_S \le \beta\|\psi\|_H\ \|\phi\|_H \le C_5(g)\|\phi\|_H$$

where $\psi = g_t \in H(Q)$ because of (2.7.8). The boundedness of forms $a_i(\phi, g)$ for $i = 3, 4, 6, 7$ may be shown in a similar manner. Then, by Riesz representation theorem [15], the ith functional is represented as the inner product

$$a_i(\phi, g) = \langle \phi, g_i \rangle_H$$

with some uniquely determined element g_i of $H(Q)$ $(i = 1, 2, \ldots, 7)$. The one-to-one correspondence $g \to g_i$ defines a linear operator A_i acting from V to $H(Q)$: $g_i = A_i g$ $(i = 1, 2, \ldots, 7)$.

Further, according to (2.4.2), the conditions $Q_i(t) \in L_2(0, T)$ $(i = 1, 2, \ldots, N)$ imply $f(r, t) \in L_2(Q)$. Therefore, for fixed functions $Q_i(t) \in L_2(0, T)$ (i.e., for a fixed forcing f), the term $a_8(f, g)$ is also a linear bounded functional (of ϕ) on $H(Q)$:

$$\left|a_8(f, g)\right| \equiv \left|\left\langle f, g_t e^{-\delta t}\right\rangle\right| \le \|f\|\ \|g_t\|\ \le C_8(f)\ \|g\|_H$$

and by Riesz representation theorem, it may be represented as $a_8(f, g) = \langle R, g \rangle_H$ with some uniquely defined element R of $H(Q)$. Thus, the operator

$$A = \sum_{i=1}^{7} A_i \tag{2.8.2}$$

acts from V to $H(Q)$. Then identity (2.7.9) can be written as

$$\langle \phi, Ag \rangle_H = \langle R, g \rangle_H \tag{2.8.3}$$

To complete the proof of the solution existence, we will need the following statement [14].

Lemma 2.8.1 *Let $G \subseteq H(Q)$ be the image of operator A, that is, A acts from V to G. Then the inverse operator A^{-1} defined on the domain G exists and is bounded.*

Proof We prove the lemma if we show that

$$\langle g, Ag\rangle_H \geq \frac{1}{2}e^{-\delta T}\langle g, g\rangle_H \tag{2.8.4}$$

for each $g \in V$ and some sufficiently large $\delta > 0$. Indeed, by (2.8.4), $Ag = 0$ implies $g = 0$, that is, inverse operator A^{-1} exists and is bounded, since

$$\left\|A^{-1}\right\|_H = \underset{w\in G}{Sup}\frac{\left\|A^{-1}w\right\|_H}{\|w\|_H} = \underset{g\in V}{Sup}\frac{\|g\|_H^2}{\|g\|_H\,\|Ag\|_H} \leq \underset{g\in V}{Sup}\frac{\langle g, g\rangle_H}{\langle g, Ag\rangle_H} \leq 2e^{\delta T}$$

where $w = Ag$. Let us now prove inequality (2.8.4). Let $g \in V$, which means $g(r, 0) = 0$. Then

$$a_1(g, g) = \left\langle g_t, g_t e^{-\delta t}\right\rangle \geq \left\|g_t e^{-\frac{\delta}{2}t}\right\|^2$$

$$a_2(g, g) = \left\langle \mu g_x, g_{tx} e^{-\delta t}\right\rangle = \tfrac{1}{2}e^{-\delta T}\int_D \mu g_x^2(r, T)dr$$
$$+\tfrac{1}{2}\delta\int\limits_Q \mu\, g_x^2\, e^{-\delta t}drdt \geq \tfrac{1}{2}\delta\alpha\left\|g_x e^{-\frac{\delta}{2}t}\right\|^2,$$

$$a_3(g, g) = \left\langle \mu g_y, g_{ty} e^{-\delta t}\right\rangle \geq \frac{1}{2}\delta\alpha\left\|g_y e^{-\frac{\delta}{2}t}\right\|^2$$

$$a_4(g, g) = \langle \sigma g, g_t e^{-\delta t}\rangle = \tfrac{1}{2}e^{-\delta T}\int\limits_D \sigma\, g^2(r, T)dr$$
$$+\tfrac{1}{2}\delta\int\limits_Q \sigma g^2 e^{-\delta t}drdt - \tfrac{1}{2}\int\limits_Q \sigma_t g^2 e^{-\delta t}drdt \geq \tfrac{1}{2}(\alpha\delta - \beta)\left\|g e^{-\frac{\delta}{2}t}\right\|^2,$$

$$a_5(g, g) \equiv \int\limits_0^T\int\limits_{S+} U_n g g_t e^{-\delta t}dSdt = \tfrac{1}{2}e^{-\delta T}\int\limits_{S+} U_n\, g^2(r, T)dS$$
$$+\tfrac{1}{2}\delta\int\limits_0^T\int\limits_{S+} U_n\left(g e^{-\frac{\delta}{2}t}\right)^2 dS \geq \tfrac{1}{2}\delta\alpha_0\left\|g e^{-\frac{\delta}{2}t}\right\|_S^2 \tag{2.8.5}$$

In the last estimate, we used the mean value theorem

$$\int\limits_0^T\int\limits_{S+} U_n\left(g e^{-\frac{\delta}{2}t}\right)^2 dS = \alpha_0\left|g e^{-\frac{\delta}{2}t}\right|_S^2$$

where $\alpha_0 > 0$, since $U_n \geq 0$ on S^+. Note that if $U_n = 0$ everywhere on S^+, then the fifth term in the l.h.s. of identity (2.7.9) is absent, and hence, the term $a_5(g, g)$ is also absent.

Now we estimate the terms $a_6(g, g)$ and $a_7(g, g)$ by using ε -inequality:

$$a_6(g, g) \equiv \left\langle ug_x, g_t e^{-\delta t}\right\rangle \leq \varepsilon \left\| g_t e^{-\frac{\delta}{2}t} \right\|^2 + \frac{\beta}{4\varepsilon} \left\| g_x e^{-\frac{\delta}{2}t} \right\|^2$$

$$a_7(g, g) \equiv \left\langle vg_y, g_t e^{-\delta t}\right\rangle \leq \varepsilon \left\| g_t e^{-\frac{\delta}{2}t} \right\|^2 + \frac{\beta}{4\varepsilon} \left\| g_y e^{-\frac{\delta}{2}t} \right\|^2$$

Let us choose ε so that $1-2\varepsilon \geq 1/2$, and take δ large enough so that $\delta\alpha - \beta/(2\varepsilon) \geq 1$ and $\delta\alpha_0 \geq 1$. Then we obtain

$$\begin{aligned}
\langle g, Ag\rangle_H &= \sum_{i=1}^{7} a_i(g, g) \geq \sum_{i=1}^{5} a_i(g, g) - \sum_{i=6}^{7} |a_i(g, g)| \\
&\geq (1 - 2\varepsilon) \left\| g_t e^{-\frac{\delta}{2}t} \right\|^2 + \tfrac{1}{2}\left(\delta\alpha - \beta(2\varepsilon)^{-1}\right) \left\{ \left\| g_x e^{-\frac{\delta}{2}t} \right\|^2 + \left\| g_y e^{-\frac{\delta}{2}t} \right\|^2 \right\} \\
&+ \tfrac{1}{2}\delta\alpha_0 \left\| g e^{-\frac{\delta}{2}t} \right\|_S^2 + \tfrac{1}{2}(\delta\alpha - \beta) \left\| g e^{-\frac{\delta}{2}t} \right\|^2 \geq \tfrac{1}{2} e^{-\delta T} \langle g, g\rangle_H
\end{aligned}$$

To derive the last inequalities, we used the norm (2.7.3) and Lemma 2.7.1. The Lemma 2.8.1 is proved.

We now proceed with the proof of Theorem 2.8.1. Let $G \subseteq H(Q)$ be a final set of the operator A. According to Lemma 2.8.1, the inverse operator A^{-1} defined on G exists and is bounded. By continuity, operator A^{-1} is defined on the closure $\overline{G}$ of the set G (in the norm (2.7.3)). Let us denote by L the orthogonal complement of G in $H(Q)$. We extend operator A^{-1} from $\overline{G}$ to $H(Q)$ by defining $A^{-1}w = 0$ for any $w \in L$. As a result, we obtain a bounded operator $\tilde{A}^{-1}$ defined on the whole space $H(Q)$: $\left\|\tilde{A}^{-1}\right\|_H \leq 2e^{\delta T}$. The adjoint $\left(\tilde{A}^{-1}\right)^*$ to $\tilde{A}^{-1}$ is uniquely defined and has the same norm as $\tilde{A}^{-1}$. Let $Ag = w$ where $w \in G$. Then, due to (2.8.3),

$$\langle \phi, w\rangle_H = \langle \phi, Ag\rangle_H = \langle R, g\rangle_H = \langle R, A^{-1}w\rangle_H \equiv \langle R, \tilde{A}^{-1}w\rangle_H \tag{2.8.6}$$

Let us look for a solution ϕ from a more restrictive than (2.8.3) condition, namely,

$$\langle \phi, w\rangle_H = \left\langle R, (\tilde{A}^{-1})w\right\rangle_H = \left\langle (\tilde{A}^{-1})^* R, w\right\rangle_H$$

for any $w \in H(Q)$. This is possible if and only if $\phi = \left(\tilde{A}^{-1}\right)^* R$. This solution also satisfies (2.8.3) and (2.8.6), and therefore original identity (2.7.9) is valid too. Thus, the existence of a weak solution is proved.

2.9 Uniqueness and Stability of a Weak Solution

We now proceed with the proof of Theorem 2.8.1. Let $\phi(r, t) \in H(Q)$ be a week solution of problem (2.4.1)–(2.4.6). Then

$$g = \int_0^t \phi(r, \tau) e^{\delta\tau} d\tau \tag{2.9.1}$$

belongs to set V. Substituting (2.9.1) into (2.7.9), we obtain after simple transformations that the weak solution satisfies the inequality

$$\frac{1}{2} \|\phi(r, T)\|^2_{L_2(D)} + \alpha\left(\|\phi_x\|^2 + \left\|\phi_y\right\|^2\right) \leq \frac{1}{2} \|\phi(r, 0)\|^2_{L_2(D)} + \langle f, \phi\rangle \tag{2.9.2}$$

where

$$\|\phi(r, t)\|^2_{L_2(D)} = \int_D \phi^2(r, t) dr$$

Inequality (2.9.2) and the initial condition $\phi(r, 0) = \phi^0(r)$ lead to

$$\begin{gathered} \tfrac{1}{2} \|\phi(r, T)\|^2_{L_2(D)} + \alpha\left(\|\phi_x\|^2 + \left\|\phi_y\right\|^2\right) \leq \\ \max_{0 \leq \tau \leq T} \|\phi(r, \tau)\|_{L_2(D)} \left(\sqrt{T}\|f\| + \tfrac{1}{2}\left\|\phi^0(r)\right\|_{L_2(D)}\right) \leq \\ |\phi|_Q \left(\sqrt{T}\|f\| + \tfrac{1}{2}\left\|\phi^0(r)\right\|_{L_2(D)}\right) \end{gathered} \tag{2.9.3}$$

where

$$|\phi|_Q \equiv \max_{0 \leq \tau \leq T} \|\phi(r, \tau)\|_{L_2(D)} + \|\phi_x\| + \left\|\phi_y\right\| \tag{2.9.4}$$

is the energetic norm of a Banach space [16]. Since each term on the left-hand side of (2.9.3) is not larger than its right-hand side, it is easy to get the estimate

$$|\phi|_Q \leq \left(\frac{2}{\sqrt{\alpha}} + \sqrt{2}\right)^2 \left(\sqrt{T}\|f\| + \frac{1}{2}\left\|\phi^0(r)\right\|_{L_2(D)}\right) \tag{2.9.5}$$

that relates the solution norm (2.9.4) to the norms of the forcing and initial condition. Note that $\phi^0(r) \equiv 0$, and therefore the last term in (2.9.5) is zero. Since the problem (2.4.1)–(2.4.6) is linear, estimate (2.9.5) implies the uniqueness of weak solution as well as its stability to variations $\delta f(r, t)$ and $\delta\phi(r, 0)$ in the forcing and initial condition, respectively:

$$|\delta\phi|_Q \le \left(\frac{2}{\sqrt{\alpha}} + \sqrt{2}\right)^2 \left(\sqrt{T}\|\delta f\| + \frac{1}{2}\left\|\delta\phi^0(r)\right\|_{L_2(D)}\right) \tag{2.9.6}$$

Theorem 2.8.1 is proved.

1. **The case of nonzero initial condition (2.4.4).** We now generalize Theorem 2.8.1 to a nonzero initial condition (2.4.4): $\phi^0(r) \neq 0$. Note that conditions (2.4.4)–(2.4.6) lead to the compatibility relations

$$\mu\frac{\partial}{\partial n}\phi^0 - U_n\phi^0 = 0 \text{ at } S^-,\ \mu\frac{\partial}{\partial n}\phi^0 = 0 \text{ at } S^+, \tag{2.9.7}$$

at the initial moment $t = 0$, and therefore problem (2.4.1)–(2.4.6) for $\phi^0(r) \neq 0$ can easily be reduced to a new problem

$$\begin{aligned} \tfrac{\partial}{\partial t}\varphi + U\cdot\nabla\varphi + \sigma\varphi - \nabla\cdot\mu\nabla\varphi &= F(r,t)\,, \\ \varphi(r,0) &= 0\,, \\ \mu\tfrac{\partial\varphi}{\partial n} - U_n\varphi = 0 \quad \text{at } S^-\,, \quad \mu\tfrac{\partial\varphi}{\partial n} &= 0 \quad \text{at } S^+ \end{aligned} \tag{2.9.8}$$

with zero initial condition, where $\varphi(r,t) = \phi(r,t) - \phi^0(r)$ and

$$F(r,t) = f(r,t) - U\cdot\nabla\phi^0 - \sigma\phi^0 + \nabla\cdot\mu\nabla\phi^0 \tag{2.9.9}$$

Therefore, if $\phi^0(r)$ is a sufficiently smooth function such that $F(r,t)$ belongs to $L_2(Q)$ then Theorem 2.8.1 is valid for problem (2.9.8), that is, there exists a unique solution $\varphi(r,t)$ that continuously depends on the initial condition and forcing. Then it follows from the formula $\phi(r,t) = \varphi(r,t) + \phi^0(r)$ that a solution $\phi(r,t)$ to problem (2.4.1)–(2.4.6) with a nonzero initial value $\phi^0(r)$ also exists, is unique and continuously depends on the initial condition and forcing.

Remark 2.9.1 It is easy to show that the same method can be used to prove Theorem 2.8.1, when both velocity $U(r,t)$ and diffusion coefficient $\mu(r,t)$ are time-dependent. However, another method is necessary to apply in this case to generalize the theorem to a nonzero initial condition.

2.10 A 3D Dispersion Problem in a Limited Area

So far we have considered the two-dimensional dispersion problem. However, more precise results require using of a three-dimensional model. One of the first analytical solutions of the 3D dispersion model describing the transport and diffusion of a passive contaminant was obtained by Berlyand [17] with the following simplifications: the motion is stationary, the wind is rectilinear with constant velocity u, the x

-axis is oriented in the direction of the wind, the vertical motions in the atmosphere are small compared to horizontal movements, the diffusion flux of pollutant in the x -direction is much less than the convective flux. Then the problem accepts the form

$$u\frac{\partial \phi}{\partial x} = \mu_y \frac{\partial^2 \phi}{\partial x^2} + \mu_z \frac{\partial^2 \phi}{\partial z^2}$$

$$u\phi = Q\,\delta(y)\,\delta(z - H) \quad \text{on the surface } x = 0$$

$$\phi = 0 \quad \text{in the limit} \quad z \to \infty, \quad |y| \to \infty$$

$$\mu_z \frac{d\phi}{dz} = 0 \quad \text{on the surface} \quad \mathrm{z} = 0$$

where Q is the intensity of the source located at point $(0, 0, H)$, $\mu_y = C\,u, C = const$, $\mu_z = const$. The Berlyand solution is

$$\phi = \frac{Q}{4\pi x\sqrt{\mu_y \mu_z}} \exp\left(-\frac{uy^2}{4\mu_y x}\right) \times \left\{\exp\left(-\frac{u(z+H)^2}{4\mu_z x}\right) + \exp\left(-\frac{u(z-H)^2}{4\mu_z x}\right)\right\}$$

1. **A 3D dispersion model.** We now consider a more general, non-stationary three-dimensional problem. Let D be a "cylindrical" domain with boundary $\partial D = S \cup S_0 \cup S_H$, where S is the lateral surface of D and S_0 and S_H are the base ($z = 0$) and top cover ($z = H$) of D.

Let us denote by $\phi(r, t)$ the concentration of a pollutant at point $r = (x, y, z)$ and moment $t \geq 0$. The evolution of pollutant concentration in the domain D and time interval $(0, T)$ is governed by the advection–diffusion equation [10]

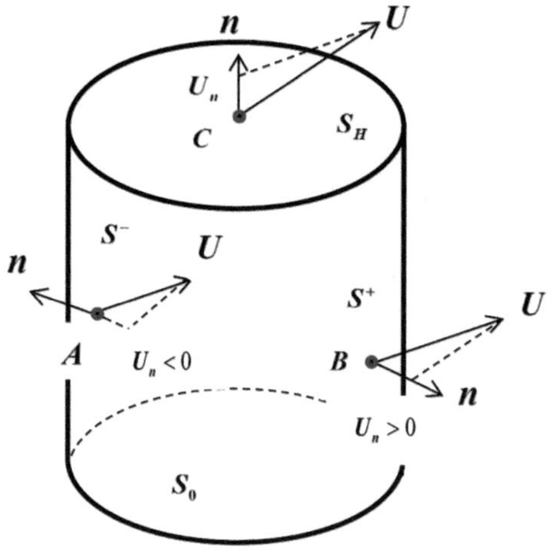

Fig. 2.7 A limited domain D. The boundary points A, B and C belong to surfaces S^-, S^+ and S_H^+, respectively;$U_n = U \cdot n$ is the normal component of velocity vector to boundary surface S

$$\frac{\partial}{\partial t}\phi + U \cdot \nabla\phi + \sigma\phi = \nabla \cdot \mu\, \nabla\phi + \frac{\partial}{\partial z}\nu\frac{\partial \phi}{\partial z} + f(r,t) \tag{2.10.1}$$

where $\mu(r,t) > 0$ and $\nu(r,t) > 0$ are the diffusion coefficients, ∇ is the two-dimensional gradient in (x, y) -direction, $\sigma(r,t) > 0$ is the parameter that characterizes the exponential decay rate of $\phi(r,t)$ because of different physical and chemical processes, and $f(r,t)$ describes the pollutant sources. Assume that flow velocity $U(r,t) = \{u(r,t), v(r,t), w(r,t)\}$ is known in D and satisfies the continuity equation

$$\nabla \cdot U = \frac{\partial u}{\partial x} + \frac{\partial v}{\partial y} + \frac{\partial w}{\partial z} = 0 \tag{2.10.2}$$

Denote by $U_n = U \cdot n$ the projection of velocity vector on the outward normal n to the boundary ∂D. Note that $U_n = 0$ (or $w = 0$) on S_0. Given the horizontal components $u(r,t)$ and $v(r,t)$ of velocity U in D, its vertical component is determined from (2.10.2) as

$$w(x,y,z,t) = -\int_0^z \left(\frac{\partial u}{\partial x} + \frac{\partial v}{\partial y}\right) dz \tag{2.10.3}$$

Therefore, $w(x, y, H, t)$ is generally non-zero on the top surface S_H, and there is a pollution flux through S_H. We divide the lateral surface S into two parts again: the part S^+ where $U_n \geq 0$ and the pollutant is removed from domain D, and the part S^- where $U_n < 0$ and the pollution is directed inside the domain D. Analogously, S_H^+ and S_H^- denote the parts of S_H where $U_n = w \geq 0$ and $U_n = w < 0$, respectively (Fig. 2.7). As the initial and boundary conditions for Eq. (2.10.1) we take

$$\phi(r,0) = \phi^0(r) \tag{2.10.4}$$

$$\mu\frac{\partial \phi}{\partial n} - U_n\phi = 0 \quad \text{on } S^- \tag{2.10.5}$$

$$\mu\frac{\partial \phi}{\partial n} = 0 \quad \text{on } S^+ \tag{2.10.6}$$

$$\nu\frac{\partial \phi}{\partial z} = \alpha\phi \quad \text{on } S_0 \tag{2.10.7}$$

$$\nu\frac{\partial \phi}{\partial z} - U_n\phi = 0 \quad \text{on } S_H^- \tag{2.10.8}$$

$$\nu\frac{\partial \phi}{\partial z} = 0 \quad \text{on } S_H^+ \tag{2.10.9}$$

where $\alpha \geq 0$ is a known function that characterizes the interaction of pollutant with the underlying surface (the process of sedimentation).

2. **Balance equations**. The integration of Eq. (2.10.1) over domain D leads to the balance equation for the total mass of pollutant:

$$\frac{\partial}{\partial t}\int_D \phi dr = \int_D f dr - \int_D \sigma\phi dr - \int_{S_0} \alpha\phi dS - \int_{S^+ \cup S_H^+} U_n \phi dS \qquad (2.10.10)$$

The operator

$$A\phi = \operatorname{div}(U\phi) + \sigma\phi - \nabla\cdot\mu\nabla\phi - \frac{\partial}{\partial z}\nu\frac{\partial\phi}{\partial z} \qquad (2.10.11)$$

of problem (2.10.1)–(2.10.9) is positive semidefinite:

$$(A\phi, \phi) = \int\limits_D \phi A\phi \, dr \geq 0 \quad \text{for any } \phi \neq 0$$

Indeed,

$$(A\phi, \phi) = \int\limits_D \left\{\sigma\phi^2 + \nu\left(\frac{\partial\phi}{\partial z}\right)^2 + \mu|\nabla\phi|^2\right\} dr + \frac{1}{2}\int\limits_{S\cup S_H} |U_n|\phi^2 dS + \int\limits_{S_0} \alpha\,\phi^2 dS \qquad (2.10.12)$$

Therefore, if we multiply Eq. (2.10.1) by $\phi(r, t)$ and integrate the result over domain D we obtain

$$\frac{1}{2}\frac{\partial}{\partial t}\int_D \phi^2 dr = \int_D f\phi dr - (A\phi, \phi) \qquad (2.10.13)$$

This equation determines the norm $\|\phi\| = \left(\int_D \phi^2 dr\right)^{1/2}$ of solution. According to (2.10.10) and (2.10.13), the total mass of pollutant $\int_D \phi \, dr$ and the solution norm $\|\phi\|$ increase due to nonzero emission rates from the sources ($f(r, t) \neq 0$) and decrease due to dissipation ($\sigma > 0$, $\mu > 0$, $\nu > 0$) and sedimentation ($\alpha > 0$) as well as due to the advective flow of pollutant through the domain boundary $S \cup S_H$. Thus, in the absence of pollution sources in domain D ($f(r, t) = 0$), both $\int_D \phi \, dr$ and $\|\phi\|$ decrease with time.

And if we consider an ideal case when in addition to $f(r, t) = 0$, there are no dissipation ($\sigma = 0$, $\mu = 0$, $\nu = 0$) and sedimentation ($\alpha = 0$), and also $U_n = 0$ on the boundary $S \cup S_H$ then both integrals are conserved in time:

$$\frac{d}{dt}\int_D \phi\,dr = 0\,,\quad \frac{d}{dt}\int_D \phi^2 dr = 0 \tag{2.10.14}$$

Although the relations (2.10.14) are valid only under artificial conditions, they are very useful for testing numerical algorithms and computer programs [10]. Note also that due to (2.10.10), the total mass $\int_D \phi\,dr$ is conserved even if $\mu > 0$ and $\nu > 0$, but $f(r,t) = 0$, $\sigma = 0$, $\alpha = 0$ and $U_n = 0$ everywhere on the boundary $S^+ \cup S_H^+$. Then, due to dissipation, the solution $\phi(r,t)$ tends to a constant value $\phi_0 \equiv \phi(r,\infty)$ in the domain D as $t \to \infty$. The constant ϕ_0 is easily be found from the equation of mass conservation:

$$\int_D \phi^0(r)\,dr = \int_D \phi_0 dr = \phi_0\,\mathrm{vol}(D)$$

where $\mathrm{vol}(D)$ is the volume of D. Thus,

$$\phi_0 = \mathrm{vol}(D)^{-1}\int_D \phi^0(r)\,dr \tag{2.10.15}$$

References

1. Godunov, S.K.: Equations of Mathematical Physics. Nauka, Moscow (1971)
2. Morton, K., Mayers, D.: Numerical Solution of Partial Differential Equations. Cambridge University Press, Cambridge (1994)
3. Skiba, Y.N.: Métodos y Esquemas Numéricos: Un Análisis Computacional. Dirección General de Publicaciones y Fomento Editorial, UNAM, México (2005)
4. Marchuk, G.I.: Mathematical Models in Environmental Problems. Elsevier, New York (1986)
5. Skiba, Y.N., Filatov, D.M.: On an efficient splitting-based method for solving the diffusion equation on a sphere. Numer. Meth. Part. Differ. Eqs. **28**(1), 331–352 (2012)
6. Skiba, Y.N., Filatov, D.M.: Splitting-based schemes for numerical solution of a nonlinear diffusion equation on a sphere. Appl. Math. Comput.Comput. **219**(16), 8467–8485 (2013)
7. Hadamard, J.: Lectures on Cauchy's Problem in Linear Partial Differential Equations. Yale University Press, New Haven (1923)
8. Marchuk, G.I., Skiba, Y.N.: Numerical calculation of the conjugate problem for a model of the thermal interaction of the atmosphere with the ocean and continents. Izv. Atmos. Ocean. Phys.. Atmos. Ocean. Phys. **12**(5), 279–284 (1976)
9. Skiba, Y.N.: Air pollution estimates. World Resour. Rev. **9**(4), 542–556 (1997)
10. Skiba, Y.N.: Balanced and absolutely stable implicit schemes for the main and the adjoint pollutant transport equations in the limited area. Rev. Int. Contam. Ambient. **9**(2), 39–51 (1993)
11. Parra-Guevara, D., Skiba, Y.N.: Elements of the mathematical modelling in the control of pollutants emissions. Ecol. Model. **167**(3), 263–275 (2003)
12. Skiba, Y.N., Davydova-Belitskaya, V.: On the estimation of impact of vehicular emissions. Ecol. Model. **166**(1–2), 169–184 (2003)
13. Reed, M., Simon, B.: Functional Analysis I. Academic Press, London (1980)
14. Skiba, Y.N., Parra-Guevara, D.: Industrial pollution transport. Part I: formulation of the problem and air pollution estimates. Environ. Model. Assess. **5**(3), 169–175 (2000)

15. Kolmogorov, A.N., Fomin, S.V.: Elements of the Theory of Functions and Functional Analysis. Graylock Press, Rochester, New York (1957)
16. Ladyzhenskaya, O.A.: The Boundary Value Problems of Mathematical Physics. Springer, New York (1973)
17. Berlyand, M.E.: Modern Problems of Atmospheric Diffusion and Pollution of Atmosphere. Gidrometeoizdat, Leningrad (1975). (in Russian)

Chapter 3
Assessment of Pollutant Concentrations

This chapter describes the methods used to estimate pollutant concentrations. The focus is on explaining the adjoint method and deriving adjoint estimates of pollutant concentrations. The method is based on the use of adjoint operators and models. Adjoint equations originally defined by Lagrange entered the golden fund of science thanks to quantum mechanics, the theory of nuclear reactors, optimal control, and helped to solve many problems of perturbation theory [1, 2]. The adjoint equations were also used to develop the perturbation theory in nonlinear equations [3–7]. To emphasize the important role of adjoint equations in science, we should mention the problem of four-dimensional analysis and the solution of inverse problems.

In meteorology and oceanography, the adjoint method was first used by Marchuk [8]. The results obtained in the study of atmospheric temperature anomalies [9, 10], sensitivity of atmospheric models [11–13], data assimilation [14–18], and the evolution of normally distributed random fields [19, 20] confirm its effectiveness.

The adjoint method makes it possible to use both direct and adjoint estimates to calculate the concentrations of pollutants in various ecologically significant areas. These dual estimates complement each other fairly well in estimating pollutant concentrations and in optimal control of emission rates from pollution sources. Indeed, direct estimates use the solution of the dispersion problem and provide comprehensive information about the ecological situation in the entire region. Unlike direct estimates, adjoint estimates show the contribution of each pollution source to the average pollutant concentration in an area. And this contribution is determined by the values of the solutions of adjoint problems. Therefore, adjoint estimates are very effective in studying the sensitivity of pollutant concentrations to variations in the number, location, and intensity of pollutant sources, as well as to variations in input data.

Y. N. Skiba and D. Parra Guevara, *Mathematical Methods for the Assessment and Control of Industrial Emissions*, Mathematical Engineering,
https://doi.org/10.1007/978-3-031-58109-0_3

Direct estimates of pollutant concentrations are defined in Sect. 3.1. The definition of the adjoint operator in Hilbert space is given in Sect. 3.2, and Sect. 3.3 contains simple examples of adjoint operators. Section 3.4 formulates the adjoint problem to the two-dimensional dispersion problem. Adjoint estimates of pollutant concentrations are obtained in Sect. 3.5. Features of dual (direct and adjoint) estimates of the pollutant concentrations and their sensitivity are discussed in Sects. 3.6 and 3.7. Section 3.8 interprets adjoint estimates in terms of the convolution operation. Two numerical examples of lead particle dispersion from a point source located in a limited area with open boundaries are given in Sect. 3.9 to illustrate the application of the adjoint method and dual estimates.

3.1 Direct Estimates of Pollutant Concentrations

The solution of the dispersion problem (2.4.1)–(2.4.6) makes it possible to estimate the impurity concentration at any point or zone in the space–time domain $D \times (0, T)$. Indeed, the level of pollution in $D \times (0, T)$ can be characterized by the value

$$J(\phi) = \int_0^T \int_D p(r,t)\phi(r,t)drdt \tag{3.1.1}$$

where $p(r, t)$ is a non-negative weight function satisfying the condition

$$\int_0^T \int_D p(r,t)drdt = 1 \tag{3.1.2}$$

In this work, we consider two particular cases of function $p(r, t)$. Let $(r_j, t_j) \in D \times (0, T)$ be a monitoring point of the pollutant concentration. Taking $p(r,t) = \delta(r - r_j)\delta(t - t_j)$ in (3.1.1) we obtain

$$J(\phi) = \phi(r_j, t_j) \tag{3.1.3}$$

i.e., $J(\phi)$ is the concentration of the pollutant at the monitoring point (r_j, t_j). Let now Ω be a monitoring zone $\Omega \subseteq D$ with an area $|\Omega|$ and $\tau \in [0, T]$. Then taking

$$p(r,t) = \begin{cases} 1/(\tau|\Omega|), & \text{if } (r,t) \in \Omega \times (T-\tau, T) \\ 0, & \text{otherwise} \end{cases} \tag{3.1.4}$$

we obtain that

$$J(\phi) = \frac{1}{\tau|\Omega|} \int_{T-\tau}^{T} \int_{\Omega} \phi(r,t)drdt \tag{3.1.5}$$

is the average concentration of the pollutant in the zone Ω and time interval $(T - \tau, T)$.

3.2 Lagrange Identity and Adjoint Operator in Hilbert Space

A Hilbert space is an abstract vector space equipped with an inner product that defines a distance function, for which it is a complete metric space. The definition of an adjoint operator in a Hilbert space generalizes the notion of an adjoint matrix in a complex Euclidean space.

Definition 3.2.1. Let H be a Hilbert space with an inner product $(\cdot, \cdot)$, and let $L : H \to H$ be a linear operator with domain $\Phi \subset H$ (Φ is a linear subspace of H). The adjoint of L is the linear operator $L^* : H \to H$ with domain $\Phi^* \subset H$ (Φ^* is a linear subspace of H that satisfies the Lagrange identity [21]:

$$(L\varphi, g) = \left(\varphi, L^* g\right) \forall \varphi \in \Phi \text{ and } \forall g \in \Phi^* \tag{3.2.1}$$

where symbol $\forall \varphi \in \Phi$ means "for all φ from the domain Φ".

1. Existence of the adjoint operator. An adjoint of operator L always exists, that is, it is always possible to construct an operator L^* which satisfies the identity (3.2.1). To this end, let us define a set Φ^* as the largest subset of H such that $g \in \Phi^*$ if and only if the linear functional $\varphi \mapsto (L\varphi,\ g)$ is bounded. The set Φ^* is non-empty, because $0 \in \Phi^*$. Further, Φ^* is a linear subspace of H because for any $g_1,\ g_2 \in \Phi^*$ and real numbers a, b we have

$$\frac{|(L\varphi, ag_1 + bg_2)|}{||\varphi||} \le \frac{|a||(L\varphi, g_1)| + |b|\,|(L\varphi, g_2)|}{||\varphi||} \le |a|C_1 + |b|C_2$$

for each $\varphi \in \Phi - \{0\}$. Moreover, if $g \in \Phi^*$ then by the Riesz representation theorem [22], there exists a unique element $\nu \in H$ such that

$$(L\varphi,\ g) = (\varphi,\ \nu)$$

Thus, the operator $L^* : \Phi^* \to H$ can be defined as $L^* g = \nu$. This operator is well defined by the uniqueness of ν, and

$$(L\varphi, g) = \left(\varphi, L^* g\right) \forall \varphi \in \Phi \text{ and } \forall g \in \Phi^*,$$

i.e., L^* is the adjoint operator to L.

2. Uniqueness of the adjoint operator. In the general case, the adjoint operator is not unique, since if the operator L^* is restricted to any subspace of Φ^*, then it also satisfies (3.2.1). To establish uniqueness, it is assumed that Φ is a dense subspace in H, and the domain Φ^* of the operator L^* is considered to be the largest subspace where (3.2.1) is satisfied. Under these conditions, if F^* is adjoint then its domain of definition Dom_{F^*} belongs to Φ^* (otherwise L^* can be extended to a subspace $Dom_{F^*} \cup \Phi^*$, and Φ^* would not be maximal), and consequently,

$$(L\varphi, g) = \left(\varphi, F^* g\right) = \left(\varphi, L^* g\right) \text{ for } g \in Dom_{F^*} \text{ and } \forall \varphi \in \Phi$$

i.e.,

$$\left(\varphi, F^* g - L^* g\right) = 0, \forall \varphi \in \Phi$$

Furthermore, if $(\varphi, h) = 0 \, \forall \varphi \in \Phi$ then $h = 0$. This statement is true, since Φ is dense in H, i.e., there is a sequence $\varphi_n \to h$ in this set, which means that if $h \neq 0$ then

$$\|\varphi_n - h\|^2 = \|\varphi_n\|^2 + \|h\|^2 \geq \|h\|^2 > 0.$$

This contradicts the convergence, therefore, $h = 0$ and $F^* g - L^* g = 0$, i.e., L^* is unique.

It should be emphasized that the linear operators L and L^* are generally different. Only in special cases, such as the Laplace operator with Dirichlet, periodic or homogeneous Neumann conditions, do we have this: $L^* = L \quad (\Phi^* = \Phi)$, in such cases L is called self-adjoint operator. The case $L^* \neq L$ but $\Phi^* = \Phi$. is important, since properties of one operator are transmitted to the other operator, for example, if L is positive definite, then L^* is also positive definite:

$$\left(g, L^* g\right) = (Lg, g) > 0, \quad \forall g \in \Phi$$

3.3 Simple Examples of Adjoint Operators

In this section, using the Lagrange identity (3.2.1), we construct adjoint operators in various Hilbert spaces [23].

1. Let H be a real Euclidean n-dimensional vector space, and let A be a real $n \times n$ matrix. The scalar product of vectors $\vec{\phi}$ and $\vec{g}$ in H is defined as $\left(\vec{\phi}, \vec{g}\right) = \vec{g}^T \vec{\phi}$ where $\vec{g}^T$ is the transpose of $\vec{g}$. then

$$\left(A\vec{\phi}, \vec{g}\right) = \vec{g}^T A\vec{\phi} = (\vec{g}^T A)\vec{\phi} = \left(A^T \vec{g}\right)^T \vec{\phi} = \left(\vec{\phi}, A^T \vec{g}\right) \qquad (3.3.1)$$

and due to (3.2.1), $A^* = A^T$. Thus, A^* is just the transpose of A. In particular, $A^T = A$ if matrix A is symmetric, and $A^T = -A$ if A is skew-symmetric.

2. Let $H = L_2(D)$ be a space of square Integrable real functions $\varphi(x)$ defined in a domain D with inner product $(\cdot, \cdot)$, and let $L\varphi = k\varphi$ where $\Phi = L_2(D)$ and k is a real number. Then $L^*g = kg$ with $\Phi^* = \Phi$, since

$$(L\varphi, g) = (k\varphi, g) = (\varphi, kg) = \left(\varphi, L^*g\right)$$

and therefore operator L is self-adjoint.

3. Let us now explain the importance of boundary conditions in the definition of the adjoint operator. Let $H = L_2[0, 1]$ be the space of square integrable real periodic functions $\varphi(x)$ on the interval $[0, 1]$ with inner product

$$(\varphi, \psi) = \int_0^1 \varphi(x)\psi(x)dx$$

A function $\varphi(x)$ is called a class $C^k(D)$ function if all its derivatives up to order k exist and are continuous. Consider the simplest advection operator $L\varphi = u\frac{d\varphi}{dx}$ with

$$\Phi = \{\varphi \in C^1[0, 1] | \varphi(0) = \varphi(1)\} \subset L_2[0, 1] \tag{3.3.2}$$

Then

$$L^*g = -u\frac{dg}{dx} \text{ with } \Phi^* = \Phi \tag{3.3.3}$$

In fact,

$$(L\varphi, g) = \int_0^1 \left(u\frac{d\varphi}{dx}\right)g\,dx = u\left[g\varphi|_0^1 - \int_0^1 \frac{dg}{dx}\varphi dx\right] = \int_0^1 \varphi\left(-u\frac{dg}{dx}\right)dx = (\varphi, L^*g)$$

and hence L is a skew-symmetric operator ($L^* = -L$).

Let now

$$\Phi = \left\{\varphi \in C^1[0, 1] | \varphi(0) = c, \varphi(1) = d\right\} \subset L_2[0, 1] \text{ and } c \neq d$$

In this case, $L^* \neq -L$ because the identity $(L\varphi, g) = (\varphi, -Lg)$ is true only if

$$\Phi^* = \left\{g \in C^1[0, 1] | g(0) = d, g(1) = c\right\} \subset L_2[0, 1]$$

Since φ and g have different boundary conditions, the domains of definition Φ^* and Φ of operators L^* and L are different.

4. Let $H = L_2[0, 1]$, and let

$$L\varphi = \mu\frac{d^2\varphi}{dx^2}, \quad \Phi = \left\{\varphi \in C^2[0,1] | \varphi(0) = 0, \frac{d\varphi}{dx}(1) = 0\right\} \subset L_2[0,1] \quad (3.3.4)$$

where $\mu = Const$. Then

$$L^*g = \mu\frac{d^2g}{dx^2} \quad \text{with} \quad \Phi^* = \Phi, \quad (3.3.5)$$

Since

$$(L\varphi, g) = \int_0^1 \left(\mu\frac{d^2\varphi}{dx^2}\right) g\,dx = \mu\left[g\varphi'|_0^1 - \left(g'\varphi|_0^1 - \int_0^1 \frac{d^2g}{dx^2}\varphi dx\right)\right]$$
$$= \int_0^1 \varphi\left(\mu\frac{d^2g}{dx^2}\right)dx = (\varphi, L^*g)$$

Thus, L is self-adjoint.

5. Let $H = L_2[0, 1]$ and let

$$L\varphi = u\frac{d\varphi}{dx} - \mu\frac{d^2\varphi}{dx^2}, \quad \varphi \in \Phi \quad (3.3.6)$$

where
$u = Const$, $\mu = Const$, and

$$\Phi = \left\{\varphi \in C^2[0,1] | \left(u\varphi - \mu\varphi'\right)(0) = 0, \mu\varphi'(1) = 0\right\} \subset L_2[0,1] \quad (3.3.7)$$

Then the adjoint of L is

$$L^*g = -u\frac{dg}{dx} - \mu\frac{d^2g}{dx^2}, \quad g \in \Phi^* \quad (3.3.8)$$

where

$$\Phi^* = \left\{\varphi \in C^2[0,1] | \mu g'(0) = 0, \quad \left(ug + \mu g'\right)(1) = 0\right\} \subset L_2[0,1] \quad (3.3.9)$$

Indeed,

$$
\begin{aligned}
(L\varphi, g) &= \int_0^1 \left(u\frac{d\varphi}{dx} - \mu \frac{d^2\varphi}{dx^2} \right) g dx = ug\varphi|_0^1 - u\int_0^1 \frac{dg}{dx}\varphi dx \\
&\quad - \mu g\varphi'|_0^1 + \mu g'\varphi|_0^1 - \mu \int_0^1 \frac{d^2 g}{dx^2}\varphi dx \\
&= -g(\mu\varphi')\big|_{x=1} - g(u\varphi - \mu\varphi')\big|_{x=0} + \varphi(ug + \mu g')\big|_{x=1} - (\mu g')\varphi\big|_{x=0} \\
&\quad -u\int_0^1 \frac{dg}{dx}\varphi dx - \mu\int_0^1 \frac{d^2 g}{dx^2}\varphi dx \\
&= \int_0^1 \varphi\left(-u\frac{dg}{dx} - \mu\frac{d^2 g}{dx^2} \right) dx = (\varphi, L^* g)
\end{aligned}
$$

Note that the boundary conditions (3.3.7) are a one-dimensional analogue of conditions (2.4.5) and (2.4.6) for the case when $u > 0$.

3.4 Adjoint Problem to the 2D Dispersion Problem

Consider again the two-dimensional dispersion model (2.4.1)–(2.4.6) to determine the concentration $\phi(r, t) \equiv \phi(x, y, t)$ of a pollutant in a time interval $(0, T)$ and in a limited atmospheric domain D (Fig. 2.5). The model operator L is

$$L\phi = \frac{\partial \phi}{\partial t} + U \cdot \nabla\phi + \sigma\phi - \nabla \cdot (\mu\nabla\phi), \tag{3.4.1}$$

$$H = L_2(Q), \quad Q \equiv \overline{D} \times [0, T] \tag{3.4.2}$$

and $\Phi \subset H$ is the set of all functions $\phi \in C^2(\overline{D}) \cap C^1[0, T])$ satisfying the initial condition (2.4.4) and the boundary conditions (2.4.5)–(2.4.6).

We define the adjoint operator L^* to the operator L using the Lagrange identity:

$$(L\phi, g) = (\phi, L^* g), \quad \forall\phi \in \Phi \text{ and } \forall g \in \Phi^* \tag{3.4.3}$$

The inner product in H is given by

$$(\phi, \psi) = \int_0^T \int_D \phi\psi \, drdt \tag{3.4.4}$$

and therefore

$$
\begin{aligned}
(L\phi, g) &= \int_0^T \int_D g\frac{\partial\phi}{\partial t} drdt + \int_0^T \int_D gU \cdot \nabla\phi drdt \\
&+ \int_0^T \int_D g\sigma\phi drdt - \int_0^T \int_D g\nabla \cdot (\mu\nabla\phi) drdt
\end{aligned}
\tag{3.4.5}
$$

where $g(r, t)$ is a test function whose characteristics will be determined in the process of constructing the adjoint operator. Integrating by parts and using the condition

$$g(r, T) = 0 \quad \text{in } D \tag{3.4.6}$$

we get

$$\int_0^T \int_D g \frac{\partial \phi}{\partial t} drdt = -\int_0^T \int_D \phi \frac{\partial g}{\partial t} drdt$$

If we use (2.4.3), (2.4.5), (2.4.6) and the Gauss formula, we obtain

$$\int_0^T \int_D gU \cdot \nabla \phi drdt = \int_0^T \int_{S^+} g\phi U_n dSdt + \int_0^T \int_{S^-} g\phi U_n dSdt - \int_0^T \int_D \phi U \cdot \nabla g drdt,$$

$$\int_0^T \int_D g\nabla \cdot (\mu \nabla \phi) drdt = \int_0^T \int_{S^-} \mu g \frac{\partial \phi}{\partial n} dSdt - \int_0^T \int_{S^+} \mu \phi \frac{\partial g}{\partial n} dSdt -$$
$$\int_0^T \int_{S^-} \mu \phi \frac{\partial g}{\partial n} dSdt + \int_0^T \int_D \phi \nabla \cdot (\mu \nabla g) drdt$$

Summing up the last two equations and using the conditions

$$\mu \frac{\partial g}{\partial n} = 0 \text{ on } S^- \quad \text{and} \quad \mu \frac{\partial g}{\partial n} + U_n g = 0 \text{ on } S^+ \tag{3.4.7}$$

on the boundary of domain D (see Fig. 2.5) we obtain

$$(L\phi, g) = \int_0^T \int_D \phi \left[-\frac{\partial g}{\partial t} - U \cdot \nabla g + \sigma g - \nabla \cdot (\mu \nabla g) \right] drdt = (\phi, L^* g),$$

i.e., the adjoint to L is

$$L^* g = -\frac{\partial g}{\partial t} - U \cdot \nabla g + \sigma g - \nabla \cdot (\mu \nabla g) \tag{3.4.8}$$

The domain of definition Φ^* of L^* consists of all functions $g \in C^2(\overline{D}) \cap C^1[0, T])$ satisfying conditions (3.4.6) and (3.4.7).

We now introduce the adjoint model in the domain D (Fig. 2.5) and the time interval $(0, T)$ [9, 24]:

$$-\frac{\partial g}{\partial t} - U \cdot \nabla g + \sigma g - \nabla \cdot (\mu \nabla g) = p(r, t) \quad \text{in } D \times (0, T) \tag{3.4.9}$$

$$g(r, T) = 0 \quad \text{in } D \tag{3.4.10}$$

$$\mu \frac{\partial g}{\partial n} = 0 \quad \text{on } S^- \tag{3.4.11}$$

$$\mu \frac{\partial g}{\partial n} + U_n g = 0 \quad \text{on } S^+ \tag{3.4.12}$$

where $p(r, t)$ is the forcing to be defined later.

Let us compare problems (2.4.1)–(2.4.6) and (3.4.9)–(3.4.12) in the case when $f(r, t) \equiv 0$ and $p(r, t) \equiv 0$. It is seen that after the replacements $t' = T - t$ and $U' = -U$ the adjoint problem (3.4.9)–(3.4.12) takes the same form as the dispersion problem (2.4.1)–(2.4.6). Thus, the boundary S^+ (or boundary S^-) for problem (2.4.1)–(2.4.6) is the boundary S^- (boundary S^+) for the adjoint problem. This explains the replacement of boundary conditions (2.4.5) and (2.4.6) by boundary conditions (3.4.11) and (3.4.12). In addition, it follows from this that the adjoint problem is well-posed according to Hadamard only if it is solved in the reverse direction of time: from $t = T$ to $t = 0$. Therefore, the existence, uniqueness and stability of a solution to the adjoint problem directly follows from Theorem 2.8.1 [25]:

Theorem 3.4.1. *Let* $p(r, t) \in L_2\left(\overline{D} \times [0, T]\right)$. *If all the coefficients of adjoint problem* (3.4.9)–(3.4.12) *satisfy the conditions of Theorem* 2.8.1, *then the adjoint problem has in the space* H(Q) *a unique weak solution that depends continuously on the forcing and initial condition.*

Remark 3.4.1. Theorems 2.8.1 and 3.4.1 can be directly generalized to the case of three dimensions.

Remark 3.4.2. The solution of adjoint problem $g(r, t)$ is nonnegative in the domain $D \times (0, T)$ if $p(r, t)$ is a nonnegative forcing.

3.5 Adjoint Estimates of Pollutant Concentrations

We will now use the principle of duality to introduce two sets of equivalent formulas for estimating the concentration of a pollutant. One set consists of the direct estimates introduced in Sect. 3.1 using the solution to the dispersion problem. The other set contains equivalent estimates based on the use of the solution of the adjoint problem, and therefore they will be called adjoint estimates. The importance of dual estimates that complement each other will be discussed below.

Suppose we are studying the concentration of a pollutant $\phi(r, t)$ in a domain D and time interval $(0, T)$. Let $\Omega \subset D$ be an ecologically sensitive zone in D, and $(T - \tau, T) \subset (0, T)$ be a subinterval of length τ. If $\omega(r, t)$is a positive function in the domain $\Omega \times (T - \tau, T)$ such that

$$\int_{T-\tau}^{T} \int_{\Omega} \omega(r, t) dr dt = 1 \tag{3.5.1}$$

then the integral

$$J(\phi) = \int_{T-\tau}^{T} \int_{\Omega} \omega(r, t)\phi(r, t) dr dt \tag{3.5.2}$$

is the direct estimate of the average concentration of the pollutant $\phi(r, t)$ in the zone Ω and time interval $(T - \tau, T)$.

To obtain the adjoint estimate, we subtract Eq. (3.4.9) pre-multiplied by $\phi(r, t)$ from Eq. (2.4.1) pre-multiplied by $g(r, t)$, and then integrate the result over domain $D \times (0, T)$. Considering the form of the forcing (2.4.2) and the boundary conditions (2.4.5)–(2.4.6) and (3.4.11)–(3.4.12), we obtain

$$\int_0^T \int_D p(r, t)\phi(r, t) dr dt + \int_D g(r, T)\phi(r, T) dr = \sum_{i=1}^{N} \int_0^T g(r_i, t) Q_i(t) dt + \int_D g(r, 0)\phi^0(r) dr$$

Considering the initial condition (3.4.10) we obtain

$$\int_0^T \int_D p(r, t)\phi(r, t) dr dt = \sum_{i=1}^{N} \int_0^T g(r_i, t) Q_i(t) dt + \int_D g(r, 0)\phi^0(r) dr \tag{3.5.3}$$

If we define the forcing of the adjoint problem as

$$p(r, t) = \begin{cases} \omega(r, t), & \text{if } (r, t) \in \Omega \times (T - \tau, T) \\ 0, & \text{otherwise} \end{cases}, \tag{3.5.4}$$

then (3.5.3) leads to

$$J(\phi) \equiv \int_{T-\tau}^{T} \int_{\Omega} \omega(r, t)\phi(r, t) dr dt = \sum_{i=1}^{N} \int_0^T g(r_i, t) Q_i(t) dt + \int_D g(r, 0)\phi^0(r) dr \tag{3.5.5}$$

In the particular case when $\omega(r, t) = 1/(\tau|\Omega|)$ where $|\Omega|$ is the area of Ω, formulas (3.5.2) and (3.5.5) give dual (equivalent) estimates of the average concentration of pollutant $\phi(r, t)$ in the zone Ω and time interval $(T - \tau, T)$ [9]: the direct estimate

$$J(\phi) \equiv \frac{1}{\tau|\Omega|}\int_{T-\tau}^{T}\int_{\Omega}\phi(r,t)drdt, \tag{3.5.6}$$

and the adjoint estimate

$$J(\phi) = \sum_{i=1}^{N}\int_{0}^{T} g(r_i,t)Q_i(t)dt + \int_{D} g(r,0)\phi^0(r)dr \tag{3.5.7}$$

3.6 Features of Dual (Direct and Adjoint) Estimates

The adjoint estimate (3.5.7) explicitly relates the average concentration $J(\phi)$ to the emission rates $Q_i(t)$ of the sources over the time interval $(0, T)$ and the initial distribution $\phi^0(r)$ of the pollutant in D. The adjoint solution $g(r, t)$ is nonnegative and does not depend on $Q_i(t)$ and $\phi^0(r)$, therefore its values $g(r_i, t)$ and $g(r, 0)$ serve in the estimate (3.5.7) as weight functions (or influence functions) showing both the contribution of each pollution source and the contribution of the initial distribution of pollution $\phi^0(r)$ in the region D.

Note that the role of the initial concentration $\phi^0(r)$ decreases with increasing the length of the interval $(0, T-\tau)$ [26]. Indeed, due to (3.5.4), $p(r, t) \equiv 0$ on the interval $(0, T-\tau)$ and due to the presence of model dissipation ($\sigma > 0$), the value $g(r, 0)$ in (3.5.7) reduces to zero as $(T-\tau) \to \infty$.

It should also be noted that the direct estimates use the solution of dispersion problem (2.4.1)–(2.4.6), which allows a comprehensive analysis of the environmental situation in the entire region D. On the other hand, the solution of the adjoint problem (3.4.9)–(3.4.12), calculated for an ecologically important zone Ω (Fig. 3.1), does not depend on the number of pollution sources N as well as on their intensity $Q_i(t)$ and location r_i. This feature makes adjoint estimates very effective for studying the N-dependence, $Q_i(t)$-dependence and/or r_i-dependence of the average concentration $J(\phi)$ in the zone Ω [9]. Sometimes adjoint estimates (3.5.7) help to solve non-trivial problems. In particular, they are necessary when solving the problem of optimal control of industrial emissions or identifying the intensity and location of unknown sources of pollution (see Chaps. 4–7). On the whole, dual estimates complement each other fairly well in solving various pollution problems.

3.7 Sensitivity of Pollution Estimates

Let us now use adjoint estimates to evaluate the sensitivity of the estimate $J(\phi)$ with respect to variations in the main parameters of the dispersion problem (2.4.1)–(2.4.6). Since the problem is linear, estimate (3.5.7) immediately leads to

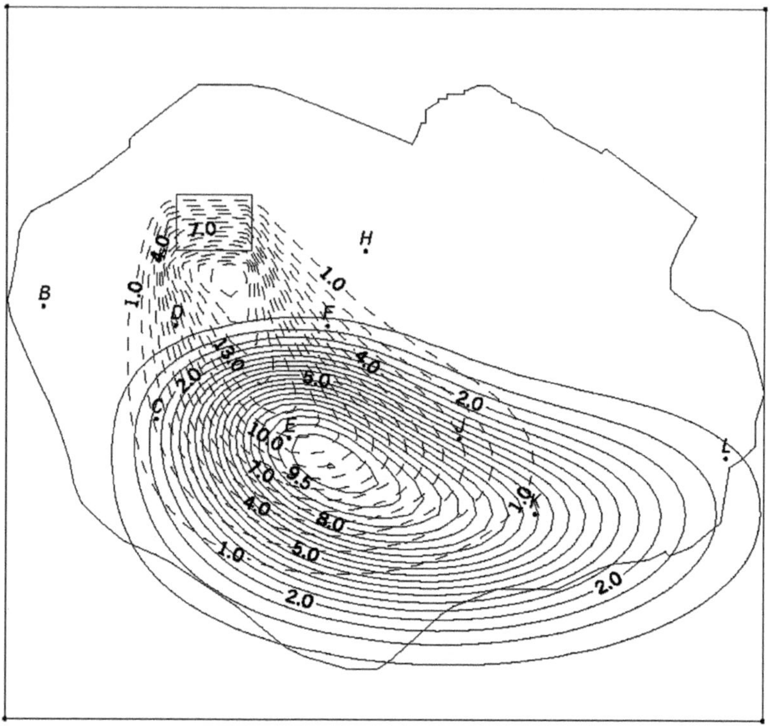

Fig. 3.1 Isolines of adjoint problem solution $g(r, t)$ calculated for a zone in Guadalajara at $t = T - 60$ min (dotted lines) and $t = T - 90$ min. (solid lines); $T = 360$ min

$$\delta J(\phi) = \sum_{i=1}^{N} \int_0^T g(r_i, t)\delta Q_i(t)dt + \int_D g(r, 0)\delta\phi^0(r)dr \tag{3.7.1}$$

where $\delta J(\phi)$ is the change in the average pollution concentration (3.5.7) in the zone Ω due to changes in the initial pollution distribution $\phi^0(r)$, the emission intensity $Q_i(t)$ and the number of sources N.

Denote by r_i and r_i' two different locations of the industrial plants. Then the r_i-dependence of $J(\phi)$ can be expressed as

$$\delta J(\phi) = \sum_{i=1}^{N} \int_0^T \left[g(r_i', t) - g(r_i, t)\right]Q_i(t)dt \tag{3.7.2}$$

It should be emphasized that (3.7.1) and (3.7.2) are valid for arbitrary variations. Finally, we now derive a general formula for analyzing the sensitivity of the average concentration $J(\phi)$ to small variations in model parameters. Let ϕ be the solution to problem (2.4.1)–(2.4.6), and let $\phi + \phi'$ be the solution to the perturbed problem

$$\frac{\partial}{\partial t}(\phi+\phi')+(U+U')\cdot\nabla(\phi+\phi')+(\sigma+\sigma')(\phi+\phi') \\ -\nabla\cdot\left[(\mu+\mu')\nabla(\phi+\phi')\right]=f+f' \tag{3.7.3}$$

$$\phi(r,0)+\phi'(r,0)=\phi^0(r)+\delta\phi^0(r) \quad \text{in } D, \tag{3.7.4}$$

$$\mu\frac{\partial}{\partial n}(\phi+\phi')-U_n(\phi+\phi')=0 \text{ at } S^- \tag{3.7.5}$$

$$\mu\frac{\partial}{\partial n}(\phi+\phi')=0 \text{ at } S^+ \tag{3.7.6}$$

where

$$f(r,t)+f'(r,t)=\sum_{i=1}^{N}\left[Q_i(t)+\delta Q_i(t)\right]\delta(r-[r_i+\delta r_i]) \tag{3.7.7}$$

For simplicity, we will assume that in this problem there are no perturbations in U and μ on the boundary S, and all perturbations $U', \phi', \mu', \sigma', \delta\phi^0, \delta r_i$ are small. Then the linearized Eq. (3.7.3) has the form

$$\frac{\partial}{\partial t}\phi'+U\cdot\nabla\phi'+\sigma\phi'-\nabla\cdot\left[\mu\nabla\phi'\right]=f-U'\cdot\nabla\phi-\sigma'\phi+\nabla\cdot\left[\mu'\nabla\phi\right].$$

Applying the adjoint method to this equation we obtain

$$\delta J(\phi)=\sum_{i=1}^{N}\int_0^T g(r_i,t)\delta Q_i(t)dt+\int_D g(r,0)\delta\phi^0(r)dr \\ +\sum_{i=1}^{N}\int_0^T\left[\nabla g(r_i,t)\cdot\delta r_i\right]Q_i(t)dt-\int_0^T\int_D g\left\{U'\cdot\nabla\phi+\sigma'\phi-\nabla\cdot\left[\mu'\nabla\phi\right]\right\}drdt \tag{3.7.8}$$

This formula shows how small disturbances $\delta Q_i(t)$ and $U', \phi', \mu', \sigma', \delta\phi^0, \delta r_i$ can change the average concentration $J(\phi)$ in the zone Ω. The first two terms on the right side of (3.7.8) coincide with (3.7.1) and show the contribution of variations $\delta Q_i(t)$ and $\delta\phi^0$, while the third term describes the effect of changing the position of the sources. The last term reflects the influence of perturbations U', σ' and μ', and this is the only term in (3.7.8) that uses the solution $\phi(r,t)$ of the unperturbed dispersion problem in the domain $D\times(0,T)$. However, if $U'=\sigma'=\mu'=0$ everywhere in D then the solution $\phi(r,t)$ is not used, and the sensitivity formula uses only the solution of the adjoint problem (3.4.9)–(3.4.12).

3.8 Dispersion of Pollutants and Convolution Operation

Systems in nature are often modeled using linear differential equations. Since there are different ways to write these equations in the form of integral equations [27], it can be assumed that the physical system has two representations that allow us to understand and analyze different aspects of the phenomenon. The differential formulation of the solution is particularly useful for predicting state variables [28], while the integral formulation, together with specific information about the state variables, allows the estimation of specific system parameters [29]. These two types of problem, i.e. prediction and parameter estimation, do not have the same mathematical properties even if they are associated with the same physical system. In particular, the integral formulation is endowed with a function called the integral kernel, which contains in a compact form information about the simulated physical process, the characteristics of which determine the mathematical properties of the integral equation. Depending on the context of the system being studied, this kernel is also known as the Green's function [27] or influence function [24, 30]. When the integral kernel takes a special form, the integral formulation of the system is called the convolution integral [27].

In mathematics (specifically functional analysis), convolution is a mathematical operation on two functions (q and h) that results in a third function $q*h$ that expresses how the shape of one function changes the shape of the other. The term convolution refers to both the result function and the process of calculating it. It is defined as the integral [31]

$$(q * h)(\tau) = \int_0^{\tau} q(t)h(\tau - t)dt, \quad \tau > 0 \tag{3.8.1}$$

This operation with an integral form appears in the second half of the eighteenth century, but it received its current name and designation until the beginning of the twentieth century [32]. Its application areas include electrical circuit analysis, signal and image processing, heat transfer, radiation therapy, and spectroscopy, among others [27, 33, 34].

This section deals with pollutants dispersed in the atmosphere. It is shown that the point concentration at the observation site, the average concentration in the control zone and the mass in the region are state variables that are expressed as convolution integrals. For this purpose, a well-posed dispersion model is considered for the stationary regime of dispersion and transformation of the pollutant in the atmosphere. At the same time, it is proven that each of the state variables is a convolution of the emission rate of the pollution source with a function obtained from the unit impulse response of the dispersion model. These results, which have an integral form, can be considered as an intermediate formulation between analytical (Gaussian) solutions of dispersion models with constant coefficients and purely numerical solutions of more complex dispersion models [28]. The importance of these formulations is that they establish an explicit relationship between certain parameters of the dispersion model and the point concentration, average concentration, and mass of the pollutant. This explicit dependence is a computational advantage for formulating and solving emission control problems [35] and parameter estimation [30].

1. Convolution formula for the total emitted mass and estimation of σ. The dispersion of a substance emitted into the atmosphere from a pollution source is modeled in region D over a time interval $(0, T)$ using a three-dimensional transport equation (see (2.10.1))

$$\frac{\partial \phi}{\partial t} + U \cdot \nabla \phi + \sigma \phi - \nabla \cdot (\mu \nabla \phi) = f(r, t) \text{ in } D \times (0, T) \tag{3.8.2}$$

This equation considers emissions f from pollution sources and several processes that influence the dispersion of matter in the atmosphere: advection, characterized by wind speed U, chemical transformation of the pollutant with the rate σ, and turbulence, determined by the diffusion coefficient μ. The vertical component of velocity U can include the deposition rate of the impurity. The function $\phi = \phi(r, t)$ is a state variable that represents the concentration of the pollutant at point $r \in D$ and time $t \in (0, T)$. In Eq. (3.8.2), the coefficients U and μ are assumed to be functions of position and time, σ is a positive constant, and U satisfies the continuity equation $\nabla \cdot U = 0$.

To show that the total mass emitted by a source can be quantified as the convolution integral, it is assumed that there is no pollutant initially, $\phi(r, 0) = 0$, and the area D is large enough to contain the pollutant emitted during the time interval $(0, T)$. Therefore, at the boundary ∂D of this region, we can assume the absence of impurities and a zero-mass flow, so the following conditions are used:

$$\phi U \cdot n = 0 \quad \text{and} \quad \nabla \phi \cdot n = 0 \quad \text{on } \partial D \tag{3.8.3}$$

where n denotes the outer unit normal vector to the boundary ∂D of the domain D. Note that on the part of the boundary that corresponds to the ground, the conditions (3.8.3) are also satisfied. Integrating Eq. (3.8.2) over the domain D and using the continuity equation, the divergence theorem, and boundary conditions (3.8.3), we obtain

$$\frac{\partial}{\partial t} \int_D \phi dr + \sigma \int_D \phi dr = \int_D f(r, t) dr \tag{3.8.4}$$

Let us denote by $m(t) = \int_D \phi dr$ the total mass of the pollutant emitted by the source until time t, with the exception of that part of the mass that was transformed during the same period, and by $Q(t) = \int_D f(r, t) dr$ the emission rate of the pollution source. Then Eq. (3.8.4) reduces to the initial value problem

$$\frac{\partial}{\partial t} m(t) + \sigma m(t) = Q(t), \quad m(0) = 0 \tag{3.8.5}$$

Integrating (3.8.5) over the interval $(0, \tau)$, where $\tau \leq T$, we obtain

$$m(\tau) = e^{-\sigma\tau} \int_0^\tau e^{\sigma t} Q(t) dt = \int_0^\tau Q(t) e^{-\sigma(\tau - t)} dt \tag{3.8.6}$$

Thus, we have the following convolution for the total mass: $m(\tau) = (Q * h)(\tau)$ where $h(t) = e^{-\sigma t}$. The function h has the following interpretation. If $Q(t) = \delta(t)$ in Eq. (3.8.6), then, using the filtering property of the Dirac function, it is obtained that the corresponding mass is $m_\delta(\tau) = e^{-\sigma\tau} = h(\tau)$. This means that the function h is a unit impulse response of the physical system under consideration.

To show the importance of this result, it should be noted that the convolution integral (3.8.6) establishes an explicit connection between the state variable m and the parameters Q and σ. Such a relationship can be used for indirect parameter estimates when additional information about the state variable is available. For example, suppose that a point and instantaneous source of emission is observed, i.e.,

$$f(r, t) = Q_e \delta(t - t_e) \delta(r - r_0) \tag{3.8.7}$$

Here, r_0 is the source location point, and Q_e and t_e are the intensity and time of emission, respectively. We will assume that these parameters are known (controlled experiment). In this case $Q(t) = Q_e \delta(t - t_e)$, and Eq. (3.8.6) reduces to

$$m(\tau) = Q_e e^{-\sigma(\tau - t_e)}, \quad t_e < \tau \tag{3.8.8}$$

Finding the parameter σ from Eq. (3.8.8) for $\tau = T$, we obtain

$$\sigma = \frac{1}{T - t_e} \text{Ln}\{Q_e / m(T)\} \tag{3.8.9}$$

Equation (3.8.9) determines the reaction coefficient of the pollutant in the atmosphere in the presence of an indirect estimate of the total mass (for a certain time) in the domain D, for example, from a satellite image [36, 37].

2. Formulation of state variables in the form of convolution integrals. Let us assume that the dispersion problem is based on Eq. (3.8.2), considered in the domain $D \times (0, T)$ with the initial condition $\phi(r, 0) = 0$ in D. Using the adjoint problem, we obtain that Eq. (3.5.3) takes the form

$$\int_0^T \int_D p(r, t) \phi(r, t) dr dt = \int_0^T \int_D g(r, t) f(r, t) dr dt \tag{3.8.10}$$

According to (3.8.10), the choice of various types of forcing $p(r, t)$ in the adjoint problem leads to integral equations representing various state variables in the dispersion process. Indeed, if $p(r, t) = \delta(r - R)\delta(t - \tau)$ then (3.8.10) represents the point concentration of the pollutant at the observation site R in D:

$$\phi(R,\tau)=\int_0^T\int_D g(r,t)f(r,t)drdt,\quad R\in D,\quad 0<\tau\le T \tag{3.8.11}$$

And if $p(r,t)=(1/|\Omega|)\,\delta(t-\tau)\chi_\Omega(r)$, where χ_Ω is the characteristic function for a certain control zone $\Omega\subset D$, then (3.8.10) represents the average concentration of the pollutant in zone Ω:

$$\overline{\phi}(\tau)=\frac{1}{|\Omega|}\int_\Omega \phi(r,\tau)dr=\int_0^T\int_D g(r,\tau)f(r,\tau)drdt,\quad 0<\tau\le T \tag{3.8.12}$$

Finally, if $p(r,t)=\delta(t-\tau)$ then (3.8.10) represents the total mass of the pollutant contained in region D:

$$m(\tau)=\int_D \phi(r,\tau)dr=\int_0^T\int_D g(r,t)f(r,t)drdt,\quad 0<\tau\le T \tag{3.8.13}$$

Note that despite the fact that the right-hand sides of Eqs. (3.8.10)–(3.8.12) have the same structure, their meaning is different, since in each case g is a solution to the adjoint problem (3.4.9)–(3.4.12) with its own forcing.

We will now show that Eq. (3.8.11) is a convolution integral. The same is true for Eqs. (3.8.12) and (3.8.13), since all three equations have the same structure. Since $p(r,t)=0$ for $\tau<t\le T$, and the adjoint problem is solved in the opposite direction of time, we obtain that $g(r,t)=0$ on the interval (τ,T). As a result, (3.8.11) reduces to

$$\phi(R,\tau)=\int_0^\tau\int_D g(r,t)f(r,t)drdt,\quad R\in D,\quad 0<\tau\le T \tag{3.8.14}$$

Without loss of generality, we will assume that $f(r,t)=Q(t)F(r)$ [38]. Therefore, Eq. (3.8.14) can be rewritten as

$$\phi(R,\tau)=\int_0^\tau Q(t)\left\{\int_D g(r,t)F(r)dr\right\}dt,\quad 0<\tau\le T \tag{3.8.15}$$

It is now assumed that the process of dispersion and transformation of impurities in the atmosphere is stationary, i.e., each coefficient of the transport Eq. (3.8.2) can be a function of position r, but does not depend on time. In this case, the adjoint function $g(r,t)$ in (3.8.15) is a translation (shift) of the *basic kernel* g_α, calculated as a solution to the adjoint problem with forcing $p(r,t)=\delta(r-R)\delta(t-T)$, i.e., $g(r,t)=g_\alpha(r,T-(\tau-t))$ when $t\le\tau$, and therefore

$$\phi(R,\tau)=\int_0^\tau Q(t)\left\{\int_D g_\alpha(r,T-(\tau-t))F(r)dr\right\}dt,\quad \tau\le T \tag{3.8.16}$$

If we define a function as

$$h(\tau - t) = \int_D g_\alpha(r, T - (\tau - t))F(r)dr \tag{3.8.17}$$

then (3.8.16) can be written as a convolution integral

$$\phi(R, \tau) = \int_0^\tau Q(t)h(\tau - t)dt = (Q * h)(\tau), \quad \tau \le T \tag{3.8.18}$$

The interpretation of the function h is as follows. If $Q(t) = \delta(t)$ then Eq. (3.8.18) takes the form

$$\phi_\delta(R, \tau) = \int_0^\tau \delta(t)h(\tau - t)dt = h(\tau) \tag{3.8.19}$$

Consequently, the function h is a partial solution ϕ_δ of the transport equation, that is, h is the response of the dispersion Eq. (3.8.2) to a single impulse. It is important to note that the function h can be calculated using Eq. (3.8.17) or by solving the dispersion model. The method used for the calculation h depends on the problem being solved. For example, if there is one emission source but many observation sites R_l then it is computationally more efficient to solve the dispersion model once to compute all functions h_l. Conversely, if there are many emission sources and only one observation site, then Eq. (3.8.17) is the most efficient option, since it involves solving the adjoint model only once.

The generalization of Eq. (3.8.18) is straightforward. We consider the dispersion model for inhomogeneous initial condition and several pollution sources

$$\frac{\partial \phi}{\partial t} + A\phi = \sum_i f_i(r, t) \text{ in } D \times (0, T), \quad \phi(r, 0) = \phi_0(r) \text{ in } D \tag{3.8.20}$$

Since the problem is linear, its solution can be written as a sum of solution functions, in particular, for the observation site R in D we have

$$\phi(R, \tau) = c_0(R, \tau) + \sum_i c_i(R, \tau), \quad 0 \le \tau \le T \tag{3.8.21}$$

where the functions $c_i (i = 0, 1, 2...)$ are solutions of the following dispersion models:

$$\frac{\partial c_0}{\partial t} + Ac_0 = \delta(t)\phi_0(r) \text{ in } D \times (0, T), \quad c_0(r, 0) = 0 \text{ in } D \tag{3.8.22}$$

$$\frac{\partial c_i}{\partial t} + Ac_i = f_i(r, t) \text{ in } D \times (0, T), \quad c_i(r, 0) = 0 \text{ in } D, \ i = 1, 2, \ldots \tag{3.8.23}$$

Note that we can consider the inhomogeneous initial condition of problem (3.8.20) as a forcing with separable structure (see Eq. (3.8.22)). Finally, it is noted that each of the problems (3.8.22) and (3.8.23) is a problem of the type

$$\frac{\partial \phi}{\partial t} + A\phi = f(r,t) \text{ in } D \times (0,T), \quad \phi(r,0) = 0 \text{ in } D \tag{3.8.24}$$

Therefore, we can apply Eq. (3.8.18) to each term (3.8.21) and obtain an expression for the point concentration of a pollutant as a sum of convolution integrals:

$$\phi(R,\tau) = c_0(R,\tau) + \sum_i c_i(R,\tau) = (\delta * h_0)(\tau) + \sum_i (Q_i * h_i)(\tau) \tag{3.8.25}$$

Note that each function $h_i (i = 0, 1, 2, ...)$ corresponds to a different forcing.

3.9 Numerical Examples

To illustrate the application of the adjoint method and dual estimates, we now consider two examples of lead particle scattering from a point source located in a limited area with open boundaries [39]. Solutions of the models (2.4.1)–(2.4.6) and (3.4.9)–(3.4.12) are calculated using the numerical scheme described in Chap. 8.

Example 3.9.1 Let $D = (0,\ 2\text{ km}) \times (0,\ 2\text{ km})$ be a square area with open boundaries, and we assume that there is a point source in D located at the point $r_1 = (1.8,\ 0.2)$ which emits lead particles at a constant rate $Q_1 = 3.8$ kg/h. Let us also assume that it starts working at the moment $t = 0$, which means that $\phi^0(r) = 0$ in D. Let us take $\sigma = 0.001 h^{-1}$ and $\mu = 0.04\text{ km}^2/\text{h}$. The velocity field $U = (u,\ v)$ was formed by a streamfunction $\psi = xy$, the streamlines of which are hyperbolas, with the wind direction from southeast to northwest:

$$u = -\frac{\partial \psi}{\partial y} = -x, \quad v = \frac{\partial \psi}{\partial x} = y$$

Obviously, $\nabla \cdot U = 0$. We will monitor the concentration of lead particles in the region $\Omega = [0,\ 0.5] \times [0.5,\ 1.0]$. The time interval $(0, T)$ is four hours, and $\tau = T$. The average concentration $J(\phi)$ of lead particles calculated according to (3.5.6) in Ω over the four-hour period is 2.11 $\mu\text{g/m}^3$, which is a violation of the sanitary standard 1.5 $\mu\text{g/m}^3$ [40].

Figure 3.2 shows the isolines of the solution $\phi(r,t)$ every hour, and a clear increase in the concentration of lead particles in Ω can be observed with the current wind direction. Figure 3.3 shows the isolines of the solution $g(r,t)$ of the adjoint problem every hour (from $t = 4$ to $t = 0$), and Fig. 3.4 contains the corresponding surfaces of $g(r,t)$. It can be seen that the region of nonzero values of the adjoint solution extends in the direction opposite to the wind velocity U, i.e., from northwest to southeast,

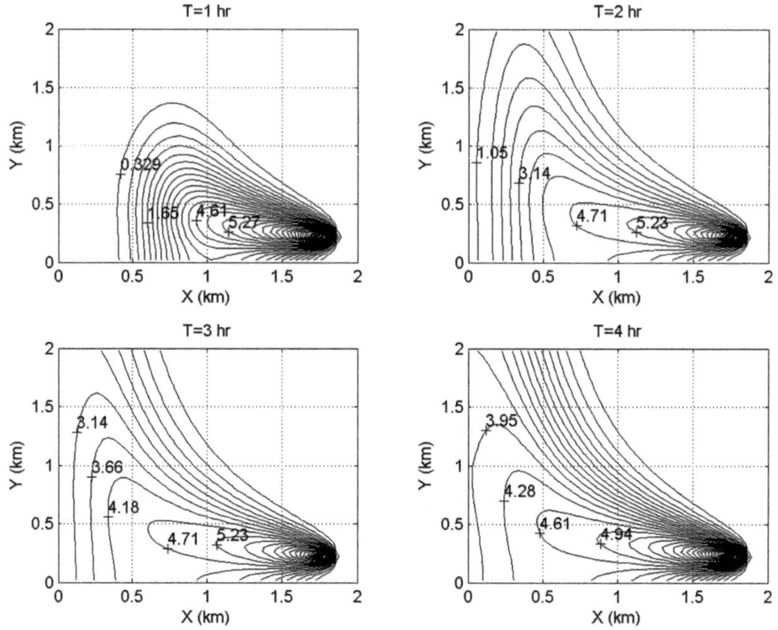

Fig. 3.2 Isolines of concentration $\phi(r, t)$ every hour: $t = 1$ h, $t = 2$ h, $t = 3$ h and $t = 4$ h

as it should be. The adjoint estimate (3.5.7) gives the same average concentration of lead particles in $\Omega : J(\phi) = 2.11\ \mu\text{g/m}^3$.

Example 3.9.2 As in the previous example, we consider in a square region $D = (0,\ 2\text{ km}) \times (0,\ 2\text{ km})$ with open boundaries a point source that emits lead particles at a constant rate $Q_1 = 4.0\,\text{kg/h}$. The source is located at the point $r_1 = (1.8,\ 1.8)$ and works only in a four-hour interval $(0, T)$. Thus, the atmosphere is free from this admixture at $t = 0$, and $\phi^0(r) = 0$ in D. As parameters of deposition and diffusion we take $\sigma = 0.001\,\text{h}^{-1}$ and $\mu = 0.03\,\text{km}^2/\text{h}$. The non-divergent velocity field $U = (u, v)$ was created by the streamfunction $\psi = y - \frac{1}{2}(x-1)^3$, whose streamlines are cubic parabolas with the wind direction from northeast to southwest, i.e.,

$$u = -\frac{\partial\psi}{\partial y} = -1,\ v = \frac{\partial\psi}{\partial x} = -\frac{3}{2}(x-1)^2$$

The dispersion problem is solved in a four-hour interval $(0, T)$, $\tau = T$ and $\Omega = [0,\ 0.5]\times[0.5,\ 1.0]$. Figure 3.5 shows the isolines of the solution $\phi(r, t)$ every hour, and one can see an increase in the concentration of lead particles in the region Ω with the current wind direction. The average concentration $J(\phi)$ of lead particles in Ω and time interval $(0, T)$, calculated according to (3.5.6) is $0.90\ \mu\text{g/m}^3$, which does

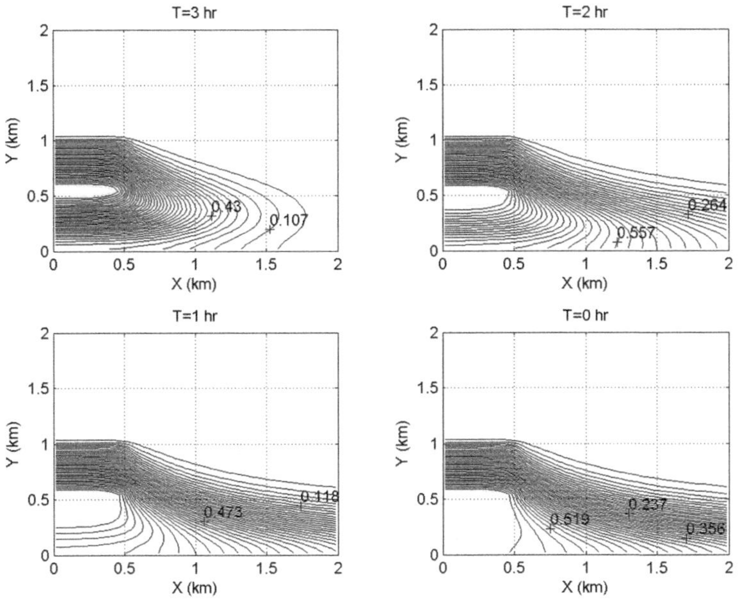

Fig. 3.3 Isolines of adjoint problem solution $g(r, t)$ every hour: $t = 3$ h, $t = 2$ h, $t = 1$ h and $t = 0$

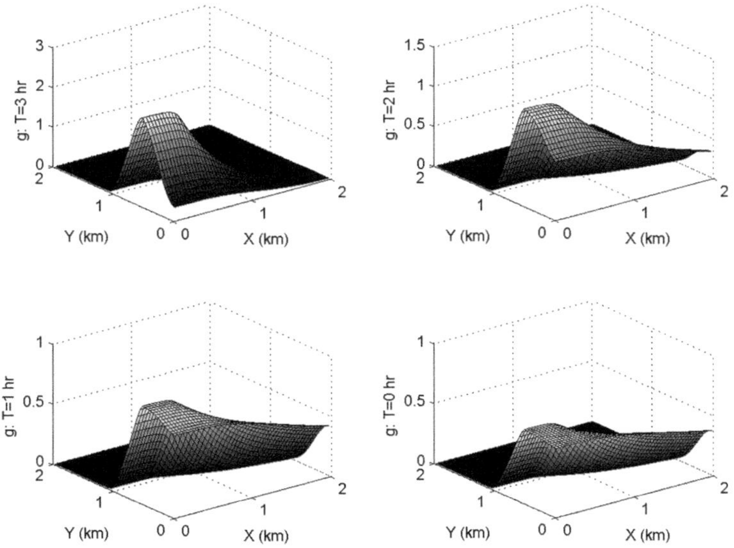

Fig. 3.4 Surfaces of adjoint problem solution $g(r, t)$ every hour: $t = 3$ h, $t = 2$ h, $t = 1$ h and $t = 0$

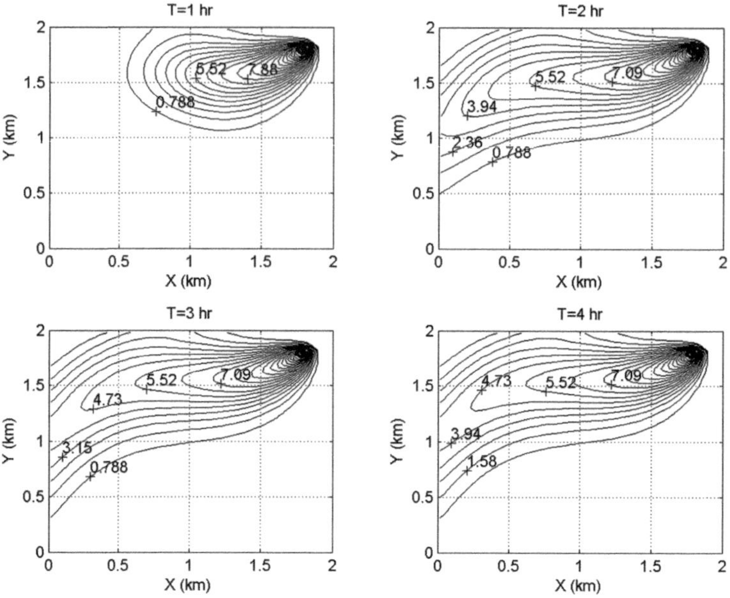

Fig. 3.5 Isolines of concentration $\phi(r, t)$ every hour: $t = 1$ h, $t = 2$ h, $t = 3$ h and $t = 4$ h

not violate the sanitary standard (1.5 $\mu g/m^3$). However, the average concentration of lead particles in a two-hour interval $(T/2, T)$ is 1.60 $\mu g/m^3$, i.e., slightly exceeds the sanitary norm. This result is explained by the fact that the atmosphere was initially clean, and it took more than an hour for the lead particles to reach the Ω zone (Fig. 3.5). Thus, the accumulation of lead particles becomes important in the second half of the four-hour period. The example shows the importance of choosing the right interval $(T - \tau, T)$ when monitoring and assessing pollutant concentrations using dual estimates (3.5.6) and (3.5.7).

Figure 3.6 shows the isolines of the solution $g(r, t)$ of the adjoint problem every hour on the interval $(0, T)$ (from $t = 4$ to $t = 0$), and Fig. 3.7 shows the corresponding surfaces of $g(r, t)$ for the case when $\tau = 4$ h. It can be seen that the region of nonzero values of the adjoint solution propagates in the direction opposite to the wind velocity U. The equivalent adjoint estimate (3.5.7) gives the same average lead particle concentrations in Ω: $J(\phi) = 0.90\,\mu g/m^3$ if $\tau = 4$, and $J(\phi) = 1.60\,\mu g/m^3$ if $\tau = 2$.

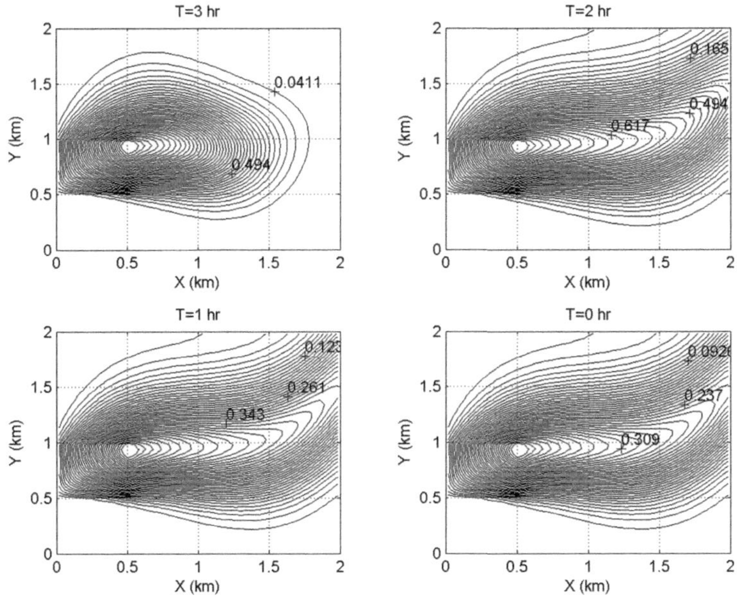

Fig. 3.6 Isolines of adjoint problem solution $g(r, t)$ every hour:$t = 3$ h, $t = 2$ h, $t = 1$ h and $t = 0$

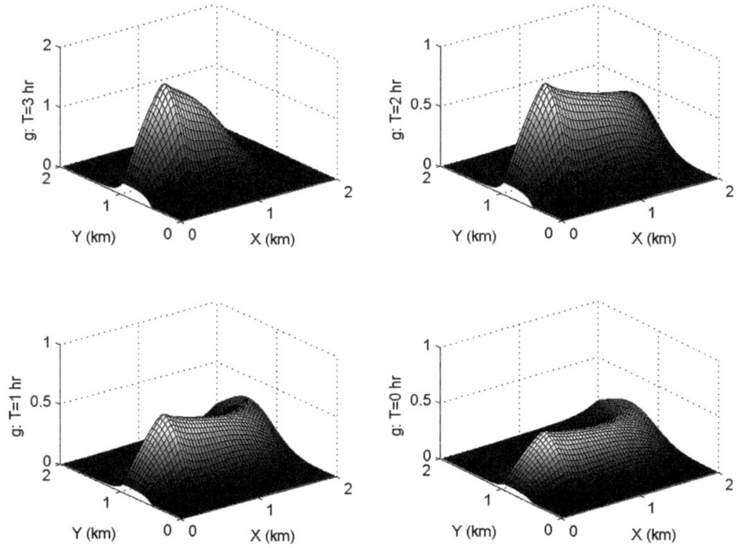

Fig. 3.7 Surfaces of adjoint problem solution $g(r, t)$ every hour:$t = 3$ h, $t = 2$ h, $t = 1$ h and $t = 0$

References

1. Lewins, J.: Importance, the Adjoint Function. Pergamon Press, Oxford (1965)
2. Marchuk, G.I., Orlov, V.V.: On the theory of adjoint functions. In: Neutron Physics, pp. 30–45. Gosatomizdat, Moscow (1961)
3. Cacuci, D.G.: Sensitivity theory for nonlinear systems. I. Nonlinear functional analysis approach. J. Math. Phys. **22**, 2794–2802 (1981). (II. Extensions to additional classes of responses. Ibid, 2803–2812).
4. Cacuci, D.G., Wever, C.F., Oblow, E.M., Marable, J.H.: Sensitivity theory for general systems of nonlinear equations. Nucl. Sci. Eng. **75**(1), 88–110 (1980)
5. Marchuk, G.I.: Adjoint Equations and Analysis of Complex Systems. Kluwer, Dordrecht (1995)
6. Marchuk, G.I., Agoshkov, V.I., Shutyaev, V.P.: Adjoint Equations and Perturbation Algorithms in Nonlinear Problems. CRC Press, Boca Raton (1996)
7. Vladimirov, V.S., Marchuk, G.I.: A definition of an adjoint operator for nonlinear problems. Dokl. Math. **61**(3), 438–441 (2000)
8. Marchuk, G.I.: Numerical Solution of Problems of the Dynamics of Atmosphere and Ocean. Gigrometeoizdat, Leningrad (1974)
9. Marchuk, G.I., Skiba, Yu.N.: Numerical calculation of the conjugate problem for a model of the thermal interaction of the atmosphere with the ocean and continents. Izv. Atmos. Ocean. Phys. **12**(5), 279–284 (1976)
10. Marchuk, G.I., Skiba, Yu.N.: Role of the adjoint equation solution in estimating the mean temperature anomalies. Atmósfera **5**(3), 119–133 (1992)
11. Hall, M.C.G., Cacuci, D.G.: Physical interpretation of the adjoint functions for sensitivity analysis of atmospheric models. J. Atmos. Sci. **40**, 2537–2546 (1983)
12. Penenko, V.V.: Computational aspects of modelling the atmospheric process dynamics and estimation of effects of various factors on the dynamics of the atmosphere. In: Actual Problems of Numerical and Applied Mathematics. , pp. 61–77. Nauka, Novosibirsk (1975)
13. Penenko, V.V.: Estimation of the parameters of discrete models of the dynamics of atmosphere and ocean. Sov. Meteorol. Hydrology **7**, 57–68 (1979)
14. Lewis, J., Derber, J.C.: The use of adjoint equations to solve a variational adjustment problem with advective constraints. Tellus **37A**, 309–322 (1985)
15. Robertson, L., Persson, C.: Attempts to apply four dimensional data assimilation of radiological data using the adjoint technique. In: Radiation Protection Dosimetry, vol. 50, pp. 333–337. Nuclear Technology Publishing (1993).
16. Shutyaev, V.P.: Solvability of the data assimilation problem in the scale of Hilbert spaces for quasilinear singularly perturbed evolutionary problems. Russ. J. Numer. Anal. Math. Modelling **12**(1), 53–66 (1997)
17. Shutyaev, V.P.: Control operators and iterative algorithms in problems of reconstructing source functions and initial data. Russ. J. Numer. Anal. Math. Modelling **14**(2), 137–176 (1999)
18. Talagrand, O., Courtier, P.: Variational assimilation of meteorological observations with the adjoint vorticity equation. I. Theory. Q. J. Roy. Meteor. Soc. **113**, 1311–1328 (1987)
19. Marchuk, G.I., Skiba, Yu.N., Protsenko, I.G.: Method of calculating the evolution of random hydrodynamic fields on the basis of adjoint equations. Izv. Atmos. Ocean. Phys. **21**(2), 87–92 (1985)
20. Marchuk, G.I., Skiba, Yu.N., Protsenko, I.G.: Application of adjoint equations to problems of estimating the state of random hydrodynamic fields. Izv. Atmos. Ocean. Phys. **21**(3), 175–180 (1985)
21. Kolmogorov, A.N., Fomin, S.V.: Elements of the Theory of Functions and Functional Analysis. Graylock Press, Rochester, New York (1957)
22. Reed, M., Simon, B.: Functional Analysis I. Academic Press, London (1980)
23. Skiba, Yu.N., Adem, J., Morales-Acoltzi, T.: Numerical algorithm for the adjoint sensitivity study of the Adem ocean thermodynamic model. Atmósfera **9**(2), 147–170 (1996)
24. Marchuk, G.I.: Mathematical Models in Environmental Problems. Elsevier, New York (1986)

25. Skiba, Yu.N., Parra-Guevara, D.: Industrial pollution transport. Part I: Formulation of the problem and air pollution estimates. Environ. Model. Assess. **5**(3), 169–175 (2000)
26. Skiba, Yu.N.: Balanced and absolutely stable implicit schemes for the main and the adjoint pollutant transport equations in the limited area. Rev. Int. Contam. Ambient. **9**(2), 39–51 (1993)
27. Rahman, M.: Integral Equations and Their Applications. WIT Press, UK (2007)
28. Seinfeld, J.H., Pandis, S.N.: Atmospheric Chemistry and Physics: From Air Pollution to Climate Change. Wiley, New York (1998)
29. Enting, I.G.: Inverse Problems in Atmospheric Constituent Transport. Cambridge University Press, Cambridge (2002)
30. Skiba, Y.N., Parra-Guevara, D.: Applications of Adjoint Equations to Problems of Dispersion and Control of Pollutants. Nova Science Publishers Inc., New York (2015)
31. Srivastava, H.M., Buschman, R.G.: Theory and Applications of Convolution Integral Equations. Springer, Dordrecht (1992)
32. Domínguez, A.: A history of the convolution operation. IEEE PULSE, 38–49 (2015). (January/February)
33. Aroda, P., Guergachi, A., Huang, H.: Application of the convolution operator for scenario integration with loss data in operational risk modeling. J. Oper. Risk **10**(4), 23–44 (2015)
34. Keats, A., Yee, E., Lien, F.S.: Bayesian inference for source determination with applications to a complex urban environment. Atmos. Environ. **41**, 465–479 (2007)
35. Parra-Guevara, D., Skiba, Yu.N.: Quadratic programming formulation for controlling the emissions of air pollution point sources. In: Clary, T.S. (ed.) Horizons in Computer Science Research, Vol. 14, pp. 207–247, Chapter 7. Nova Science Publishers, New York (2017)
36. Park, S., Shin, M., Im, J., Song, C.-K., Choi, M., Kim, J., Lee, S., Park, R., Kim, J., Lee, D.-W., Kim, S.-K.: Estimation of ground-level particulate matter concentrations through the synergistic use of satellite observations and process-based models over South Korea. Atmos. Chem. Phys. **19**, 1097–1113 (2019)
37. Saleh, S., Hasan, G.: Estimation of PM10 concentration using ground measurements and Landsat 8 OLI satellite image. J. Geophys. Remote Sens. **3**(2), 1–6 (2014)
38. Parra-Guevara, D., Skiba, Yu.N.: Elements of the mathematical modelling in the control of pollutants emissions. Ecol. Model. **167**(3), 263–275 (2003)
39. Parra-Guevara, D.: Modelación Matemática y Simulación Numérica en el Control de Emisiones Industriales, Ph. D. Thesis, Centro de Ciencias de la Atmósfera, UNAM, México, pp. 130, (2001)
40. Seinfeld, J.H.: Atmospheric Chemistry and Physics of Air Pollution. Wiley, New York (1992)

Chapter 4
Examples of Application of Dual Estimates

It is well known that adjoint estimates serve as one of the main tools in the sensitivity study of dispersion models with respect to variations in their parameters, as well as in the development of optimal strategies for controlling pollutant emissions [1]. In particular, they are very important to control the air quality in the industrial regions. For example, timely application of pollution control strategies provides quantitative information about the required reduction in the emission rates of pollutants from sources in order to prevent violations of health standards within the next few days. These control strategies take into account the positions and intensities of existing pollution sources and use complicated models that describe the processes of transformation and dispersion of pollutants in the atmosphere.

In this chapter, some applications of direct and adjoint estimates are considered. The role of dual estimates in the study of the oil spill effects is shown in Sects. 4.1 and 4.2. Two-thirds of the carbon monoxide emissions were found to come from transportation sources (due to incomplete combustion of hydrocarbons in automobiles). In urban areas, the percentage exceeded 90%. This problem is analyzed in Sect. 4.3. A method of detecting the industrial plants that violate the emission rates prescribed by a control strategy is described in Sect. 4.4. This method can be used to detect the infringers and apply sanctions against them. A method that determines a suitable place for the construction of a new industrial plant in the region is given in Sect. 4.5. Its location is considered to be optimal in the sense that the operation of the new industrial plant will not violate health standards in environmentally important areas. The method is illustrated with a simple example that considers only two types of climatic winds, eleven industrial plants and three environmentally important areas.

Y. N. Skiba and D. Parra Guevara, *Mathematical Methods for the Assessment and Control of Industrial Emissions*, Mathematical Engineering,
https://doi.org/10.1007/978-3-031-58109-0_4

4.1 Simple Direct and Adjoint Oil Spill Problems

The direct and adjoint estimates (3.4.6) and (3.4.7) can be applied to diverse pollution problems, for example, to the problem of oil spill [2–4]. Maritime oil transportation has increased enormously in recent decades. A heavy oil spill caused by an accident involving tanker often has dramatic impact on the environment. The problem is of great practical importance [5–7], and many studies have been devoted lately to the oil slick movement and spreading [2, 8, 9]. The mathematical formulation of the model is complicated by the fact that the oil being released into marine environment is subjected to various weather processes, such as spreading and drift, advection and dissolution, evaporation and sinking, etc. [10]. Nethertheless, simulation of the oil movement modeling can give additional insights into the dynamical processes that influence such a motion. Obviously, the knowledge of solution to the oil transport equation allows us to get a comprehensive idea of the oil spill consequences in the whole area, and in particular, to evaluate average oil concentrations in a few ecologically important zones (fishery region, tourist coastal zone, beaches, etc.).

We now consider a very simple one-dimensional example of a pollution problem to elucidate the role of the adjoint solution in estimating average oil concentrations in ecologically significant zones. Suppose that interval $D = (a, b)$ of the x-axis represents a straight part of the oil tanker route, x_0 is the accident point, $Q(t)$ is the oil spill rate from the damaged tanker, and $\Omega = [y, y + h]$ is an ecologically important zone in D (Fig. 4.1). For simplicity, assume that the oil velocity $U(x, t) = U$ is constant and axially directed. Then the Eq. (2.4.1) of the oil slick motion in (a, b) and time interval $(0, T)$ is reduced to

$$\frac{\partial \phi}{\partial t} + U \cdot \frac{\partial \phi}{\partial x} = Q(t)\delta(x - x_0) \tag{4.1.1}$$

provided that all dissipative processes are neglected ($\mu = \sigma = 0$). We also assume that solution is periodic in $(a, b) : \phi(b, t) = \phi(a, t)$, and satisfy the initial (free from oil) condition $\phi(x, 0) = 0$ at $t = 0$. We will estimate the average oil concentration

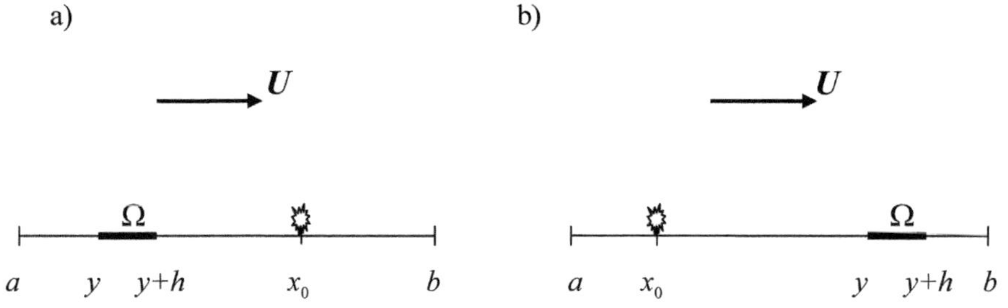

Fig. 4.1 One-dimensional oil propagation problem. Two possible variants of movement of the oil slick: **a** a harmless situation for zone Ω, **b** a dangerous situation for zone Ω

$$J(\phi) = \frac{1}{h}\int_{y}^{y+h} \phi(x, T)dx \tag{4.1.2}$$

in the zone Ω at the moment $t = T$. Introducing the inner product of two functions in (a, b) by

$$(\phi, g) = \int_a^b \phi(x)g(x)dx \tag{4.1.3}$$

and using the Lagrange identity $(A\varphi, g) = (\varphi, A^* g)$ to define the adjoint operator, it is easy to show that operator $A\phi \equiv U\partial\phi/\partial x$ of problem is skew-symmetric: $A^* = -A$ (see operator (3.3.3)). Indeed, using the periodic conditions $g(b, t) = g(a, t)$ we obtain

$$(A\varphi, g) = \int_a^b (U\frac{d}{dx}\varphi)g\, dx = \int_a^b \varphi(-U\frac{d}{dx}g)\, dx = (\varphi, -Ag) = (\varphi, A^* g)$$

Following the adjoint approach, we now formulate in the domain $(a, b) \times (0, T)$ the adjoint transport problem

$$-\frac{\partial g}{\partial t} - U\frac{\partial g}{\partial x} = 0 \tag{4.1.4}$$

based on the adjoint operator $A^* = -A$ with "initial" condition at the final moment $t = T$:

$$g(x, T) = q(x) \equiv \begin{cases} 1/h & \text{if } x \in [y, y + h] \\ 0, & \text{otherwise} \end{cases} \tag{4.1.5}$$

The exact solution of problem (4.1.4) and (4.1.5) in $(a, b) \times (0, T)$ is

$$g(x, t) = q(x + U(T - t)) \tag{4.1.6}$$

Thus, it is the step function (4.1.5) that moves with velocity U in the direction opposite to the vector U. Subtracting the Eq. (4.1.4), pre-multiplied by $\phi(x, t)$, from the Eq. (4.1.1) pre-multiplied by $g(x, t)$, then integrating the result obtained over the domain $(a, b)\times(0, T)$, and taking account of the Lagrange identity, initial conditions, and solution (4.1.6) we obtain the adjoint estimate equivalent to (4.1.2):

$$J(\phi) = \int_0^T g(x_0, t)Q(t)dt = \int_0^T q(x_0 + U(T - t))Q(t)dt \tag{4.1.7}$$

This estimate explicitly links the average concentration $J(\phi)$ of oil in the zone Ω with the rate of oil spill $Q(t)$ through the value of the adjoint problem solution $g(x_0, t)$ at the point of the accident [2].

Evidently, two situations are possible. The first one is quite safe for the zone Ω (Fig. 4.1a), while the other is dangerous (Fig. 4.1b). In fact, if the oil velocity vector U is pointed away from the zone Ω (Fig. 4.1a) then the adjoint solution $g(x_0, t) = q(x_0 + U(T - t))$ in the integral (4.1.7) is zero in $(0, T)$ (at least if $(b - x_0) + (y - a) > UT$), and therefore $J(\phi) = 0$ in the zone Ω at $t = T$, that is, the zone Ω will not be contaminated by oil during time interval $(0, T)$. On the contrary, let oil velocity vector U be pointed towards the zone Ω (Fig. 4.1b). Then $g(x_0, t) = q(x_0 + U(T - t))$ is nonzero and equal to $1/h$ only within time interval $[\tau, \tau + \Delta\tau]$ where $\tau = (y - x_0)/U$ is the time required for the oil to reach the zone, and$\Delta\tau = h/U$. Therefore, if$\tau_f = \min\{\tau + \Delta\tau, T\}$, then (4.1.7) and (4.1.5) lead to the nonzero value

$$J(\phi) = \frac{1}{h} \int_{\tau}^{\tau_f} Q(t)\, dt \tag{4.1.8}$$

of the average oil concentration in zone Ω at the moment $t = T$. It is of interest to note that if oil spill from the damaged tanker stopped at the moment $t = \tau$ ($Q(t) = 0$ for $t \geq \tau$) then average oil concentration $J(\phi)$ in Ω is zero again at $t = T$.

Thus, the adjoint problem solution is a weight function, or informative time–space function, which helps us to understand the mechanism of formation of average oil concentrations in ecologically significant zones (i.e., the response of the model solution to variations in the accident site x_0 and oil spill rate $Q(t)$).

Adjoint estimation and oil spill model response. Suppose that an accident takes place with an oil tanker in a two-dimensional sea area D. Let $r = (x, y)$ be a point of domain D, and let $r_0 = (x_0, y_0)$ denote the accident site in D. We now consider the two-dimensional dispersion model (2.4.1)–(2.4.6) to describe the oil transport in the domain D and time interval $(0, T)$:

$$\frac{\partial}{\partial t}\phi + U \cdot \nabla\phi + \sigma\phi - \nabla \cdot \mu\nabla\phi = f(r, t) = Q(t)\delta(r - r_0) \tag{4.1.9}$$

$$\nabla \cdot U = 0 \tag{4.1.10}$$

$$\phi(r, 0) = 0 \tag{4.1.11}$$

$$\mu\frac{\partial\phi}{\partial n} - U_n\phi = 0 \ \text{ at } S^- \tag{4.1.12}$$

$$\mu\frac{\partial\phi}{\partial n} = 0 \ \text{ at } S^+ \tag{4.1.13}$$

where $\phi(r, t)$ is the oil concentration in $D \times (0, T)$, and $Q(t)$ is the oil discharge rate from the oil tanker. Then the adjoint problem accepts in the domain $D \times (0, T)$ the following form (see (3.4.9)–(3.4.12)):

$$-\frac{\partial}{\partial t}g - U\cdot\nabla g + \sigma g - \nabla\cdot(\mu\nabla g) = p(r,t) \tag{4.1.14}$$

$$g(r,T) = 0 \ \text{ in } D \tag{4.1.15}$$

$$\mu\frac{\partial g}{\partial n} = 0 \ \text{ at } S^- \tag{4.1.16}$$

$$\mu\frac{\partial g}{\partial n} + U_n g = 0 \ \text{ at } S^+ \tag{4.1.17}$$

Let us estimate the average concentration of oil in a space–time domain $\Omega_\tau = \Omega\times(T-\tau,T)$ where Ω is a zone in D and $0\le\tau\le T$. It is easy to see that in our case identidad (3.5.3) is reduced to

$$\int_0^T\int_D p(r,t)\phi(r,t)drdt = \int_0^T g(r_0,t)Q(t)dt \tag{4.1.18}$$

If we take

$$p(r,t) = \begin{cases} 1/(\tau|\Omega|), \text{if}(r,t)\in\Omega_\tau \\ 0, \qquad \text{otherwise}\end{cases} \tag{4.1.19}$$

where $|\Omega|$ is the area of the zone Ω then

$$\int_{T-\tau}^T\int_\Omega p(r,t)drdt = 1 \tag{4.1.20}$$

and functional

$$J(\phi) = \int_0^T\int_D p(r,t)\phi(r,t)drdt = \frac{1}{\tau|\Omega|}\int_{T-\tau}^T\int_\Omega \phi(r,t)drdt \tag{4.1.21}$$

can be considered as the average concentration of oil in the domain Ω_τ [4, 11, 12].

Due to (4.1.18), the adjoint estimate equivalent to (4.1.21) has the form

$$J(\phi) = \int_0^T g(r_0,t)Q(t)dt \tag{4.1.22}$$

where $g(r_0,t)$ is the value of the solution of the adjoint problem at the accident site r_0. To use the adjoint estimate (4.1.22), it is necessary solve the problem (4.1.14)–(4.1.17) taking (4.1.19) as the forcing. Note that the solution $g(r,t)$ is non-negative, because $p(r, t)$ is non-negative, and therefore $g(r_0, t)$ serves in (4.1.22) as a weight function that represents the dependence of the average oil concentration $J(\phi)$ on r_0 and $Q(t)$.

In the particular case when $Q(t) = Q = Const$ (oil discharge rate does not change in the time interval $(0, T)$), formula (4.1.22) is further simplified:

$$J(\phi) = Q \int_0^T g(r_0, t) dt \tag{4.1.23}$$

Due to the linearity of problems (4.1.9)–(4.1.13) and (4.1.14)–(4.1.17), we have

$$\delta J(\phi) = \int_0^T g(r_0, t) \delta Q(t) dt \tag{4.1.24}$$

where $\delta J(\phi)$ is a variation in the average oil concentration in Ω_τ, and $\delta Q(t)$ is a variation in the oil discharge rate from the damaged tanker. Therefore, the adjoint estimate (4.1.22) determines the response of the model to a specific accident site r_0 and oil discharge rate $Q(t)$ from the damaged tanker, and (4.1.24) is useful in studying the sensitivity of the estimate $J(\phi)$ to variations in r_0 and $Q(t)$ [3].

4.2 Application of Dual Estimates in Oil Spill Problems

Direct and adjoint methods complement each other in the study of oil spills [3, 4, 13, 14]. In some situations, one of the estimates (4.1.21) or (4.1.22) is preferrable. The direct estimate (4.1.21) utilizes the solution $\phi(r, t)$ of problem (4.1.9)–(4.1.13), and hence, depends on the main oil spill parameters: the oil spill rate $Q(t)$ and accident site r_0. Therefore, the direct estimate is preferable if a comprehensive information should be obtained on the oil concentration in the whole area D or in many its parts, and also if the time required to take precautions against polluting a zone Ω is assessed ([3], example 3). By contrast, the adjoint estimate (4.1.22) employs the solution $g(r_0, t)$ of adjoint problem (4.1.14)–(4.1.17). Being independent of both $Q(t)$ and r_0, the adjoint problem can be solved for each ecologically sensitive zone Ω irrespective of a concrete incident. The following examples show that in some applications this property can be very useful.

1. Search for the most dangerous point of tanker route. Assume that the ecologically sensitive zone Ω, time interval $(0, T)$, and the value τ are fixed, and the oil velocity U, the oil spill rate $Q(t)$ from the damaged tanker are known [3]. Then the adjoint approach is efficient for studying the r_0-sensitivity of the estimate (4.1.22) (Fig. 4.2). In particular, its application is convenient to answer the question: which point of the oil tanker route is the most dangerous in the sense that accident in this point will result in the highest value of (4.1.22) in the zone Ω?

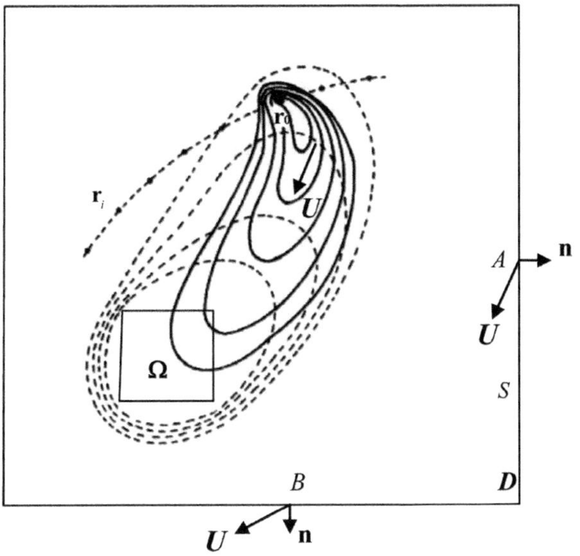

Fig. 4.2 The location of the incident r_0 and the zone Ω in domain D. Isolines of solutions of main and adjoint problems are schematically given by solid and dashed lines, respectively; r_i are the grid points lying along the tanker route (dash-dotted line); points A and B belong to parts S^- and S^+ of boundary S, respectively

Let us use a grid in the domain D and suppose that there are K grid nodes r_i lying along the oil tanker route in D ($i = 1, 2, ..., K$; Fig. 4.2). Following the direct method, the K solutions ϕ_i of problem (4.1.9)–(4.1.13) must be calculated with the forcing $Q(t)$ successively located at each of these K points r_i, and the integral (4.1.21) must be taken with each solution ϕ_i ($i = 1, 2, ..., K$). Then the most dangerous point of the oil tanker route is determined by the maximum of these integrals.

However, this method is laborious and time-consuming, and the application of the adjoint method is preferable here. Indeed, it does not require solving the problem (4.1.9)–(4.1.13), and the adjoint problem (4.1.14)–(4.1.17) must be solved for the zone Ω only once. Then for each point r_i of the tanker route the integral

$$J_i(\phi) = \int_0^T g(r_i, t)Q(t)dt \tag{4.2.1}$$

or its approximate value

$$J_i(\phi) \cong \Delta t \sum_{n=1}^{N} g(r_i, t_n)Q(t_n) \tag{4.2.2}$$

is calculated ($i = 1, 2, ..., K$). Note that $t_n = n\Delta t$ is a node of the regular grid with a sufficiently small mesh size Δt in the time interval $(0, T)$, $t_0 = 0$, $t_N = T$, and $g(r_i, t_n)$ and $Q(t_n)$ are the mean values of functions $g(r_i, t)$ and $Q(t)$ in the interval(t_{n-1}, t_n). The maximum of these K values $J_i(\phi)$ determines the most dangerous point r_i of the route. Thus, the adjoint method is K times faster than the direct method [15]. If the number K of the route points is large then the difference between two methods is essential. This difference is especially large in the case of using a three-dimensional pollution transport problem [16].

2. The case of stationary velocity. The adjoint method is especially effective in the case when the oil transport velocity U is time-invariant, for example, U is a seasonal or monthly climatic velocity [3]. Indeed, in contrast to the oil transport problem (4.1.9)–(4.1.13), the adjoint problem (4.1.14)–(4.1.17) is not linked with the accident site r_0 and the oil spill rate $Q(t)$. Its solution $g(r, t)$ depends only on the zone Ω and can be calculated in advance for each zone and each type of velocity U regardless of any concrete accident with an oil tanker. Moreover, estimate (4.1.22) requires the values of adjoint solution $g(r_i, t)$ solely at the accident site. Therefore, due to (4.2.2), it is sufficient to keep in the computer only the values $g(r_i, t_n)$ of adjoint solution at the grid points r_i that are on the path of the oil tanker (Fig. 4.2). Indeed, any of such grid points (and only those grid points) may be a possible accident site r_0. And if the oil spill happens, one can use (4.1.22) or (4.1.23) to give a preliminary estimate of the average concentration of oil in the zone Ω provided the tanker accident site r_0 and its oil spill rate $Q(t_n)$ are known. All one has to do is to take out of the computer the adjoint solution corresponding to the zone Ω, and then calculate the sum (4.2.2) using the values $g(r_i, t_n)$ at the grid point r_i which is nearest to the actual accident site r_0. Evidently, the distance between such two points (r_i and r_0) can be made arbitrary small by reducing the mesh size in the domain D. The assessment may be improved if a more accurate solution of the adjoint problem (4.1.14)–(4.1.17) is used in (4.2.2). An example is the adjoint solution found with the actual velocity U rather than with the climatic one.

3. $Q(t)$-dependence of oil concentration estimate. Let us assume that velocity of oil U and accident site r_0 are known, and zone Ω, time interval $(0, T)$ and value τ are fixed. Suppose that one needs to study we use a set of oil discharge rates $Q_i(t)$ $(i = 1, 2, ..., K)$ to study the effect of variations in the oil spill rate $Q(t)$ on the oil concentration $J(\phi)$ in the zone Ω. It can be done by comparing the values $J_i(\phi)$ obtained for different oil spill rates $Q_i(t)$ $(i = 1, 2, ..., K)$. Following the direct method, the transport problem (4.1.9)–(4.1.13) must be solved K times (for each $Q_i(t)$), and the obtained solutions should be used for calculating and comparing K integrals (4.1.21):

$$J_i(\phi) = \frac{1}{\tau|\Omega|} \int_{T-\tau}^{T} \int_{\Omega} \phi_i(r, t)drdt\,, \quad (i = 1, 2, ..., K)$$

However, if K is large enough, the use of adjoint estimate (4.1.22) (or the sum (4.2.2)) is more efficient (economical). Indeed, there is no need to solve the problem (4.1.9)–(4.1.13) at all. Instead, the adjoint problem (4.1.14)–(4.1.17) is solved only once, and then its solution $g(r, t)$ is repeatedly used to take K simple integrals

$$J_i(\phi) = \int_0^T g(r_0, t) Q_i(t) dt$$

for each $Q_i(t)$ ($i = 1, 2, ..., K$). Moreover, the adjoint estimate

$$J(\phi) = \int_0^{t_f} g(r_0, t) Q(t) dt \tag{4.2.3}$$

can be used to study the relationship between the oil concentration $J(\phi)$ in the zone Ω and the oil spill time $t_f \leq T$.

4.3 Vehicular Emission Impact Assessment

Emissions from automobiles and trucks operating on public roads represent a major portion of the air pollutants included in the emission inventories. The roadways are one of the main places of human exposure to carbon monoxide (CO). Hobbs [17] notes that more than one third of the carbon in vehicle emissions is now a highly poisonous gas CO. It is relatively insoluble, colorless, odorless and very stable gas with the residence time in the atmosphere approximately from 2 to 4 months [18]. CO is the most abundant air pollutant in the lower atmosphere with the exception of CO_2. It forms from combustion of hydrocarbon fuel. Since the principle source of carbon monoxide in urban areas is motor vehicle exhaust, CO concentrations closely correlate with traffic volume [19]. Increasing traffic volume produces high levels of carbon monoxide in urban areas affecting the health of human beings and animals. Diseases of the respiratory system, like bronchitis or pulmonary emphysema are generally correlated with level of CO [20, 21]. It does bind strongly to hemoglobin (the iron-containing protein that carries oxygen in blood). Thus, relatively small quantities of CO restrict the transport of oxygen by blood. Concentration of CO in city traffic often reaches 50 ppmv (parts per million by volume), and it can rise to 140 ppmv compared to approximately 0.05 ppmv in clean, unpolluted air [17].

Many recent works are devoted to the assessment [22–25], prediction [26, 27] and control [28, 29] of pollutants coming from vehicle emissions. There are various air pollution models, both statistical [30] and dynamical [31–33] allowing to produce air pollution forecast from global to local scales. Obviously, a dispersion model is appropriate only if it is able to predict pollution levels with a good precision. However, apart from a good model, an effective short-term control strategy is also required to avoid dangerous levels of pollutants predicted by the model. In this connection, new approaches to the problem of control of vehicle emissions should be found [34].

Vehicle emissions along the main roads can be described as line sources, and the adjoint approach allows evaluating the contribution of any part of highway to the contamination of any urban area. As we know, the adjoint method is especially effective in the study of the sensitivity of pollution concentrations in a zone with respect to variations in the positions of sources and their emission rates.

1. Dispersion problem. Let us consider again in a domain D and time interval $0 \leq t \leq T$ the two-dimensional pollution dispersion model (2.4.1)–(2.4.6):

$$\frac{\partial}{\partial t}\phi + U \cdot \nabla\phi + \sigma\phi - \nabla \cdot \mu\nabla\phi = f(r,t) \tag{4.3.1}$$

$$f(r,t) = \begin{cases} Q_i(r,t) \text{ if } r \in R_i \\ 0 \text{ if } r \notin R_i \end{cases} \quad (i = 1, 2, ..., L) \tag{4.3.2}$$

$$\nabla \cdot U = 0 \tag{4.3.3}$$

$$\phi(r,0) = \phi^0(r) \tag{4.3.4}$$

$$\mu\frac{\partial\phi}{\partial n} - U_n\phi = 0 \text{ at } S^- \tag{4.3.5}$$

$$\mu\frac{\partial\phi}{\partial n} = 0 \text{ at } S^+ \tag{4.3.6}$$

where $\phi(r,t)$ is the CO concentration at the point $r = (x, y)$ of domain D and the time moment t $(0 \leq t \leq T)$; $\mu(r,t) > 0$ is the diffusion coefficient, and term $\sigma\phi$ describes the exponential decay of CO due to deposition and destruction. Numerical experiments with Eq. (4.3.1) showed that the parameterization $\sigma\phi$ was rather good in the case of such contaminants as carbon monoxide, sulfur dioxide, lead, and carbon [19]. In general, the coefficients $\mu(r,t)$ and $\sigma(r,t)$ are rather complicated functions, and their exact values are often unknown. In practice, to simplify the problem, these functions are usually approximated by constant values: $\mu(r,t) = \mu = Const > 0$, $\sigma(r,t) = \sigma = Const > 0$ [33, 35, 36]. In forcing (4.3.2), $Q_i(r,t)$ is a continuous function that describes the time-dependent CO emission rate along the i-th road R_i, $i = 1, 2, ..., L$ (Fig. 4.3).

The model (4.3.1)–(4.3.6) is applied to the Metropolitan Area of Guadalajara City (MAGC), Mexico [37]. Since the most part of Guadalajara is located in the valley, we ignore the domain topography. Nevertheless, in the case of a 3D model, the topography can be very important. Then it can be taken into account by introducing the σ-coordinate [38, 39] or s-coordinate [40], which are widely used in atmospheric and oceanic models.

Integrating Eq. (4.3.1) over D and taking into account (4.3.3), (4.3.5) and (4.3.6) we obtain the CO balance equation

$$\frac{\partial}{\partial t}\int_D \phi\, dr = \sum_{i=1}^{L}\int_{R_i} Q_i(r,t)dl - \int_D \sigma\phi\, dr - \int_{S^+} U_n\phi\, dS \tag{4.3.7}$$

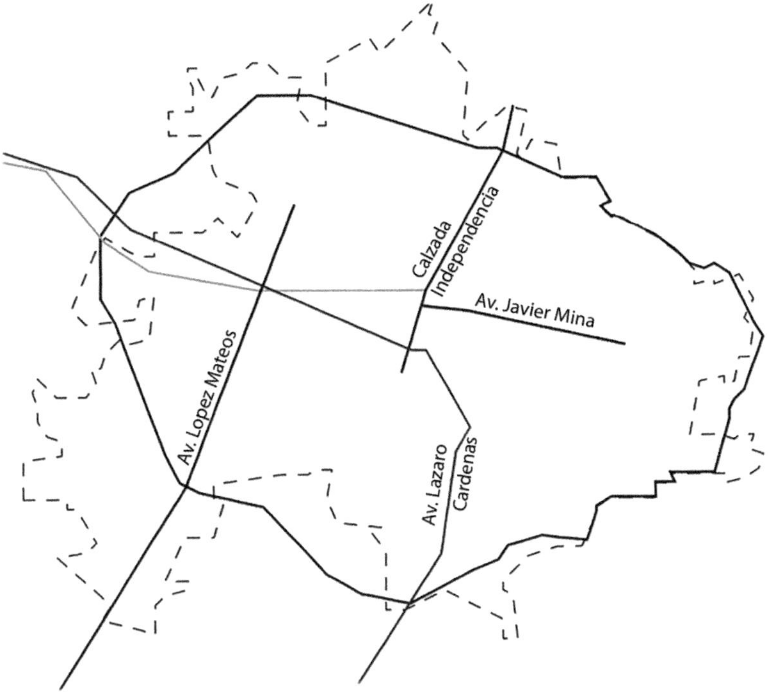

Fig. 4.3 The main roads in Guadalajara

Vehicular emissions of the i-th road R_i contribute to (4.3.7) by means of the line integral $\int_{R_i} Q_i(r,t)dl$ where dl is a line element of the road R_i [37].

According to (4.3.7), the total mass $\int_D \phi dr$ of carbon monoxide in the domain D increases because of vehicular emissions and decreases by reason of both deposition ($\sigma > 0$) and pollution outflow across the boundary part S^+.

Multiplying Eq. (4.3.1) by $\phi(r,t)$, integrating the result over D, and using again (4.3.3), (4.3.5) and (4.3.6) we obtain

$$\frac{\partial}{\partial t}\int_D \phi^2 dr = 2\sum_{i=1}^{L}\int_{R_i} Q_i(r,t)\phi(r,t)dl - 2\int_D (\sigma\phi^2 + \mu|\nabla\phi|^2)dr - \int_S |U_n|\phi^2 dS \tag{4.3.8}$$

This balance equation describes the evolution of the norm $\|\phi\| = (\int_D \phi^2 dr)^{1/2}$ of solution. One can see that this norm increases due to vehicular emissions and decreases due to deposition ($\sigma > 0$), diffusion ($\mu > 0$) and pollution flow across the boundary S of domain D.

If problem (4.3.1)–(4.3.6) is solved by a finite-difference method, the domain D is replaced by a grid. Suppose that the mesh size is sufficiently fine, and r_{ij} denotes the j-th grid point belonging to the road R_i ($i = 1, 2, ..., L; j = 1, 2, ..., N_i$). Then for each i, the emission rate $Q_i(r, t)$ of the line source is approximated by a set of point sources with emission rates $Q_{ij}(t)$ located at points r_{ij} (Fig. 4.4). In other words, $Q_{ij}(t)$ is the emission rate of the j-th line element of the road R_i ($i = 1, 2, ..., L; j = 1, 2, ..., N_i$). Thus, the discrete form of the source (4.3.2) can be represented as

$$f(r, t) \equiv \sum_{i=1}^{L} \sum_{j=1}^{N_i} Q_{ij}(t)\, \delta(r - r_{ij}) \tag{4.3.9}$$

Numerical experiments were carried out for the MAGC, with the Historical Centre was chosen as the zone Ω (*HC* zone in Fig. 4.4). It is the principle tourist and commercial zone that contains many historical monuments and is characterized by a high density of population. Due to serious air pollution problems suffers not only the health of people. Also, the historical monuments are subjected to serious corrosion

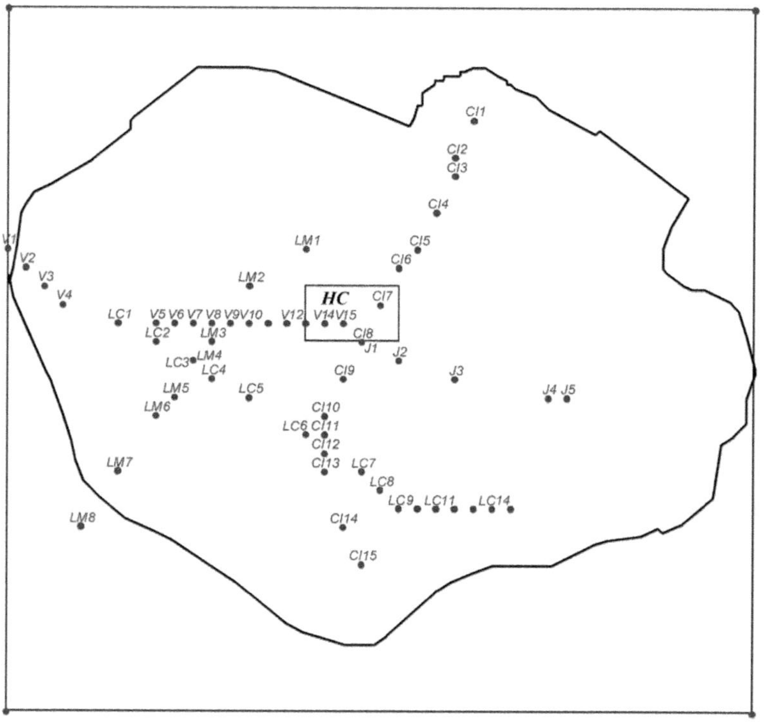

Fig. 4.4 Approximation of the road sources by the point sources in the Metropolitan Area of Guadalajara City

Fig. 4.5 Isolines of the solution to the pollution transport problem (4.3.1)–(4.3.6) calculated at $t = 180$ min with the rainy season wind (**a**) and dry season wind (**b**). *LC*: Av. Lazaro Cardenas; *CI*: Av. Calzada Independencia; *V*: Av. Vallarta; *J*: Av. Javier Mina

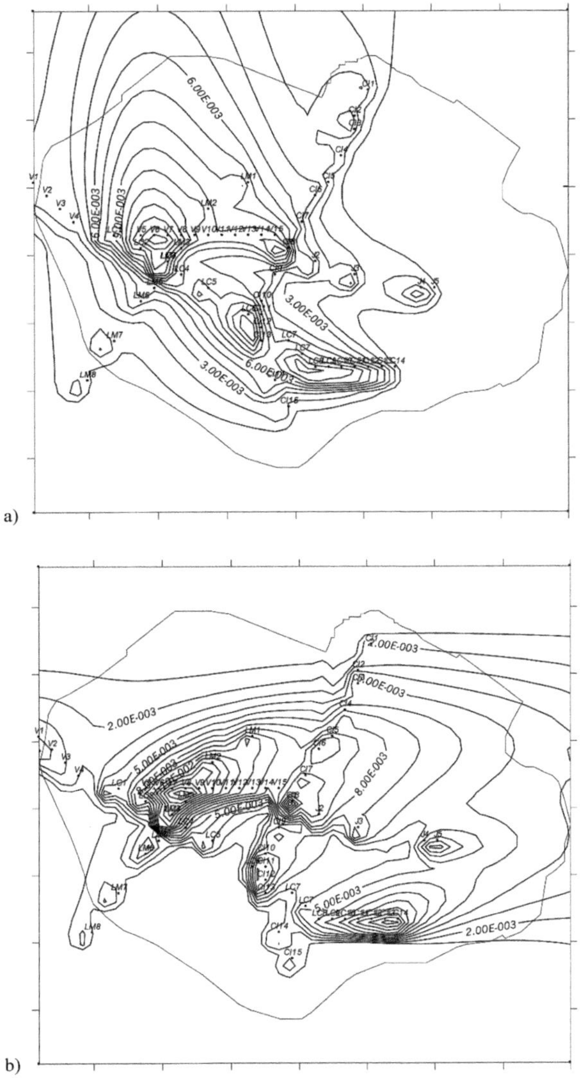

processes and require significant financial support for the preservation and restoration. Therefore, the protection of this zone from pollution is a problem of great social and economic importance. In particular, with the aim to control vehicular emissions, it is important to determine the parts of the roads that pollute the Historical Centre most of all.

To show the role of wind velocity U in the CO concentration, two types of climatic wind were used in all numerical experiments with the direct and adjoint pollution transport models in the MAGC: Wind I and Wind II, characteristic of the rainy season (June–September) and dry season (October–May), respectively. Both the winds were defined using a climatological archive covering the period 1881–1999 [41]. Also, in order to illustrate the possibilities of the direct and adjoint methods, we used just five main roads in the MAGC: Avenida Vallarta, Avenida Lopez Mateos, Calzada Independencia, Avenida Lazaro Cardenas, and Avenida Javier Mina (Fig. 4.3).

The CO emission rates for various parts of the roads sections were calculated using the methodology developed in MOBILES by the US Environmental Protection Agency (EPA-450/4-81-026d, 1992) for high altitudes, which takes into account local statistical information of the State Commission of Ecology [22, 23] on the average speed and mileage of the three types of the vehicles: (1) the light duty gasoline powered vehicls for mileage up to 50,000 mile; (2) the light duty gasoline powered vehicles for mileage greater than 50,000 mile; and (3) the heavy duty gasoline powered vehicles, light and heavy duty diesel powered vehicles and trucks. The total CO emission concentration values are in the interval from 17.8 to 28.9 g/s, having the maximum at the Minerva traffic circle (point $V9$ in Fig. 4.4) and along the Lazaro Cardenas avenue where the most heavy-duty gasoline and diesel traffic is observed.

At $t = 180$ min, isolines of the solution $\phi(r, t)$ of pollution transport problem (4.3.1)–(4.3.6) are given in Fig. 4.5a and b for Wind I and Wind II, respectively. In the figures, points on the main roads are marked by the corresponding letters and numbers (for example, $V5$, $V6$ for Av. Vallarta, $CI7$, $CI8$ for Av. Calzada Independencia, etc.).

The isolines with the largest values, initially located near the main roads, lengthen with time in the direction of the local wind U and also extend due to diffusion ($\mu(r, t) \neq 0$). One can see that vehicule emissions affect the HC zone during both dry and rainy seasons.

However, in practice it is often very importance to have quantitative information on the contribution of each road, or each its part, to the mean CO concentration value

$$J(\phi) \equiv \frac{1}{\tau|\Omega|} \int_{T-\tau}^{T} \int_{\Omega} \phi(r, t) dr dt \tag{4.3.10}$$

in a zone Ω (HC zone). It is rather difficult to get such quantitative information using the model (4.3.1)–(4.3.6), since this approach requires solving the model (4.3.1)–(4.3.6) repeatedly for each single source $Q_i(r, t)$ of the road R_i, or even for each source $Q_{ij}(t)$ of the road part. Thus, this way is laborious, since it requires time-consuming calculations.

However, it can easily be determined with the adjoint estimate (3.5.7). Indeed, for each type of wind, it is easy to calculate the solution of the adjoint model (4.1.14)–(4.1.17) for the HC zone, and then just apply the adjoint estimate (3.5.7) for various combinations of emission rates $Q_{ij}(t)$.

2. Adjoint problem. We recall that the adjoint problem (4.1.14)–(4.1.17) is run backward in time (from $t = T$ to $t = 0$). Its solution $g(r, t)$ is also subjected to the processes of advection ($U \neq 0$) and diffusion ($\mu \neq 0$). It should be noted that the isolines with the largest values of $g(r, t)$ are located near the HC zone as long as the forcing (4.1.20) is nonzero (i.e., in the time interval $(T-\tau, T)$. As t decreases from $t=T-\tau$ to $t=0$, $p(r,t)\equiv 0$, and the plume of the largest values of adjoint solution $g(r,t)$ moves in the direction opposite to that of velocity U and expands due to diffusion ($\mu \neq 0$). Figure 4.6a and b show the adjoint solutions calculated at $t=T-60$ min (dotted lines) and $t=T-120$ min (solid lines) for the HC zone with Wind I and Wind II, respectively. One can see that the number of the road points appearing in the region of large values of $g(r,t)$ essentially depends on the wind direction [37].

3. Influence functions. If we use the adjoint model (4.1.14)–(4.1.17) then the adjoint estimate (3.5.7) can be approximated as

$$J(\phi)=\sum_{i=1}^{L}\sum_{j=1}^{N_i}\delta l_{ij}\int_0^T g(r_{ij},t)Q_{ij}(t)dt+\int_D g(r,0)\phi^0(r)dr \tag{4.3.11}$$

where δl_{ij} is a small line element corresponding to the point r_{ij} of the road R_i ($i=1,2,...,L; j=1,2,...,N_i$), and $g(r_{ij},t)$ is the non-negative value of adjoint solution at road point r_{ij} and moment t. Therefore, $g(r_{ij},t)$ serves as an influence function that can intensify or weaken the role of the emission rate $Q_{ij}(t)$ in the pollution of the HC zone, and therefore its temporal behavior is of great importance. This behavior is shown in Figs. 4.7 and 4.8 for the Wind I and Wind II, respectively. It is seen that the values of $g(r_{ij},t)$ vary with time and are nonzero only within certain time subintervals [37]. As a result, the contribution of the source located at r_{ij} to the average concentration $J(\phi)$ in the HC zone is significant only if both $g(r_{ij},t)$ and $Q_{ij}(t)$ are large enough in the time interval $(0,T)$.

We now analyze the influence functions $g(r_{ij},t)$ for various parts of the roads. First, we note that values of these functions are always zero at the points of Av. Lopez Mateos for Wind I (Fig. 4.6a), and at the points of Av. Javier Mina for Wind II (Fig. 4.6b). Thus, these avenues do not pollute the HC zone during the corresponding seasons. At the same time, values of $g(r_{ij},t)$ at the points V14 and V15 of Av. Vallarta are large for both types of wind, i.e., these parts of the road pollute the HC zone throughout the year.

In particular, for Wind I, functions have maximum values at the points J1 and J2 of Av. Javier Mina (Fig. 4.7d) and have significant values at the points V15 and V14 of Av. Vallarta (Fig. 4.7b) and points CI8 and CI7 of Av. Calzada Independencia (Fig. 4.7c). Thus, according to (4.3.11), any traffic jam at these parts of the roads will result in significant polluting the HC zone. At the same time, Fig. 4.7 shows that the weights $g(r_{ij},t)$ for points of Av. Lazaro Cardenas are small compared to the corresponding weights for points of Av. Javier Mina and Av. Vallarta, i.e., in formula (4.3.11), the time integrals corresponding to the points of Av. Lazaro Cardenas can be neglected.

Obviously, the contribution of various points of the roads depends on the wind direction. For example, CO emissions at the point $V12$ of Av. Vallarta contribute considerably in the pollution of the HC zone during the dry season (Fig. 4.8b), but

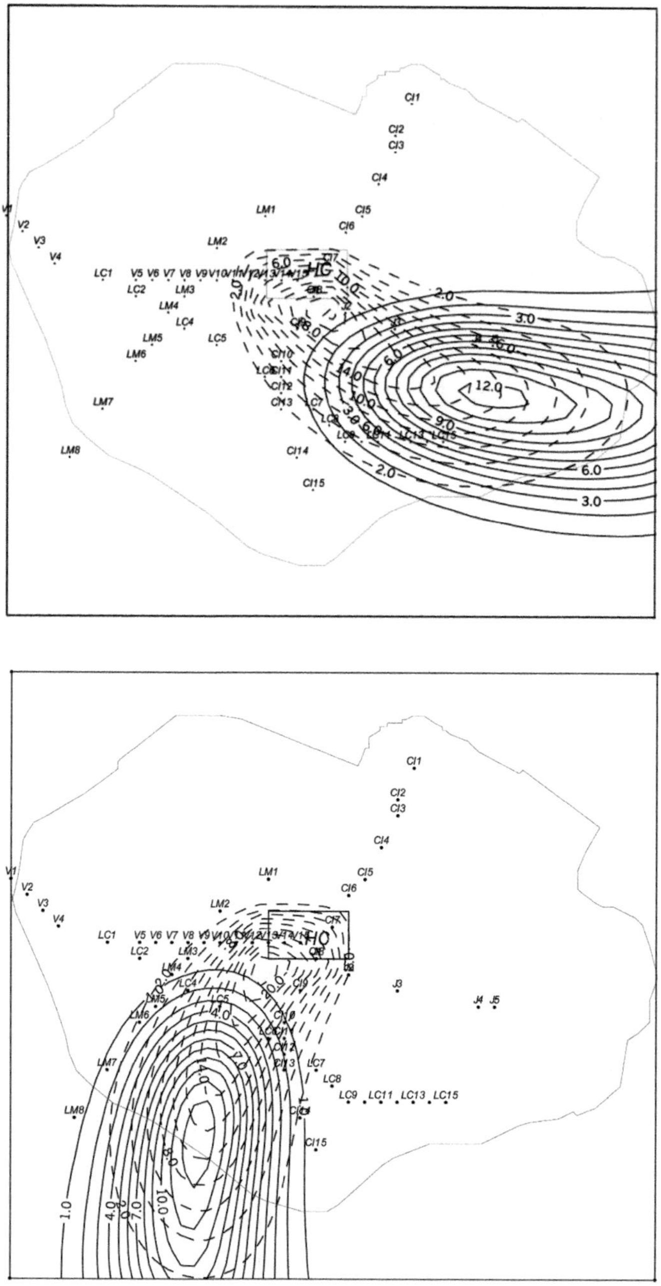

a)

b)

Fig. 4.6 Isolines of solution $g(r, t)$ of adjoint problem (4.1.14)–(4.1.17) calculated for the *HC* zone for two moments: $t = T - 60$ min (dotted lines) and $t = T - 120$ min (solid lines); $T = 180$ min, $\tau = 60$ min. Wind I (**a**) and Wind II (**b**)

Fig. 4.7 Temporal behavior of functions $g(r_{ij}, t)$ calculated with Wind I for *HC* zone at points r_{ij} of the roads ($T = 180$ min, $\tau = 60$ min). **a** Av. Lazaro Cardenas (*LC*); **b** Av. Vallarta (*V*); **c** Av. Calzada Independencia (*CI*), **d** Av. Javier Mina (*J*)

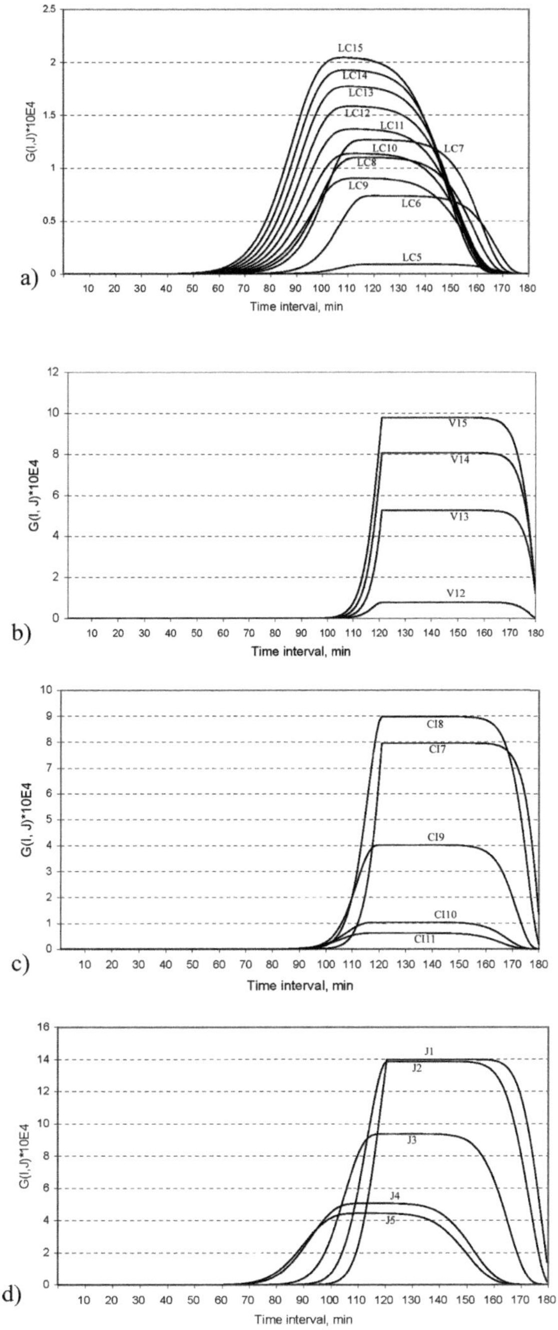

Fig. 4.8 Temporal behavior of functions $g(r_{ij}, t)$ calculated with Wind II for *HC* zone at points r_{ij} of the roads ($T = 180$ min, $\tau = 60$ min). **a** Av. Lazaro Cardenas (*LC*); **b** Av. Vallarta (*V*); **c** Av. Calzada Independencia (*CI*), **d** Av. Lopez Mateos (*LM*)

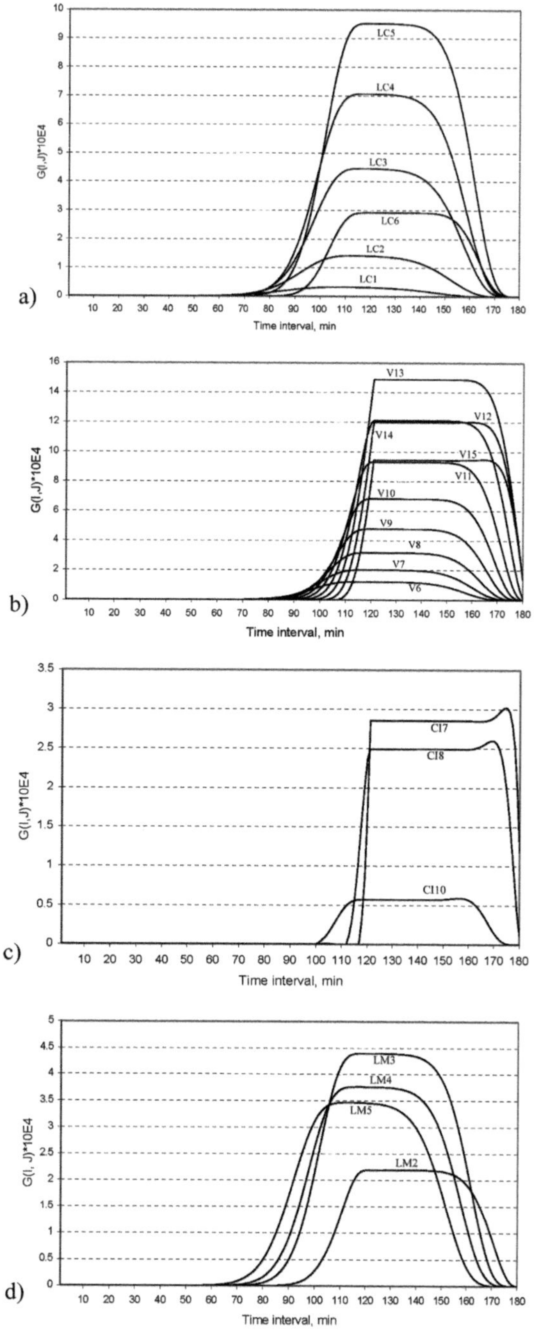

practically have no influence on this zone within the rainy season (Fig. 4.7b). In general, for Wind II, the weights $g(r_{ij}, t)$ are large for the points $V11$–$V15$ of Av. Vallarta (Fig. 4.8b) and the points $LC5$ and $LC4$ of Av. Lazaro Cardenas (Fig. 4.8a) while the impact of Av. Calzada Independencia and Av. Lopez Mateos is much weaker (Fig. 4.8c and d).

Since there is a non-zero distance between a road point r_{ij} and the HC zone, it takes some time for the pollutant to reach the zone. That is why each function $g(r_{ij}, t)$ in Figs. 4.7 and 4.8 is zero within its own initial period $(0, t_{ij})$ where t_{ij} depends on the wind and mutual location of the road point r_{ij} and the HC zone.

4.4 Detection of Industrial Plants that Violate Prescribed Emission Rates

Suppose that in a two-dimensional limited area D, there are N industrial plants in operation. Denote by $r_i = (x_i, y_i)$ and $Q_i(t)$ the position and emission rate of the i-th industrial plant, respectively ($i = 1, \ldots, N$). We now assume that a dispersion model (for example, the model (2.4.1)–(2.4.6)) is used in the domain D and time interval $(0, T)$ with emission rates $Q_i(t)$ and predicts a violation of air quality standard J_0 in a zone $\Omega \subset D$:

$$J(\phi) \equiv \frac{1}{\tau|\Omega|} \int_{T-\tau}^{T} \int_{\Omega} \phi(r, t) dr dt > J_0 \tag{4.4.1}$$

Suppose also that a control strategy has been applied and some reduced emission rates $\tilde{Q}_i(t)$ have been prescribed to all industrial plants to ensure the compliance with health standard in the zone Ω:$J(\phi) \leq J_0$. Thus, according to the control results, all sources must lower their emission rates from $Q_i(t)$ to $\tilde{Q}_i(t)$ in order to prevent the excessive pollution of zone Ω.

Let us consider a hypothetical situation when some industrial plants ignore this requirement and continue to work with unknown emission rates $\overline{Q}_i(t)$ that exceed the prescribed values $\tilde{Q}_i(t)$: $\overline{Q}_i(t) > \tilde{Q}_i(t)$. As a result, the measurements show again that $J(\phi) > J_0$. This situation raises the question of detecting and penalizing the infringers. How can one detect such rule-breakers? Though this problem is not trivial for time-varying emission rates $Q_i(t)$ (a partial solution will be given at the end of this section), we now show that it can easily be solved for invariable emission rates Q_i. Note that the requirement of constant emission intensity is not a strong constraint if the time interval $(0, T)$ is sufficiently small.

1. Permanent emission rates. We now describe the method of solution of the problem [42]. So, suppose that the i-th industrial plant usually operates at a constant emission rate Q_i if no control is used, had to operate within interval $(0, T)$ with a constant rate $\tilde{Q}_i$ prescribed by the control, but in fact operated with an unknown constant rate $\overline{Q}_i$. We denote $\delta Q_i = Q_i - \overline{Q}_i, i = 1, ..., N$. Let $J(\phi)$ and $\overline{J}(\phi)$ be the mean pollutant concentrations in the zone Ω forecasted by the model before control and measured after the control, respectively. Thus,

$$\delta J(\phi) = J(\phi) - \overline{J}(\phi) \tag{4.4.2}$$

is a known value. Let us choose K zones Ω_k in the domain D ($K \geq N$), and denote

$$\delta J_k(\phi) = J_k(\phi) - \overline{J}_k(\phi)$$

the known value (4.4.2) obtained for the k-th zone Ω_k ($k = 1, 2, \ldots, K$). Assuming that $\overline{J}_k(\phi) > J_0$ at least for one k, let us detect the industrial plants which are responsible for excessive pollution of the zone(s). Denote by $g_k(r, t)$ the solution of the adjoint problem (3.4.9)–(3.4.12) for the zone Ω_k with forcing

$$p_k(r,t) = \begin{cases} \dfrac{1}{\tau|\Omega_k|}, & \text{if } (r,t) \in \Omega_k \times (T-\tau, T) \\ 0, & \text{otherwise} \end{cases} \tag{4.4.3}$$

and by $g_k(r_i, t)$ its value in the position r_i ($i = 1, ..., N$) of the i-th source.

Since $\delta\phi^0(r) \equiv 0$ and δQ_i is constant for each i, the adjoint estimates (3.5.7) for each of K zones Ω_k accept the form

$$\delta J_k(\phi) = \sum_{i=1}^{N} \delta Q_i \int_0^T g_k(r_i, t)\, dt \quad (k = 1, 2, \ldots, K) \tag{4.4.4}$$

It is a linear system.

$$A\vec{x} = \vec{b} \tag{4.4.5}$$

or

$$\sum_{i=1}^{N} a_{ki}\delta Q_i = b_k, \quad (k = 1, 2, ..., K) \tag{4.4.6}$$

where

$$a_{ki} = \int_0^T g_k(r_i, t)dt \tag{4.4.7}$$

is the non-negative element of the $N \times K$ matrix A ($k = 1, ..., K;\ i = 1, ..., N$), $K \geq N$, and

$$b_k = \delta J_k(\phi) \tag{4.4.8}$$

The solution of system (4.4.6) determines δQ_i and then one can calculate values $\overline{Q}_i = Q_i - \delta Q_i$ $(i = 1, ..., N)$. Thus, if for some i, $\overline{Q}_i > \tilde{Q}_i$ then the i-th industrial plant broke the agreement and exceeded the emission rate $\tilde{Q}_i$ prescribed by the control strategy.

Remark 4.4.1 If $K = N$ then A is a square matrix. In this case it is important to estimate its condition number $\nu(A) = \|A\| \, \|A^{-1}\|$ [43]. In special simple cases when $\|B\| = \|E - A\| < 1$ in some norm (E is the identity matrix), the condition number can be evaluated as.

$$\nu(A) = \|A\| \left\|A^{-1}\right\| \leq (1 + \|E\|)/(1 - \|B\|) \tag{4.4.9}$$

(see for example [43]). The inequality is useful to check if the condition number $\nu(A)$ is moderate. Note, however, that in many cases, inequality $\|B\| < 1$ does not hold. To test if the condition number is large, sometimes it is helpful to use the following inequality [43]:

$$\|A\| \, \|A^{-1}\| \geq \frac{\| \vec{\varepsilon} \| / \| \vec{x} \|}{\| \vec{\delta} \| / \| \vec{b} \|} \tag{4.4.10}$$

where $\| \vec{\varepsilon} \|$ and $\| \vec{\delta} \|$ are the errors in the solution $\vec{x}$ and vector $\vec{b}$, respectively. If the system (4.4.5) is ill-conditioned ($\nu(A)$ is too large) then a regularization method should be applied to resolve this system with good accuracy [43].

Remark 4.4.2 One of the regularization methods leading to a well-conditioned square matrix A consists in choosing the zones Ω_k in such a way that they do not intersect each other, besides, each of them contains just one source ($K = N$) [42]. If, in addition, we consider the problem in a small interval $(0, T)$ and take $\tau = T$ in (4.4.3), then isolines of each weight function $g_k(r, t)$ calculated for the zone Ω_k will be concentrated just near the zone Ω_k (Fig. 4.9), and A will be a diagonally dominant square matrix:

$$|a_{ii}| > \sum_{j \neq i} |a_{ij}| \quad \text{for every } i \ (i = 1, ..., N)$$

Remark 4.4.3 In some cases, it is preferable to choose $K > N$. Then A is a rectangular matrix, and system (4.4.5) can be solved by the method of least squares [44]:

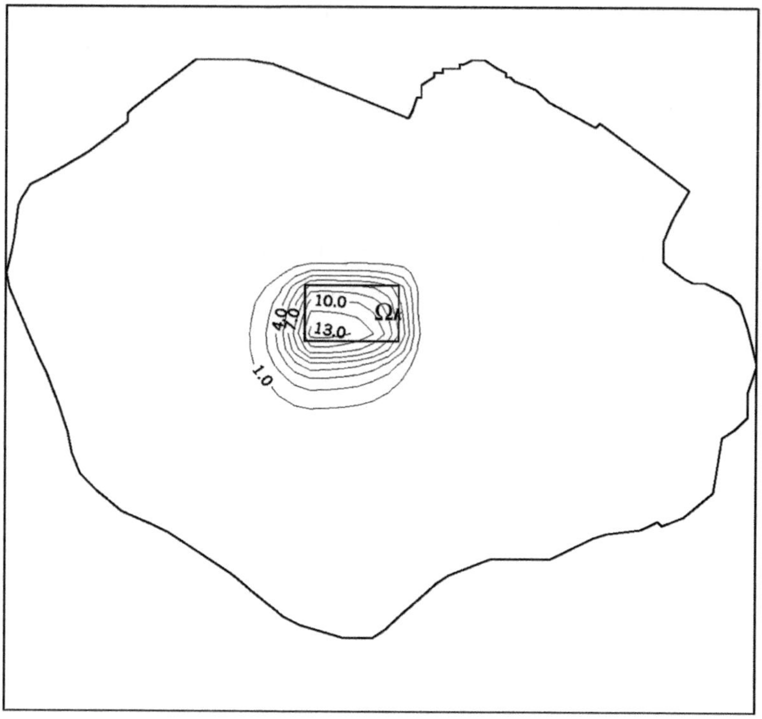

Fig. 4.9 Isolines of the adjoint solution $g_k(r, t)$ calculated for a zone Ω_k at a moment $t \in (T-\tau, T)$ which is close to $t = T$. The solution is non-zero only nearby the zone Ω_k

$$\vec{x} = (A^T A)^{-1} A^T \vec{b} \tag{4.4.11}$$

The solution (4.4.11) is less sensitive with respect to errors in the system parameters and measurements.

Remark 4.4.4 Sometimes, instead of average pollution concentrations $J_k(\phi)$ in the zones Ω_k, it is simpler to get mean values

$$J_k(\phi) = \frac{1}{\tau} \int_{T-\tau}^{T} \phi(c_k, t)dt \tag{4.4.12}$$

of $\phi(r, t)$ in K points c_k of the domain D ($k = 1, \ldots, K$). In this case, each function $g_k(r, t)$ appearing in (4.4.7) is the solution of the adjoint problem (3.4.9)–(3.4.12) with the forcing

$$p_k(r,t) = \begin{cases} \frac{1}{\tau}, & \text{if } r = c_k \text{ and } t \in (T-\tau, T) \\ 0, & \text{otherwise} \end{cases}, \quad k = 1, \ldots, K$$

In particular, if $c_k = \boldsymbol{r}_i$ where $\boldsymbol{r}_i$ is the position of the i-th industrial source then the pollution estimate (4.4.12) is transformed into the end-of-pipe control.

2. Time-varying emission rates. Let us now give a partial solution in the case when emission rates $Q_i(t)$ are time-varying. Indeed, the sensitivity formulas (4.4.4) take the form

$$\delta J_k(\phi) = \sum_{i=1}^{N} \int_0^T g_k(r_i, t)\, \delta Q_i(t)\, dt\,, \quad (k = 1, 2, \ldots, K) \tag{4.4.13}$$

where $\delta Q_i(t) = Q_i(t) - \overline{Q}_i(t)$ and $g_k(r_i, t)$ are continuous and non-negative weight functions. Therefore, according to the second mean value theorem for integrals, we have

$$\int_0^T g_k(r_i, t)\delta Q_i(t)dt = \delta Q_i^* \int_0^T g_k(r_i, t)dt$$

and hence (4.4.13) accept the same form as (4.4.4):

$$\delta J_k(\phi) = \sum_{i=1}^{N} \delta Q_i^* \int_0^T g_k(r_i, t)dt, \quad (k = 1, 2, \ldots, K) \tag{4.4.14}$$

where δQ_i^* is the mean value of function $\delta Q_i(t)$ in the interval $(0, T)$. Since the mean values δQ_i^*, Q_i^* and $\tilde{Q}_i^*$ can be calculated, the method for the permanent emission rates can also be applied for time depending emission rates with the aim to control the mean emission rates within interval $(0, T)$.

4.5 Optimal Location of a New Industrial Plant

We now consider an important problem. Assume that in addition to N industrial plants located at points r_i of a region D and operating with emission rates $Q_i(t)$ $(i = 1, \ldots, N)$, it is required to construct in D one more industrial plant, whose emission rate $Q_0(t)$ does not exceed a value Q_0:

$$Q_0 = \max_{0 \le t \le T} Q_0(t) \tag{4.5.1}$$

How a suitable location r_0 of the new plant in D can be found? The solution depends on the criteria used and is not unique [45, 46]. Let us select K zones Ω_k in

D $(k = 1, \ldots, K)$ and find the site r_0 of the new plant so that

$$J_k \leq J_s \,, \quad (k = 1, \ldots, K) \tag{4.5.2}$$

where J_k is the mean concentration $\phi(r, t)$ of pollutant in Ω_k, and J_s is the sanitary norm. For simplicity, we neglect here the initial distribution of the pollutant in D: $\phi^0(r) \equiv 0$. Then

$$J_k = \sum_{i=0}^{N} \int_0^T g_k(r_i, t) Q_i(t) dt \tag{4.5.3}$$

is the adjoint estimate (3.4.7) of the pollutant concentration in zone Ω_k when $N + 1$ plants (including the new one) are in operation, and

$$\tilde{J}_k = \sum_{i=1}^{N} \int_0^T g_k(r_i, t) Q_i(t) dt \tag{4.5.4}$$

is the analogous estimate obtained for the same zone, but in the case when the new plant is not taken into account. Then from (4.5.2) we get the following restriction for $Q_0(t)$:

$$\int_0^T g_k(r_0, t) Q_0(t) dt \leq J_0 - \tilde{J}_k = \overline{J}_k \tag{4.5.5}$$

Taking into account (4.5.1), we find that (4.5.2) is fulfilled if

$$\int_0^T g_k(r_0, t) dt \leq \overline{J}_k / Q_0 \,, \quad (k = 1, ..., K) \tag{4.5.6}$$

For each k, one can draw isolines of the function

$$F_k(x, y) \equiv F_k(r) = \int_0^T g_k(r, t) dt \,, \quad (k = 1, \ldots, K) \tag{4.5.7}$$

The isoline $F_k(x, y) \equiv \overline{J}_k / Q_0$ will separate a subdomain D_k where (4.5.6) holds from its complement $D \backslash D_k$ where $F_k(x, y) > \overline{J}_k / Q_0$. As a result, every point of the set $\tilde{D} = \cap_{k=1}^{K} D_k$, being the intersection of all subdomains D_k is suitable for locating the new industrial plant. These calculations can be repeated for typical (climatic) winds in the domain D for each season, and the intersection of domains $\tilde{D}$ obtained for different winds gives a narrower domain for the new plant. It is evident that the larger the number K of zones Ω_k, the narrower is the set $\tilde{D}$. Of course, other criteria

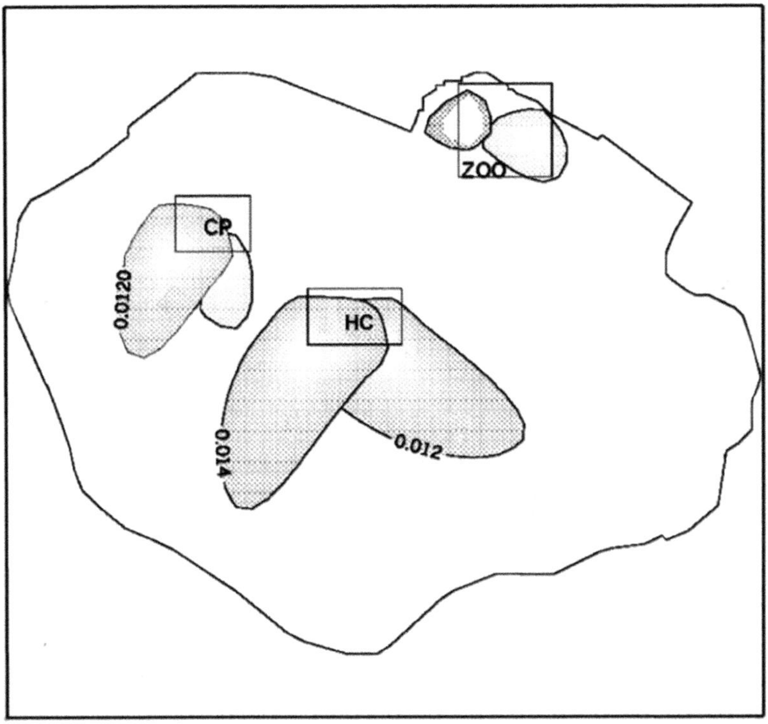

Fig. 4.10 The intersection $\tilde{D}$ of the six subdomains D_k defined by (4.5.6) for two types of climatic wind, $K = 3$ and $N = 11$. The non-shaded area $\tilde{D}$ is suitable for locating the new industrial plant. The field $F_k(x, y)$ was normalized

(social, economic, etc.) should also be taken into account to specify the best location of the new industrial plant in $\tilde{D}$. The unshaded area in Fig. 4.10 shows the area $\tilde{D}$ suitable for the locating the new industrial plant according to the criteria (4.5.2). This regine was obtained as the intersection of the six subdomains D_k defined by (4.5.6) for the case when only 11 industrial plants ($N = 11$), two climatic winds (Wind I and Wind II) and three zones (Zoo, Historical Center (HC) and Colomos park (CP)) were taken into account in Guadalajara [34]. Besides, all the plants had the same SO_2 emission rate: $Q_i(t) = 60g/s$ ($i = 1, ..., 11$).

References

1. Skiba, Yu.N., Parra-Guevara, D.: Pollution control methods. Int. J. Appl. Math. **25**(5), 675–710 (2012)
2. Proctor, R., Elliott, A.J., Flather, R.A.: Forecast and hindcast simulations of the Braer oil spill. Mar. Pollut. Bull. **28**(4), 219–229 (1994)

3. Skiba, Yu.N.: Dual oil concentration estimates in ecologically sensitive zones. Environ. Monitor. Assess. **43**(2), 139–151 (1996)
4. Skiba, Yu.N.: Direct and adjoint oil spill estimates. Environ. Monit. Assess. **59**(1), 95–109 (1999)
5. Kennicutt, M.C., II., McDonald, S.J., Sweet, S.T.: The Bahia Paraiso spill in Arthur Harbor. Anvers Island. Antarct. J. **27**, 331–340 (1992)
6. Schneider, D.: Oil spill has long-term effects. Water Envir. Technol. **5**, 30–32 (1993)
7. Wolff, G.A., Preston, M.R., Harriman, G., Rowland, J.S.: Some preliminary observations after the wreck of the oil tanker Braer in Shetland. Mar. Pollut. Bull. **26**, 567–571 (1993)
8. Elliott, A.J.: Shear diffusion and the spread of oil in the surface layers of the North Sea. Deutsche Hydrographische Zeitung **39**, 113–137 (1986)
9. Elliott, A.J., Dale, A.C., Proctor, R.: Modelling the movement of pollutants in the UK shelf seas. Mar. Pollut. Bull. **24**, 614–619 (1992)
10. Doerffer, J.W.: Oil Spill Response in the Marine Environment. Pergamon Press, Oxford (1992)
11. Marchuk, G.I., Skiba, Yu.N.: Numerical calculation of the conjugate problem for a model of the thermal interaction of the atmosphere with the ocean and continents. Izv. Atmos. Ocean. Phys. **12**(5), 279–284 (1976)
12. Parra-Guevara, D., Skiba, Yu.N.: Adjoint approach to estimate the non-steady emission rate of a point source. Int. J. Eng. Res. Appl. **3**(6), 763–776 (2013)
13. Dang, Q.A., Ehrhardt, M., Tran, G.L., Le, D.: Mathematical modelling and numerical algorithms for simulation of oil pollution. Environ. Model. Assess. **17**(3), 275–288 (2012)
14. Skiba, Y.N., Parra-Guevara, D.: Applications of Adjoint Equations to Problems of Dispersion and Control of Pollutants. Nova Science Publishers Inc., New York (2015)
15. Ramsay, M.A., Swannell, R.P.J., Shipton, W.A., Duke, N.C., Hill, R.T.: Effect of bioremediation on the microbial community in oiled mangrove sediments. Mar. Pollut. Bull. **41**, 413–419 (2000)
16. Skiba, Yu.N.: The derivation and applications of the adjoint solutions of a simple thermodynamic limited area model of the atmosphere–ocean–soil system. World Resour. Rev. **8**(1), 98–113 (1996)
17. Hobbs, P.V.: Introduction to Atmospheric Chemistry. Cambridge University Press, Cambridge (2000)
18. Wark, K., Warner, C.F.: Air Pollution, its Origin and Control. Harper & Row (1999)
19. Seinfeld, J.H.: Atmospheric Chemistry and Physics of Air Pollution. J. Wiley & Sons, New York (1992)
20. Seinfeld, J.H., Pandis, S.N.: Atmospheric Chemistry and Physics: From Air Pollution to Climate Change. Wiley, New York (1998)
21. Venn, A.J., Lewis, S.A., Cooper, M., Hubbard, R., Britton, J.: Living near a main road and the risk of wheezing illness in children. Am. J. Respir. Crit. Care Med. **164**(12), 2177–2180 (2001)
22. COESE, Factores de Emisión para Vehículos Automotores. Comisión Estatal de Ecología, Gobierno del Estado de Jalisco (1995)
23. EPA-450/4-81-026d, Procedures for Emission Inventory Preparation, Volume IV: Mobile Sources. United States Environmental Protection Agency, 1992.
24. Fu, L.X., Hao, J.M., He, D.Q., He, K.B., Li, P.: Assessment of vehicular pollution in China. J. Air Waste Manag. Assoc. **51**, 658–668 (2001)
25. Parrish, D.D., Trainer, M., Hereid, D., Williams, E.J., Olszyna, K.J., Harley, R.A., Meagher, J.F., Fehsenfeld, F.C.: Decadal change in carbon monoxide to nitrogen oxide ratio in US vehicular emissions. J. Geophys. Res. Atmos. **107**(D12): art. no. 4140 (2002)
26. Hassan, A.A., Crowther, J.M.: A simple model of pollutant concentrations in a street canyon. Environ. Monitor. Assess. **52**, 269–280 (1998)
27. Yu, L.E., Hildemann, L.M., Ott, W.R.: A mathematical model for predicting trends in carbon monoxide emissions and exposures on urban arterial highways. J. Air Waste Manag. Assoc. **46**, 430–440 (1996)
28. Goddard, H.C.: Promoting urban sustainability: the case for a tradable supplementary licence system for vehicle use. Urban Stud. **36**, 2317–2331 (1999)

29. Rusco, F.W., Walls, W.D.: Vehicular emissions and control polices in Hong-Kong. Contemp. Econ. Policy **13**, 50–61 (1995)
30. Murphy-Klimova, E., Fisher, B.E.A., Sokhi, R.: Treatment of urban areas within a regional transport model of sulphur and nitrogen oxides. Environ. Monitor. Assess. **52**, 213–224 (1998)
31. Brandt, J., Christensen, J.H., Frohn, L.M., Zlatev, Z.: Operational air pollution forecast modelling using the THOR system. Phys. Chem. Earth (B) **26**(2), 117–122 (2001)
32. Liu, M., Carroll, J.J.: A high-resolution air pollution model suitable for dispersion studies in complex terrain. Mon. Wea. Rev. **124**, 2396–2409 (1996)
33. Zakarin, E.A., Mirkarimova, B.M.: GIS-based mathematical modeling of urban air pollution. Izv. Atmos. Ocean. Phys. **36**, 334–342 (2000)
34. Skiba, Yu.N., Davydova-Belitskaya, V.: Air pollution estimates in Guadalajara City. Environ. Model. Assess. **7**, 153–162 (2002)
35. Gavrilov, V.P., Gormatyuk, Yu.K.: Pollutant dispersion from stationary sources in the surface layer of the atmosphere. Sov. Meteorol. Hydrol. **2**, 28–37 (1989)
36. Ku, J.-Y., Trivikrama Rao, S., Shankar Rao, K.: Numerical simulation of air pollution in urban areas: model development. Atmos. Environ. **21**(1), 201–212 (1987)
37. Skiba, Yu.N., Davydova-Belitskaya, V.: On the estimation of impact of vehicular emissions. Ecol. Model. **166**(1–2), 169–184 (2003)
38. Kantha, L.H., Clayson, C.A.: Numerical Models of Oceans and Oceanic Processes. Academic Press, San Diego (2000)
39. Phillips, N.A.: A coordinate system having some special advantages for numerical forecasting. J. Meteorol. **14**, 184–185 (1957)
40. Gerdes, R.: A primitive equation ocean circulation model using a general vertical coordinate transformation. 1. Description and testing of the model. J. Geophys. Res. Oceans 98(C8), 14683–14701 (1993)
41. Davydova-Belitskaya, V., Skiba, Yu.N.: Climate of Guadalajara city (Mexico); its variation and change within latest 120 years. World Res. Rev. **11**(2), 258–270 (1999)
42. Skiba, Yu.N.: On a method of detecting the industrial plants which violate prescribed emission rates. Ecol. Model. **159**(2–3), 125–132 (2003)
43. Skiba, Yu.N.: Métodos y Esquemas Numéricos: Un Análisis Computacional. Dirección General de Publicaciones y Fomento Editorial, UNAM, México (2005)
44. Lawson, Ch.L., Hanson, R.J.: Solving Least Squares Problem. Prentice-Hall, Englewood Cliffs, New Jersey (1974)
45. Marchuk, G.I.: Mathematical issues of industrial effluent optimization. J. Meter. Soc. Japan **60**, 481–485 (1982)
46. Marchuk, G.I.: Mathematical Models in Environmental Problems. Elsevier, New York (1986)

Chapter 5
Emission Control

The use of optimization methods in problems related to the control of air quality begins in the 60s of XX century. The central goal of such techniques has been the development of optimal strategies for the operation of polluting sources, i.e., the minimal-cost strategies that permit to satisfy a set of air quality standards in urban and industrial centers [1].

All optimization tasks in the environmental protection can be divided into two groups. The first group focuses on controlling the emissions from known sources (industrial plants [2–12], motor roads [13, 14], oil spill [15–17], and so on). The existence and uniqueness of solution to the pollution dispersion problems of this type in a limited area was demonstrated in [18]. The second group is dedicated to the identification of the location and intensity of unknown sources. These problems arise from accidents at the nuclear or chemical plants, explosions, terrorist attacks, etc. [19–23]. The uniqueness of the solution of the problem of identification of basic parameters of an unknown source was proved in [21], and the solution is usually obtained by adjusting the source parameters so as to achieve the best match between the model output and the observations (by minimizing a cost function).

In this chapter, different strategies of control of emission rates from pollution sources are suggested. The strategies are based on the application of the adjoint method. An example of the long-term and short-term strategies of control of emissions in the case of a simple box model is considered in Sect. 5.1. The main features of these two strategies of control of industrial emissions are discussed in Sect. 5.2. The purpose of control is to prevent a situation in which the average concentration of a pollutant in the selected monitoring zones will exceed the established sanitary norm. Control is called optimal if, as a result of its application, the average concentration of a pollutant in the selected zones is exactly equal to the sanitary norm. Thus, optimal control minimizes the limitation on emissions of pollution sources. Section 5.3 describes the optimal control of total emissions. A general strategy of optimal control of emissions using damping parameters is given in Sect. 5.4. Section 5.5 proves the solvability of the general problem of optimal emission control.

Y. N. Skiba and D. Parra Guevara, *Mathematical Methods for the Assessment and Control of Industrial Emissions*, Mathematical Engineering,
https://doi.org/10.1007/978-3-031-58109-0_5

All strategies use solutions of adjoint problems and ensure compliance with the air quality health standard in a zone Ω: $J(\phi) \leq J_0$ [7, 24, 25]. The numerical solution of the optimal control problem using the algorithm of successive orthogonal projections is discussed in Sect. 5.6. Section 5.7 considers an example of optimal control of lead particulate emissions. New control strategies are constructed by using convex linear combinations of existing control strategies in Sect. 5.8, and the use of linear programming to control multiple air pollutants is described in Sect. 5.9.

5.1 Emission Control Using a Simple Box Model

Let D be a three-dimensional limited domain of volume V. Suppose that D contains all the sources emitting a primary pollutant A. Then, using a mass balance, one can write the following system of equations:

$$\frac{d\alpha}{dt} = \frac{Q}{V} - \left(\sigma_1 + \frac{r}{V}\right)\alpha, \tag{5.1.1}$$

$$\alpha(0) = \alpha_0, \tag{5.1.2}$$

$$\frac{d\beta}{dt} = \sigma_1\alpha - \left(\sigma_2 + \frac{r}{V}\right)\beta, \tag{5.1.3}$$

$$\beta(0) = \beta_0 \tag{5.1.4}$$

where α and β are the concentrations of the primary pollutant A and secondary pollutant B, respectively, and Q represents the total emission rate of the primary pollutant [26]. For simplicity, we suppose here a mechanism of first-order transformation with the rates σ_1 and σ_2:

$$A \rightarrow^{\sigma_1} B \rightarrow^{\sigma_2} C$$

Since all pollutant sources are located in the domain D, the concentrations of primary and secondary pollutants are independent of external factors. However, the mixture of pollutants is removed from the domain by the wind U at a speed

$$r = \int_{S^+} U \cdot n \, dS$$

Here n is the outer unit normal to the part S^+ of boundary where $U \cdot n > 0$. Finally, α_0 and β_0 represent the initial concentrations of pollutants in D.

As a first approximation of a complex phenomenon, we assume that the pollutants are dispersed in the domain quickly and homogenously. This hypothesis is applied in some box models U.S. EPA (UAM, EKMA) where each box has a size of a few tens of kilometers [27, 28].

In the case when U and Q are invariant in time, the solution to problem (5.1.1) and (5.1.2) is

$$\alpha = \alpha_e + (\alpha_0 - \alpha_e) \cdot \exp\Big[-\Big(\sigma_1 + \frac{r}{V}\Big)t\Big],$$

and hence

$$\alpha_e = \frac{Q}{r + \sigma_1 V}$$

is a globally asymptotically stable equilibrium solution. The solution of problem (5.1.3) and (5.1.4) is

$$\beta = \beta_e + (\beta_0 - \beta_e) \cdot \exp\Big[-\Big(\sigma_2 + \frac{r}{V}\Big)t\Big] + \frac{\sigma_1}{\sigma_2 - \sigma_1}(\alpha_0 - \alpha_e) \cdot \Big\{\exp\Big[-\Big(\sigma_1 + \frac{r}{V}\Big)t\Big] - \exp\Big[-\Big(\sigma_2 + \frac{r}{V}\Big)t\Big]\Big\}, \quad (\sigma_1 \neq \sigma_2)$$

$$\beta = \beta_e + (\beta_0 - \beta_e) \cdot \exp\Big[-\Big(\sigma_1 + \frac{r}{V}\Big)t\Big] + \sigma_1 t(\alpha_0 - \alpha_e) \cdot \exp\Big[-\Big(\sigma_1 + \frac{r}{V}\Big)t\Big], \quad (\sigma_1 = \sigma_2)$$

and

$$\beta_e = \frac{\sigma_1 V}{r + \sigma_2 V} \cdot \alpha_e$$

is also a state of equilibrium, which is globally asymptotically stable.

Thus, the concentration of primary and secondary pollutants is stabilized at large times, and therefore the health standards will be satisfied at large times if Q is bounded so that the following inequalities are fulfilled:

$$\alpha_e < N_A \quad \text{and} \quad \beta_e < N_B$$

where N_A and N_B are the health standards for primary and secondary pollutants, respectively. The long-term control is achieved by considering

$$Q < \min\Big\{N_A(r + \sigma_1 V),\ N_B \frac{(r + \sigma_1 V)(r + \sigma_2 V)}{\sigma_1 V}\Big\}$$

It should be noted that quotas $N_A(r + \sigma_1 V)$ and $N_B(r + \sigma_1 V)(r + \sigma_2 V)/(\sigma_1 V)$ are sufficient to gain control of two pollutants. Since they are different values, the long-term control of the primary pollutant does not necessarily imply the control of the secondary pollutant, and vice versa, so one need to take the minimum of these two quotas. It will be shown later that the same conclusion is also valid for the short term control.

Also note that long-term control does not imply the short-term control. Indeed, if $\alpha_0 > N_A > \alpha_e$ then solution of problem (5.1.1) and (5.1.2), though it decreases, does not meet the air quality standards over finite time interval $(0, \tau)$, where

$$\tau = \frac{V}{r + \sigma_1 V} \cdot \ln\left(\frac{\alpha_0 - \alpha_e}{N_A - \alpha_e}\right)$$

In order to limit the intensity Q and realize a short-term control of the primary pollutant, we consider the mean concentration

$$\overline{\alpha} = \frac{1}{T}\int_0^T \alpha dt = \alpha_e\left(1 - \frac{\theta_1}{T}\right) + \alpha_0\frac{\theta_1}{T}$$

of this pollutant, where θ_i $(i = 1, 2)$ is defined as

$$\theta_i = \frac{1 - \exp\{-(\sigma_i + r/V)T\}}{\sigma_i + r/V}$$

If $0 < \theta_1/T < 1$ and $0 < 1 - \theta_1/T < 1$ for each $T > 0$ then the short-term control is not always possible and its realization depends on the initial distribution of the pollutant. Indeed, there are three cases:

(1) if $\alpha_0\frac{\theta_1}{T} > N_A$ then the sanitary norm is violated at any $Q \geq 0$, that is, $\overline{\alpha} > N_A$ even at full stop of industrial activity $(Q = 0)$;
(2) if $\alpha_0\frac{\theta_1}{T} = N_A$ then the only possible control is the complete cessation of production activity $(Q = 0)$ in the time interval $[0, T]$;
(3) if $\alpha_0\frac{\theta_1}{T} < N_A$ then short-term control of the primary pollutant will be successful if

$$Q < \left(N_A - \alpha_0\frac{\theta_1}{T}\right)(r + \sigma_1 V)/\left(1 - \frac{\theta_1}{T}\right)$$

This analysis allows us to conclude that the initial distribution of contaminants is a critical value in the case of short-term control of industrial activity. Unlike this, the initial distribution of pollutants is not so critical in the case of long-term control, where the large-time behavior of the solution is more important.

The short-term control of the secondary pollutant gives a similar conclusion, since for $\sigma_1 \neq \sigma_2$ we get

$$\overline{\beta} = \frac{1}{T}\int_0^T \beta dt = \alpha_e\left[\left(1 - \frac{\theta_2}{T}\right)\frac{\sigma_1 V}{r + \sigma_2 V} - \frac{\sigma_1(\theta_1 - \theta_2)}{T(\sigma_2 - \sigma_1)}\right] + \beta_0\frac{\theta_2}{T} + \alpha_0\frac{\sigma_1(\theta_1 - \theta_2)}{T(\sigma_2 - \sigma_1)}$$

If

$$\frac{\theta_1 - \theta_2}{T(\sigma_2 - \sigma_1)} > 0 \text{ for any } T > 0$$

then we obtain the following three cases:

(1) if $\beta_0 \frac{\theta_2}{T} + \alpha_0 \frac{\sigma_1(\theta_1 - \theta_2)}{T(\sigma_2 - \sigma_1)} > N_B$ then the sanitary norm is violated for any $Q \geq 0$, that is, $\overline{\beta} > N_B$ even at full stop of industrial activity ($Q = 0$);
(2) if $\beta_0 \frac{\theta_2}{T} + \alpha_0 \frac{\sigma_1(\theta_1 - \theta_2)}{T(\sigma_2 - \sigma_1)} = N_B$ then the only possible control is a complete cessation of production activity ($Q = 0$) in the time interval $[0, T]$;
(3) if $\beta_0 \frac{\theta_2}{T} + \alpha_0 \frac{\sigma_1(\theta_1 - \theta_2)}{T(\sigma_2 - \sigma_1)} < N_B$ then the short-term control of secondary pollutant will be successful if

$$Q < \left[N_B - \beta_0 \frac{\theta_2}{T} + \alpha_0 \frac{\sigma_1(\theta_1 - \theta_2)}{T(\sigma_2 - \sigma_1)} \right] (r + \sigma_1 V) / \left\{ \left(1 - \frac{\theta_2}{T} \right) \frac{\sigma_1 V}{r + \sigma_2 V} - \frac{\sigma_1(\theta_1 - \theta_2)}{T(\sigma_2 - \sigma_1)} \right\}$$

In the case $\sigma_1 = \sigma_2$, we get

$$\overline{\beta} = \frac{1}{T} \int_0^T \beta dt = \alpha_e \frac{\sigma_1 V}{r + \sigma_1 V} \left(1 - \frac{\theta_1}{T} - \Theta_1 \right) + \beta_0 \frac{\theta_1}{T} + \alpha_0 \frac{\sigma_1 V}{r + \sigma_1 V} \Theta_1$$

where

$$\Theta_1 = \frac{\theta_1}{T} - \exp\left[-\left(\sigma_1 + \frac{r}{V} \right) T \right] > 0 \text{ for any } T > 0$$

Thus, the following conditions of control are obtained [26]:

(1) if $\beta_0 \frac{\theta_1}{T} + \alpha_0 \frac{\sigma_1 V}{r + \sigma_1 V} \Theta_1 > N_B$ then the sanitary norm is violated for any $Q \geq 0$, i.e., $\overline{\beta} > N_B$ even at full stop of industrial activity ($Q = 0$);
(2) if $\beta_0 \frac{\theta_1}{T} + \alpha_0 \frac{\sigma_1 V}{r + \sigma_1 V} \Theta_1 = N_B$ then the only possible control is a complete cessation of production activity ($Q = 0$) in the time interval $[0, T]$;
(3) if $\beta_0 \frac{\theta_1}{T} + \alpha_0 \frac{\sigma_1 V}{r + \sigma_1 V} \Theta_1 < N_B$ then the short-term control of the secondary pollutant will be successful if

$$Q < \left[N_B - \beta_0 \frac{\theta_1}{T} - \alpha_0 \Theta_1 \frac{\sigma_1 V}{r + \sigma_1 V} \right] (r + \sigma_1 V)^2 / \left[\sigma_1 V \left(1 - \frac{\theta_1}{T} - \Theta_1 \right) \right]$$

It should be noted that there is a continuous connection between the quotas on the emission rates Q provided in this example by the short-term and long-term control strategies. Indeed, the emission rate Q provided by the long-term control can also be obtained if we proceed to the limit as $T \to \infty$ in the emission rate provided by the short-term control and take into account that

$$\lim_{T \to +\infty} \frac{\theta_i}{T} = 0 \quad (i = 1, 2) \text{ and } \lim_{T \to +\infty} \Theta_1 = 0$$

This suggests that in case of a more complicated model, such limit transition to the limit can also serve a way of building a long-term strategy to control emissions.

5.2 Emission Control Strategies

The control of industrial emissions is one of the most important inverse problems in the air pollution modeling. This control provides quantitative criteria to restrict industrial emissions and reduce the air pollution concentrations. Such quantitative criteria should be developed for all industrial sources, taking into account the complexity of the processes of dispersion and transformation of pollutants in the atmosphere.

As discussed in Chap. 1, the goal of any air pollution control is to keep the maximum concentrations of pollutants below sanitary norms, or at least to minimize the number of hours or days when these norms are violated. All control strategies can be classified into two types: long-term and short-term control [29]. These control programs supplement each other, but differ in the goals and in the strategies used to achieve these goals.

A long-term control (ranging from several months to several years) is usually applied to large-scale regions, from urban to global. The control consists in implementing the strategies that reduce the total mass of pollutant emissions throughout the period. Besides, the central objective of long-term control is to reduce the number of emergency days per year when, due to a weak and slow dispersion, the air quality standards are violated and the concentration of pollutants reaches dangerous levels. In other words, its application does not protect the region from the emergency days. This flexibility is inherent in the nature of long-term control that takes into account not only the air quality standards, but also cost–benefit criteria.

On the other hand, the short-term control (ranging from several hours to several days) is generally used for small areas (from urban to regional) during the events of atmospheric stability (atmospheric inversions, calm), which favor the accumulation of pollutants. Typical actions for this type of control are those that immediately lead to a reduction in pollutant emissions, up to complete stop of some industrial plants. The specific objective of the control during the events of atmospheric stability is to maintain the concentration of contaminants below the health standards.

We now give the general formulation of the problem of short-term control. Let $\boldsymbol{M}$ be a dispersion model which is used for short-term forecasting the concentration ϕ of a pollutant in a limited area $D \subset \mathbb{R}^m$ ($m = 2$ or $m = 3$) in a time interval $[0, T]$:

$$\boldsymbol{M}\colon \vec{Q} \mapsto \phi_{\vec{Q}}$$

where $\overrightarrow{Q} = (Q_1, Q_2, ..., Q_N)$ is the vector defined by non-negative and non-stationary emission rates Q_i of N pollutant sources located at points r_i of the domain D. Let $\Omega \subset D$ be an ecologically important zone in D whose area (or volume) is $|\Omega|$. The mean concentration of contaminant in zone Ω and time interval $[T - \tau, T]$ of length τ is defined by the functional (3.5.6):

$$J\left(\phi_{\vec{Q}}\right) \equiv \frac{1}{\tau|\Omega|} \int_{T-\tau}^{T} \int_{\Omega} \phi_{\vec{Q}}(r,t)drdt \tag{5.2.1}$$

Suppose now that J_0 is a sanitary norm of the concentration of pollutant in Ω, but the result obtained with a forecast model is negative: $J(\phi_{\vec{Q}}) > J_0$. It means that the emission rates $\vec{Q}$ are excessive, and the control problem consists in determining such new reduced emission rates $\vec{Q}^*$ that their use in $[0, T]$ will prevent the zone Ω from excessive pollution. In other words, the forecast with new rates $\vec{Q}^*$ must lead to

$$J\left(\phi_{\vec{Q}^*}\right) \leq J_0 \tag{5.2.2}$$

Hereinafter, for simplicity, the index $\vec{Q}$ is omitted and $J(\phi)$ is used instead of $J(\phi_{\vec{Q}})$. Also, every time we write $J(\phi)$, the area $\Omega \subset D$ and the length τ of interval $[T-\tau, T]$ are supposed to be given and fixed.

Note that each emission rate Q_i^* should be non-negative and bounded from above. In general, this inverse problem is ill-posed, since it can have many solutions or no solution, depending on the initial distribution of contaminants ϕ^0 [25, 30]. In order to pose the problem well, a regularization technique should be applied. In a sense, each control strategy is defined by its own regularization method.

In order to develop a short-term control strategy, the dispersion model (2.4.1)–(2.4.6) and its adjoint model (3.4.9)–(3.4.12) will be used. As mentioned before, the adjoint models are a very powerful tool in the inverse modeling. The adjoint method covers a wide range from models of "zero" dimension (so called box models) to sophisticated general circulation models. This variety in the use of adjoint equations is related with the duality principle (3.5.5), which, as we will show later, is also used in the development of different control strategies.

We will assume that the results obtained with a forecast model are unfavorable in the sense that the predicted mean pollution concentration $J(\phi)$ exceeds the sanitary standard J_0 in a zone Ω: $J(\phi) > J_0$. In this case, a control strategy should be applied to reduce emissions rates and prevent excessive pollution of zone Ω. In the following sections, we consider strategies that differ from each other in the ways of limiting the emission rates of the sources. They represent the control of the total mass of a pollutant emitted during a time interval $(0, T)$, and an optimal control using damping parameters, respectively.

5.3 Control of the Total Mass of Emissions

1. **The case of a single point source**. We first consider the case when there is only one point source of a pollutant in a domain D, which is located at a point r_0 and whose emission rate is $Q(t)$. Later this case will be generalized to several point sources. With this simplification, Eq. (3.5.7) takes the form

$$J(\phi) = \int_0^T g(r_0, t)Q(t)dt + \int_D g(r, 0)\phi^0(r)dr \tag{5.3.1}$$

where $J(\phi)$ is the mean concentration of the pollutant in zone Ω and time interval $[T - \tau, T]$. The second integral in (5.3.1) represents the atmospheric pollution accumulated as a result of industrial activity in the previous time period and shows the contribution to $J(\phi)$ of the initial distribution of the pollutant in D.

We now suppose that $J(\phi) > J_0$. The impact of ϕ^0 on the concentration $J(\phi)$ cannot be avoided. Therefore, in order to satisfy the condition

$$J(\phi) \le J_0 \tag{5.3.2}$$

the emission rate Q should be restricted so that

$$\int_0^T g(r_0, t)Q(t)dt \le \alpha \tag{5.3.3}$$

where

$$\alpha = J_0 - \int_D g(r, 0)\, \phi^0(r)\, dr \tag{5.3.4}$$

is the parameter that characterizes different possibilities to control emissions [7, 25]. Indeed, if $\alpha = 0$ then the only possibility to satisfy the condition (5.3.2) is to put $Q \equiv 0$ in the whole interval $(0, T)$, which means a full stop of industrial activity. If $\alpha < 0$, then the problem of control has no solution, since $J(\phi) > J_0$ even in the case when industrial activity is completely stopped ($Q \equiv 0$ in $(0, T)$). Note, however, that stopping industrial activity is still useful, as this action will reduce dangerous levels of pollution.

Inequality $\alpha > 0$ corresponds to a more interesting case, when the pollution concentration ϕ^0 accumulated due to industrial activity in the previous period of time is relatively small. Then the goal is to find $Q \in L_2(0, T)$ so that

$$\int_0^T g(r_0, t)Q(t)dt = (g(r_0), Q) \leq \alpha \tag{5.3.5}$$

where $(\cdot, \cdot)$ is the inner product in the Hilbert space $L_2(0, T)$. Note that there exists an infinity of solutions to this problem, and each emission rate that satisfies the inequality (5.3.5) defines a sufficient control strategy.

Since any emission rate Q that satisfies the inequality $(g(r_0), Q) < \alpha$ can significantly limit industrial activity, it is better to look for such Q that

$$(g(r_0), Q) = \alpha \tag{5.3.6}$$

The latter problem also has an infinite number of solutions, and some of them, despite retaining acceptable concentrations in the zone Ω, can lead to high concentrations of pollutant near the source. To avoid such a situation, it is advisable to choose among all control strategies satisfying (5.3.6), a strategy that will minimize the total mass of pollutant released from the source during the time interval $(0, T)$. As will be shown below, this problem has a unique solution $Q_{\min}$.

Let us define a functional $F: L_2(0, T) \to \mathbb{R}$ by

$$F(Q) = \|Q\|^2 \tag{5.3.7}$$

where $\|Q\| = \sqrt{\int_0^T Q^2(t)dt}$ is the norm induced in the space $L_2(0, T)$ by the inner product. The functional $F(Q)$ measures the total mass of the pollutant emitted during time interval $(0, T)$ by the source of intensity $Q(t)$.

Thus, the first control strategy represents the following optimization problem:

$$\begin{gathered} \min F(Q) = \|Q\|^2, \\ \text{subject to (5.3.6): } (g(r_0), Q) = \alpha \end{gathered} \tag{5.3.8}$$

It has a simple geometric interpretation. Equation (5.3.6) defines in the space $L_2(0, T)$ a hypersurface with a normal vector $g(r_0, t)$, and $F(Q)$ represents the distance of each point of the hypersurface from the origin. Therefore, the optimization problem is to find the point on the hypersurface with minimum distance from the origin.

We now remind the classical results on the convexity which are necessary to prove the uniqueness of a solution to problem (5.3.8).

Definition 5.3.1 A set D is convex if $\lambda x + (1 - \lambda)y \in D$ for any $x, y \in D$ and every $\lambda \in [0,1]$ (Fig. 5.1).

Definition 5.3.2 A function $f(x)$ defined in a convex set D is convex if

$$f(\lambda x + (1 - \lambda)y) \leq \lambda f(x) + (1 - \lambda)f(y)$$

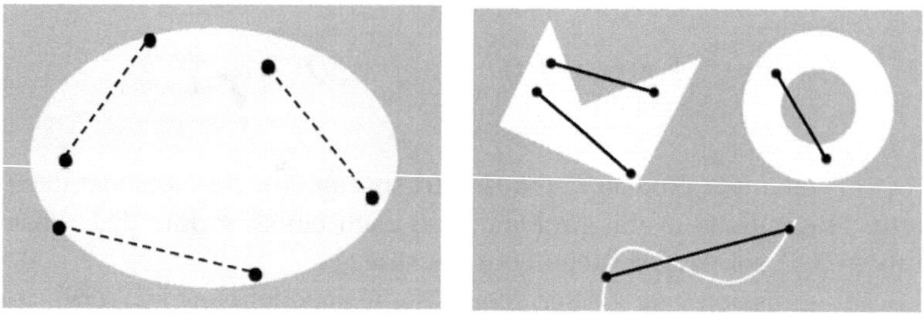

Fig. 5.1 Convex sets (**a**) and non-convex sets (**b**)

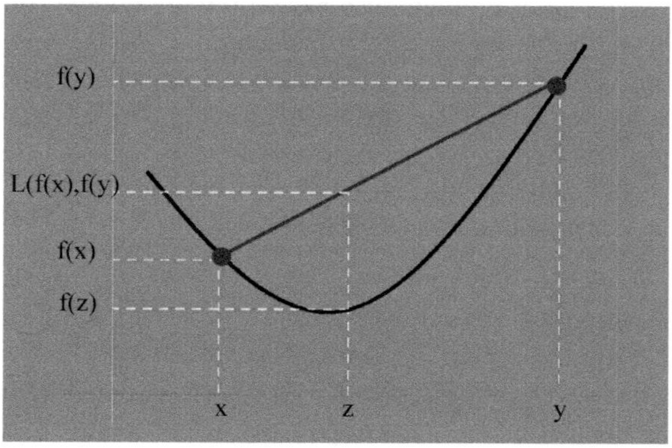

Fig. 5.2 Graphic explanation of the convexity of functions; $z = \lambda x + (1 - \lambda)y$; $L(f(x), f(y)) = \lambda f(x) + (1 - \lambda)f(y)$

for any $x, y \in D$ and every $\lambda \in [0,1]$. If the inequality is strict then the function is called strictly convex (Fig. 5.2).

Lemma 5.3.1 *A set*

$$C = \{Q \in L_2(0, T) | (g(r_0), Q) = \alpha\} \tag{5.3.9}$$

is convex, and $F(Q)$ *is a strictly convex functional in* C.

Proof Let Q_1 and Q_2 be two different elements in C, and $\lambda \in (0,1)$. Evidently, $\lambda Q_1 + (1 - \lambda)Q_2 \in L_2(0, T)$ and

$$(g(r_0), \lambda Q_1 + (1 - \lambda)Q_2) = \lambda(g(r_0), Q_1) + (1 - \lambda)(g(r_0), Q_2) = \lambda\alpha + (1 - \lambda)\alpha = \alpha$$

Therefore, C is convex. On the other hand,

$$(\lambda Q_1 + (1 - \lambda)Q_2)^2 = \lambda Q_1^2 + (1 - \lambda)Q_2^2 - \lambda(1 - \lambda)(Q_1 - Q_2)^2 < \lambda Q_1^2 + (1 - \lambda)Q_2^2$$

for any t. Integrating the last inequality over $(0, T)$, we obtain that F is strictly convex.

Lemma 5.3.2 *Let F be a strictly convex function defined on a convex set C of a normed space. Then the following two statements are valid:*

(i) *if there is a local minimum of F on C, then it must be a global minimum on C;*
(ii) *if there is a global minimum of F on C then it must be unique.*

Proof To prove the first assertion, let's assume that $Q^* \in C$ is a local minimum of F. It means that there exists such $\varepsilon > 0$ that $F(Q^*) \le F(Q)$ for any $Q \in C$ from the set $\|Q - Q^*\| < \varepsilon$. Let $q \in C$ and $\|q - Q^*\| \ge \varepsilon$. If we assume that $F(q) < F(Q^*)$ then

$$\lambda_0 = 1 - \frac{\varepsilon}{2\|q - Q^*\|} < 1$$

Moreover, $\varepsilon/(2\|q - Q^*\|) \le \frac{1}{2} < 1$, since $\|q - Q^*\| \ge \varepsilon$, and hence, $\lambda_0 > 0$. Using the strict convexity of F we obtain

$$F\big(\lambda_0 Q^* + (1 - \lambda_0)q\big) < \lambda_0 F\big(Q^*\big) + (1 - \lambda_0)F(q) < F\big(Q^*\big)$$

On the other hand,

$$\big\|\lambda_0 Q^* + (1 - \lambda_0)q - Q^*\big\| = (1 - \lambda_0)\big\|q - Q^*\big\| = \frac{\varepsilon}{2} < \varepsilon$$

Since $\lambda_0 Q^* + (1-\lambda_0)q \in C$ (due to convexity), the last two inequalities contradict the fact that Q^* is the local minimum. Thus, $F(Q^*) \le F(q)$.

To prove the second assertion, assume that Q_1^* and Q_2^* are different global minima, and $F(Q_1^*) = F(Q_2^*)$. Moreover, if $0 < \lambda < 1$ then

$$F\big(\lambda Q_1^* + (1 - \lambda)Q_2^*\big) < \lambda F\big(Q_1^*\big) + (1 - \lambda)F\big(Q_2^*\big) = F\big(Q_1^*\big)$$

This contradicts the fact that Q_1^* is the global minimum. Consequently, $Q_1^* = Q_2^*$. The lemma is proved.

More general results on the convexity can be found in [31, 32]. It follows from Lemmas 5.3.1 and 5.3.2 that if the optimization problem formulated above has a minimum in C, then this minimum is global and unique.

Theorem 5.3.1 *Let*

$$Q_{\min} = \frac{\alpha g(r_0, t)}{\|g(r_0)\|^2} \tag{5.3.10}$$

where

$$\|g(r_0)\| = \left(\int_0^T |g(r_0, t)|^2 dt \right)^{1/2}$$

Then the functional F attains its minimum at point $Q_{\min}$ on C.

Proof Since $J(\phi) > J_0$ for the initial emission rate, then by (5.3.1) and (5.3.4),

$$0 < \alpha < \int_0^T g(r_0, t)Q(t)\, dt \le \|g(r_0)\| \|Q\|,$$

and therefore $\|g(r_0)\| > 0$, that is, $Q_{\min}$ is well defined. On the other hand, $Q_{\min} \in L_2(0, T)$ and

$$(g(r_0), Q_{\min}) = \left(g(r_{0,}t), \alpha \frac{g(r_{0,}, t)}{\|g(r_{0,})\|^2} \right) = \alpha$$

i.e., $Q_{\min} \in C$. Let $Q \in C$ and $Q \ne Q_{\min}$. Define $\delta Q = Q - Q_{\min}$. Then for any $\lambda \in \mathbb{R}, (g(r_{0,}), Q_{\min} + \lambda\delta Q) = \alpha,$ and hence, $(g(r_{0,}), \delta Q) = 0$.

The last equation together with (5.3.10) implies

$$(Q_{\min}, \delta Q) = 0$$

It is easy to show that the function $h(\lambda) = F(Q_{\min} + \lambda\delta Q) = \int_0^T (Q_{\min} + \lambda\delta Q)^2 dt$ reaches its minimum at $\lambda = 0$. In fact, $\lambda = 0$ is a critical point of $h(\lambda)$, since

$$h'(\lambda) = 2 \int_0^T (Q_{\min} + \lambda\delta Q)\delta Q dt \text{ and } h'(0) = \int_0^T Q_{\min}\delta Q dt.$$

Besides,

$$h''(\lambda) = 2 \int_0^{\mathrm{T}} |\delta Q|^2 dt > 0.$$

Since $h(\lambda)$ is the restriction of F on one line in C, this function is strictly convex, and by Lemma 5.3.2 the point $\lambda = 0$ is the unique global minimum. Therefore, $h(0) < h(1)$, or $F(Q_{\min}) < F(Q)$. The theorem is proved.

Theorem 5.3.2 *Let us assume that $\alpha > 0$, Ω is an ecologically significant zone, and $Q(t) \ge 0$ is the emission rate of the pollutant source located at the point $\mathbf{r}_0$. If*

$$\|Q\| \le \|Q_{\min}\| = \alpha / \|g(r_0)\| \tag{5.3.11}$$

then the average concentration of the pollutant in the zone Ω does not exceed the maximum permissible concentration J_0.

Proof The application of Schwarz' inequality [33] leads to

$$(g(r_0), Q) \le \|g(r_0)\| \|Q\| \le \|g(r_0)\| \frac{\alpha}{\|g(r_0)\|} = \alpha.$$

Using (5.3.4) we obtain

$$J(\phi) = \int_0^T g(r_0, t) Q(t) dt + \int_D g(r, 0)\, \phi^0(r)\, dr \le J_0.$$

The theorem is proved.

It is important to note that this is a sufficient (non-optimal) strategy that imposes limitation (5.3.11) on the total amount of a pollutant emitted by the industrial source during time interval $(0, T)$, and therefore this strategy does not restrict the temporal behavior of emission rate $Q(t)$ in the interval $(0, T)$.

1. **Generalization to a few point sources.** To extend this strategy to the case of a few sources located at points $r_k \in D$, one must first eliminate from the consideration those sources for which $\int_0^T g(r_k, t) Q_k(t) dt = 0$, because they do not pollute the zone Ω, and therefore there is no need to restrict their operation.

Let r_i be a position of the i-th point source polluting the zone Ω ($i = 1, ..., N$). In the case when $J(\phi) > J_0$, the emission rates $Q_i^*(t)$ should be chosen so that

$$J(\phi) = \sum_{i=1}^{N} \int_0^T g(r_i, t) Q_i^*(t) dt + \int_D g(r, 0)\, \phi^0(r)\, dr \le J_0 \tag{5.3.12}$$

As before, assume that the parameter α defined in (5.3.4) is positive. Let $m_j = F(Q_j)$ the denote the total mass of pollutant, emitted by the j-th source during time interval $(0, T)$ when it operates without control. Then

$$\gamma_i = m_i \left(\sum_{j=1}^{N} m_j \right)^{-1}, \quad i = 1, \ldots, N \tag{5.3.13}$$

is the fraction of the total mass of pollutant, emitted by the i-th source, and therefore

$$\gamma_1 + \gamma_2 + \cdots + \gamma_N = 1 \tag{5.3.14}$$

Thus, γ_i characterizes the power of each source. For example, if there are only two sources in D, and the emission rate of the source located at r_2 is twice as much than that of the source located at r_1 then their weights are defined as $\gamma_1 = 1/3$ and $\gamma_2 = 2/3$.

Given the weights γ_i, the new emission rates prescribed by the control are

$$Q_i^* = \gamma_i \alpha\, g(r_i,\ t)/\|g(r_i)\|^2, \quad (i = 1, \ldots, N) \tag{5.3.15}$$

where

$$\|g(r_i)\| = \left(\int_0^T |g(r_i, t)|^2 dt \right)^{1/2}$$

Then

$$\sum_{i=1}^{N} \left(g(r_i),\ Q_i^*\right) = \sum_{i=1}^{N} \gamma_i \alpha \ = \alpha,$$

and the application of (5.3.4) and (5.3.12) leads to $J(\phi) \le J_0$. It should be noted that $\|g(r_i)\| > 0$ for all i, since $\int_0^T g(r_i, t) Q_i(t) dt > 0$, i.e., all new emission rates Q_i^* are well defined.

5.4 Optimal Emission Control Using Damping Parameters

We consider separately the cases of permanent and time-varying emission rates.

1. **Stationary emission rates**. Suppose there are N industrial sources operating in a domain D. Let δ_i be the maximum possible emission rate of the i-th source located at r_i ($i = 1, \ldots, N$). Suppose that $J(\phi) > J_0$ in a zone Ω when all sources operate at full capacity. Then it is necessary to lower the emission rates of sources. We now consider a control strategy, the main idea of which is to establish the maximum possible stationary emission rates $Q_i \le \delta_i$ for each industrial plant so that the condition $J(\phi) \le J_0$ is satisfied in the zone Ω.

Let again $\alpha > 0$, and let $\gamma_1, \gamma_2, \ldots, \gamma_N$ be positive constants (5.3.13) that characterize the power of industrial sources. Taking into account (5.3.4) and the constancy of rates Q_i in $[0, T]$, one can rewrite (5.3.12) as $\sum_{i=1}^{N} a_i Q_i \le \alpha$, where $a_i = \int_0^T g(r_i, t) dt > 0$. To minimize the capacity of each industrial plant, we will seek such constant emission rates Q_i that minimize the values $\delta_i - Q_i$ and satisfy the condition

$$\sum_{i=1}^{N} a_i Q_i = \alpha \tag{5.4.1}$$

Therefore, this strategy can be formulated as the following optimization problem:

$$\begin{cases} \min H(Q_1, Q_2, \ldots, Q_I) = \sum_{i=1}^{N} \frac{1}{\gamma_i}(Q_i - \delta_i)^2, \\ \text{subject to } \sum_{i=1}^{N} a_i Q_i = \alpha \end{cases} \tag{5.4.2}$$

Using the Lagrange multipliers [34], we obtain

$$\begin{aligned} \lambda &= \left(\alpha - \sum_{i=1}^{N} a_i\delta_i\right)\left(\sum_{i=1}^{N} \frac{a_i^2\gamma_i}{2}\right)^{-1} \\ Q_i &= \delta_i + \lambda\frac{a_i\gamma_i}{2}, \quad i = 1, \ldots, N \end{aligned} \tag{5.4.3}$$

Due to (5.4.1), the factor λ is negative, and hence, $Q_i \leq \delta_i$ for any i. Also, $Q_i \approx \delta_i$ for small γ_i. Thus, this strategy imposes restrictions mainly on industrial sources of high power.

Note that instead of (5.4.2), the new emission rates Q_i can also be defined as $Q_i = \beta_i\delta_i$ where β_i are specific damping parameters that have to minimize the values $(1 - \beta_i)\delta_i$ and satisfy the constraint $\sum_{i=1}^{N} a_i\beta_i = \alpha$ where $a_i = \delta_i \int_0^T g(r_i, t)dt > 0$. However, in case of stationary emission rates, formulation (5.4.2) is preferred due to its simplicity, and the damping parameters will be used in case of non-stationary emission rates.

2. **Time-varying emission rates**. We now consider the general case, when emission rates are non-stationary, and apply the optimal strategy based on using the damping parameters:

$$Q_i^*(t) = \beta_i\, Q_i(t) \tag{5.4.4}$$

The advantage of this approach lies in the fact that it does not change the graphs of functions $Q_i(t)$ in $(0, T)$, reducing only their intensity. It may be of interest to the industrial plants. The set of positive damping parameters β_i $(0 \leq \beta_i \leq 1,\ i = 1, 2, \ldots, N)$ is the solution to the following optimization problem [35, 36]:

$$\begin{cases} \min \quad H(\beta_1, \beta_2, \ldots, \beta_N) = \sum_{i=1}^{N} \|Q_i\|^2(1 - \beta_i)^2 \\ \quad \text{subject to: } \sum_{i=1}^{N} a_i\beta_i = \alpha \end{cases} \tag{5.4.5}$$

In other words, we want to minimize the objective function H that estimates the difference between the initial emission rates and the emission rates (5.4.4) prescribed by the control with damping. Note that $a_i = \int_0^T g(r_i, t)Q_i(t)dt \geq 0$ for all i, besides, a positive value a_i means that the i-th industrial source has a nonzero impact on the zone Ω.

The control is optimal in the sense that all parameters β_i are searched as close as possible to unity, so that industrial plants can satisfy condition (5.3.12) with minimum material losses. It makes this strategy less restrictive and therefore more attractive for the plants. The optimization problem (5.4.5) can be written in the equivalent form as

$$\begin{cases} \min \quad H(x_1, x_2, \ldots, x_N) = \sum_{i=1}^{N} \left(x_i - x_i^0\right)^2 \\ \text{subject to}: \sum_{i=1}^{N} b_i x_i = \alpha \\ x_1 \geq 0, \ldots, x_N \geq 0 \end{cases} \tag{5.4.6}$$

where $x_i = \|Q_i^*\|$, $x_i^0 = \|Q_i\| > 0$ and $b_i = a_i / \|Q_i\| > 0$. Besides, $\sum_{i=1}^{N} b_i x_i^0 > \alpha$, and the feasible space **F** is determined by the constraints of problem (5.4.6). This problem has unique solution. Indeed, the existence of solution immediately follows from the fact that objective function H is continuous, and feasible space **F** is compact (Weierstrass theorem [37]). The uniqueness is a consequence of the facts that the H is a strictly convex function and set **F** is also convex [37]. Moreover, due to convexity, the optimal values β_i^* of damping parameters β_i satisfy the conditions

$$\beta_i^* \leq 1, \quad i = 1, \ldots, N \tag{5.4.7}$$

Parra-Guevara [26], and therefore some industrial sources must reduce their activity:

$$Q_i^*(t) \leq Q_i(t), i = 1, \ldots, N, \quad 0 \leq t \leq T \tag{5.4.8}$$

Finally, we point out that the formulation (5.4.6) is a quadratic programming problem whose solution can be found by the algorithm of successive orthogonal projections described in Sect. 5.6.

5.5 General Problem of Optimal Control

We now develop an optimal strategy without using the condition (5.4.4). To this end, we introduce a regularization method that makes the inverse problem well-posed (that is, the optimal solution exists, is unique, and is stable). Suppose that there are N point sources located at points r_i of a limited area D with emission rates $Q_i(t)$ of a pollutant ($i = 1, \ldots, N$). Denote $\vec{Q}(t) = \{Q_1(t), Q_2(t), \ldots, Q_N(t)\}$.

Let $\boldsymbol{M}$ be a dispersion model used for short-term prediction of pollutant concentrations in D and finite time interval $(0, T)$ (for example, the model (2.4.1)–(2.4.6)). Suppose that with emission rates $\vec{Q}(t)$, the model $\boldsymbol{M}$ (coupled with a weather forecast model) predicts an unfavorable result: $J(\phi_{\vec{Q}}) > J_0$, that is, the mean concentration of pollutant in an ecologically important zone $\Omega \subset D$ and time interval $[T - \tau, T]$ exceeds the health standard J_0. To avoid excessive pollution in the zone Ω, an optimal short-term control strategy should be applied by prescribing more appropriate emission rates in interval $(0, T)$. In other words, it is necessary to determine reduced emission rates $\vec{q}(t) = \{q_1(t), q_2(t), ..., q_N(t)\}$, optimal in some sense, in order to satisfy the health standard in zone Ω: $J(\phi_{\vec{Q}}) \le J_0$.

As it was mentioned before, this inverse problem is ill-posed, since it may have many solutions or have no solution, depending on the initial distribution of the contaminant (see (2.4.4)). Therefore, we now apply a regularization method.

We denote by $\boldsymbol{H}$ a space of vector functions $\vec{q}(t) = \{q_1(t), q_2(t), ..., q_N(t)\}$ with the norm

$$\|\vec{q}\| \equiv \left(\sum_{i=1}^{N} \|q_i\|^2_{L_2[0,T]}\right)^{1/2} = \left(\sum_{i=1}^{N} \int_0^T q_i^2(t)dt\right)^{1/2}$$

and define functional

$$F(\vec{q}) = \left\|\vec{Q} - \vec{q}\right\|^2 \equiv \sum_{i=1}^{N} \|Q_i - q_i\|^2_{L_2[0,T]} = \sum_{i=1}^{N} \int_0^T (Q_i - q_i)^2 dt \tag{5.5.1}$$

in the set

$$\Theta = \{q_i(t) \in L_2[0, T]; \quad q_i(t) \ge 0, \quad (i = 1, \dots, N) | J\left(\phi_{\vec{Q}}\right) \le J_0\} \tag{5.5.2}$$

where components $q_i(t)$ of vector $\vec{q}(t) = \{q_1(t), q_2(t), ..., q_N(t)\}$ represent new emission rates which guarantee the compliance with health regulations in the zone Ω: $J(\phi_{\vec{q}}) \le J_0$.

The problem of optimal control is to determine such $\vec{Q}^* \in \Theta$ that

$$F\left(\vec{Q}^*\right) = \inf_{\vec{q} \in \Theta} F(\vec{q}). \tag{5.5.3}$$

Components $\vec{Q}^*_i(t)$ of vector $\vec{Q}^*$ represent the least restriction imposed by the control on the industrial activity. Such variational problem is solved by an iterative optimization method that applies a successive evaluation of the dynamic model $\boldsymbol{M}$. In general, this process is not very efficient, because it requires a lot of calculations due to the complexity of the model $\boldsymbol{M}$. We now use the adjoint method, which allows solving the optimal control problem without successive evaluations of model $\boldsymbol{M}$.

Note again that the solution of optimal control problem (5.5.3) critically depends on α. If $\alpha < 0$ then due to (5.3.4), this problem has no solution, since $J\left(\phi_{\vec{q}}\right) > J_0$ even if $\vec{q}(t) \equiv 0$ in interval $(0, T)$.

Theorem 5.5.1 *If* $\alpha = 0$ *then the unique solution* $\vec{Q}^* \in \Theta$ *of optimal control problem (5.5.3) is*

$$Q_i^*(t) = \begin{cases} 0, & if\ t \in I_i \\ Q_i(t), & if\ t \in I_i^c \end{cases} \tag{5.5.4}$$

where $I_i = \{t \in [0, T] | g(r_i, t) > 0\}$ *and* $I_i^c = [0, T] \backslash I_i$ *is the complement of set* I_i *to* $[0, T]$.

Proof Without loss of generality, we can assume that $|I_i| > 0$ for any i. Indeed, if $|I_j| = 0$ for some index j then

$$\int_0^T g(r_j, t) Q_j(t)\, dt = 0,$$

and the j-th source does not pollute the area Ω and should be excluded from the consideration. Note that $\alpha = 0$ and (5.3.4) imply

$$J_0 = \int_D g(r, 0)\, \phi^0(r)\, dr$$

Further, set Θ consists of functions

$$q_i(t) = \begin{cases} 0, & if\ \ t \in I_i \\ p_i(t), & if\ \ t \in I_i^c \end{cases}$$

where $p_i(t) \in L_2[0, T]$ and $p_i(t) \geq 0$. Therefore,

$$F(\vec{q}) = \sum_{i=1}^{N} \left\{ \int_{I_i} Q_i^2 dt + \int_{I_i^c} (Q_i - p_i)^2 dt \right\} \geq \sum_{i=1}^{N} \int_{I_i} Q_i^2 dt$$

and $p_i(t) = Q_i(t)$ in I_i^c for the optimal solution. The theorem is proved.

The case $\alpha > 0$ is more interesting and complicated, since the optimum emission rates depend not only on $g(r_i, t)$, but also on the difference $J(\phi_{\vec{Q}}) - J_0$. The optimal solution is calculated numerically. We now show its existence and uniqueness [30].

Theorem 5.5.2 *The solution of optimal control problem (5.5.3) is unique if* $\alpha > 0$.

Proof First, note that the feasible space Θ in $\boldsymbol{H}$ can be defined as

$$\Theta = \{q_i(t) \in L_2[0, T];\ q_i(t) \geq 0,\ (i = 1, .., N) | \sum_{i=1}^{N} \int_0^T g(r_i, t) q_i(t) dt \leq \alpha\} \tag{5.5.5}$$

Also note that if $\vec{q}_1, \vec{q}_2 \in \Theta$ and $\lambda \in [0,1]$ then

$$\sum_{i=1}^{N} \int_0^T g(r_i, t)\{\lambda q_{1i}(t) + (1-\lambda)q_{2i}(t)\}dt \le \alpha,$$

and hence, Θ is convex. Moreover, the functional $F(\vec{q}) = \sum_{i=1}^{N} \|Q_i - q_i\|^2_{L_2[0,T]}$ is also strictly convex in Θ, since each term of this sum is strictly convex in Θ. These properties ensure the uniqueness of solution of problem (5.5.3) [38]. The theorem is proved.

To show the existence of solution of problem (5.5.3) we need the following results:

Lemma 5.5.1 *The feasible space Θ is closed in $\boldsymbol{H}$.*

Proof We must show that $\Theta = \overline{\Theta}$. Let $\vec{q}_0$ be an element of $\overline{\Theta}$. Then there exists a sequence $\{\vec{q}_k\}_{k=1}^{\infty}$ in Θ such that $\|\vec{q}_k - \vec{q}_0\|^2 \to 0$ as $k \to \infty$.

To prove that $q_{0i} \ge 0$ for all i ($i = 1, ..., N$) we apply the method of proof by contradiction. Assume that for some j, $q_{0j} < 0$ in a set $I \subset [0, T]$ of positive measure: $|I| > 0$. Then

$$\|\vec{q}_k - \vec{q}_0\|^2 = \sum_{i=1}^{N} \int_0^T [q_{ki}(t) - q_{0i}(t)]^2 dt \ge \int_I [q_{kj}(t) - q_{0j}(t)]^2 dt$$
$$\ge \int_I q_{0j}^2(t)dt = l > 0$$

for any k. The last inequality contradicts the convergence of $\{\vec{q}_k\}_{k=1}^{\infty}$. Therefore, the norm $\|g(r_i, t)\|_{L_2[0,T]}$ is limited, and $\|q_{ki}(t) - q_{0i}(t)\|_{L_2[0,T]} \to 0$ as $k \to \infty$. Then

$$\sum_{i=1}^{N} \int_0^T g(r_i, t)q_{ki}(t)dt \to \sum_{i=1}^{N} \int_0^T g(r_i, t)q_{0i}(t)dt$$

due to the Schwarz inequality, and by proceeding to the limit in the inequality

$$\sum_{i=1}^{N} \int_0^T g(r_i, t)q_{ki}(t)dt \le \alpha \text{ as } k \to \infty$$

we obtain $\sum_{i=1}^{N} \int_0^T g(r_i, t)q_{0i}(t)dt \le \alpha$, i.e., $\vec{q}_0 \in \Theta$. The lemma is proved.

Definition 5.5.1 A uniformly convex space is a normed vector space if for each $\varepsilon > 0$ there exists $\delta > 0$ such that for any pair of vectors x, y with $\|x\| = 1$ and $\|y\| = 1$, the inequality $\|x + y\| > 2 - \delta$ implies $\|x - y\| < \varepsilon$.

Lemma 5.5.2 [31]. *A closed convex set in uniformly convex Banach space has a unique point closest to a given point.*

Theorem 5.5.3 *The optimal control problem (5.5.3) has a solution.*

Proof Since $\boldsymbol{H}$ is a Hilbert space, it is a uniformly convex Banach space [38]. Moreover, by Theorem 5.5.2 and Lemma 5.5.1, Θ is a closed convex set in $\boldsymbol{H}$, and $\vec{Q} \notin \Theta$ because

$$\sum_{i=1}^{N} \int_{0}^{T} g(r_i, t) Q_i(t) dt > \alpha \tag{5.5.6}$$

Therefore, by Lemma 5.5.2, there exists a point $\vec{Q}^{*} \in \Theta$ that minimizes the distance between the set Θ and the element $\vec{Q}$. This point is a solution of problem (5.5.3). The theorem is proved.

We now obtain an explicit form of the solution to control problem (5.5.3) in the simplest case, when there is only one pollution source in the domain D.

Theorem 5.5.4 *Let $Q(t)$ be an emission rate of the single point source located at r_0 and let $J(\phi_{\vec{Q}}) > J_0$. Then*

$$Q^*(t) = Q(t) - \beta g(r_0, t), \quad \beta = \left(J\left(\phi_Q\right) - J_0\right) / \int_{0}^{T} g^2(r_0, t) dt \tag{5.5.7}$$

is the solution of the optimal control problem (5.5.3), provided that $Q^*(t) \geq 0$ in $[0, T]$.

Proof The function (5.5.7) is obtained as the first variation of the Lagrangian $L(q) = F(q) - \vartheta(\int_0^T g(r_0, t) q(t) dt - \alpha)$. We now show that (5.5.7) is the solution of the optimal control problem (5.5.3). Indeed, $Q^*(t) > 0$ and $Q^*(t) \in \Theta$, since

$$\int_{0}^{T} g(r_0, t) Q^*(t) dt = J_0 - \int_{D} g(r, 0)\, \phi^0(r)\, dr = \alpha$$

Let now $q(t) = Q^*(t) + \delta q(t)$ be any element of set Θ different from $Q^*(t)$. Then

$$-\int_{0}^{T} g(r_0, t) \delta q(t) dt \geq \int_{0}^{T} g(r_0, t) Q^*(t) dt - \alpha$$

and

$$F(q) - F\left(Q^*\right) = -2\beta \int_0^T g(r_0, t)\delta q(t)dt + \int_0^T \left[\delta q(t)\right]^2 dt > 0.$$

Therefore, (5.5.7) is the solution of problem (5.5.3). The theorem is proved.

Remark 5.5.1 Due to (5.5.7), $Q^*(t) = Q(t)$ for any $t \in [0, T]$ such that $g(r_0, t) = 0$, that is, the optimum emission rate coincides with the initial rate $Q(t)$ whenever the source located at r_0 has no impact on the zone Ω. Unfortunately, if the difference $J(\phi_Q) - J_0$ or value $g(r_0, t)$ is large enough then the optimal rate Q^* given by (5.5.7) may be negative. In this case, (5.5.7) is not the optimal solution, and a numerical method should be used to approximate the optimal rate Q^* [30].

Example 5.5.1 Suppose that at the point $r_1 = (1.8, 0.2)$ of the square domain $D = (0, 2\,\text{km}) \times (0, 2\,\text{km})$ there is only one source that emits lead particles with a non-stationary rate [30]. For simplicity, assume that the initial distribution of lead is insignificant: $\phi^0(r) = 0$ in D. The coefficients of deposition σ and diffusion μ are equal to $0.001\,\text{h}^{-1}$ and $0.04\,\text{km}^2\,\text{h}^{-1}$, respectively. The wind velocity $U = (u, v)$ is nondivergent and is determined by the streamfunction $\psi = xy$:

$$u = -\frac{\partial \psi}{\partial y} = -x, \quad v = \frac{\partial \psi}{\partial x} = y \tag{5.5.8}$$

Thus, streamlines have the form of hyperbolas with the direction from southeast to northwest. The dispersion model and its adjoint model are considered in the four-hour interval $(0, T)$. The mean concentration of lead particles $J(\phi)$ in the zone $\Omega = [0{,}0.5] \times [0.5{,}1.0]$ is monitored during the same time period (i.e., $\tau = T = 4h$). The health standard J_0 is equal to $1.5\,\mu\text{g}\,\text{m}^{-3}$ [29]. The four types of emission rates Q (see Fig. 5.5) were considered in the units kg/h:

$$Q_1(t) = 3.8 \quad , \quad Q_2(t) = \begin{cases} 4, & 0 \le t < 1 \\ 5 - t, & 1 \le t \le 4 \end{cases}, \tag{5.5.9}$$

$$Q_3(t) = \cos \pi t + 3.5\,, \quad Q_4(t) = \begin{cases} 3, & 0 \le t < 1 \\ \frac{2}{3}(t-1) + 3, & 1 \le t \le 4 \end{cases} \tag{5.5.10}$$

Thus, Q_1 is constant, Q_2 is constant for the first hour and then decreases linearly, Q_3 is a periodic function with the two-hour period, and Q_4 is invariable during the first hour, and then increases linearly. Isolines of the dispersion model solution ϕ calculated using Q_1 are shown in Fig. 5.3 every hour (from $t = 1$ to $t = 4$). One can see a steady increase in the concentration of lead in Ω due to wind.

Isolines of the solution g of the adjoint model are shown in Fig. 5.4 also every hour (from $t = 4$ to $t = 0$). Under the action of the forcing $p(r, t)$, the adjoint solution has the form of a step function at $t = 4$, which then moves in the direction opposite to the wind U, i.e., from northwest to southeast, as it should be. The mean

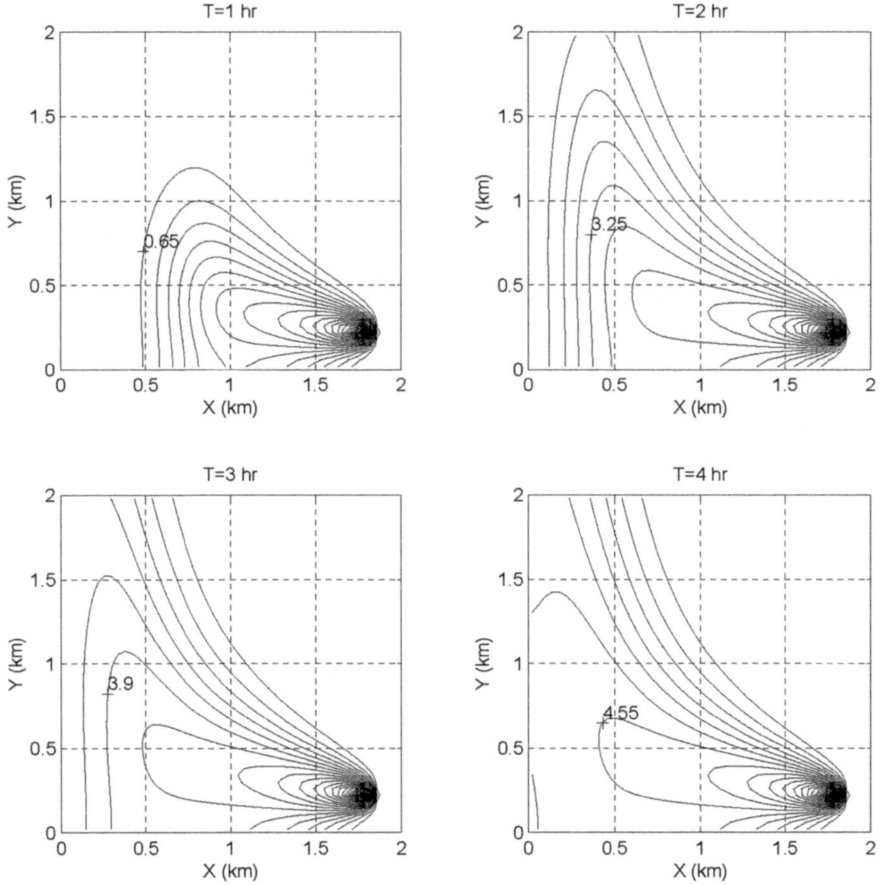

Fig. 5.3 Isolines of the solution of dispersion model $\phi(r, t)$ at $t = 1h$, $t = 2h$, $t = 3h$ and $t = 4h$

concentrations of lead particles $J(\phi)$ (see (3.1.5)) calculated in the zone Ω with each of the emission rates (5.5.9) and (5.5.10) are 2.11 $\mu g\, m^{-3}$, 2.02 $\mu g\, m^{-3}$, 1.97 $\mu g\, m^{-3}$ and 1.81 $\mu g\, m^{-3}$, respectively.

All the results are not satisfactory, since they exceed the health standard $J_0 = 1.5\, \mu gm^{-3}$, and therefore we now apply the optimal control method (5.5.7). The optimal emission rates $q_{opti}(t)$ given by the control are shown in Fig. 5.5. As a result, the mean pollution concentrations $J(\phi)$ obtained with optimal emission rates coincide with the health standard J_0, as it should be. In total agreement with theory, $q_{opti}(t) \leq Q_i(t)$ for each $t \in (0,4)$, $i = 1, \ldots, 4$.

Figure 5.5 also shows that $q_{opti}(t) = Q_i(t)$ during the last hour ($3 \leq t \leq 4$), i.e., the emission rates, optimal and original, matching each other during the period in which the solution of adjoint model $g(r_1, t)$ is zero, and therefore the emissions give no contribution to polluting the zone Ω. Finally, it is interesting to note that in the interval in which these values do not match ($0 \leq t < 3$), the temporal behavior of

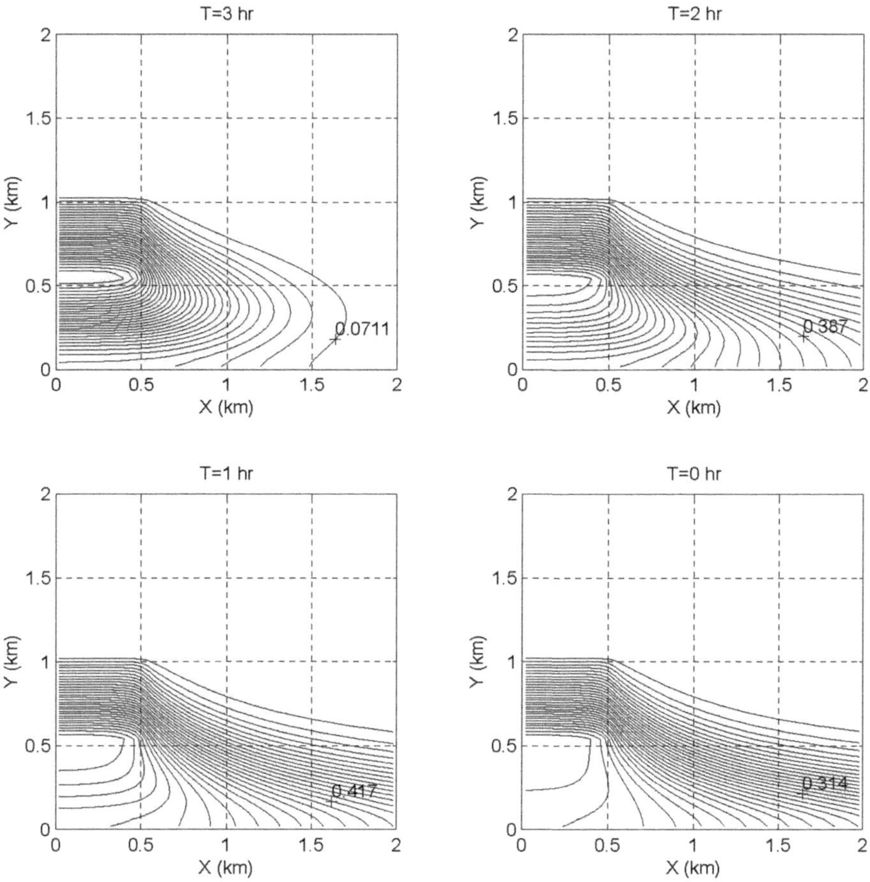

Fig. 5.4 Isolines of the solution $g(r, t)$ of adjoint model at $t = 3h$, $t = 2h$, $t = 1h$ and $t = 0h$

each optimum emission rate $q_{\text{opti}}(t)$ is similar to that of the corresponding original rate $Q_i(t)$. This is something useful for the industrial plants, because they should not drastically change its routine work.

5.6 Numerical Solution of Optimal Control Problem

This section begins with a few results that characterize the optimal solution and are very useful for its numerical approximation.

Lemma 5.6.1 *Let $\overrightarrow{Q}^* \in \Theta$ be a solution of optimal control problem (5.5.3). Then*

$$Q_i^*(t) \leq Q_i(t)\,, \quad 0 \leq t \leq T, \quad 1 \leq i \leq N. \tag{5.6.1}$$

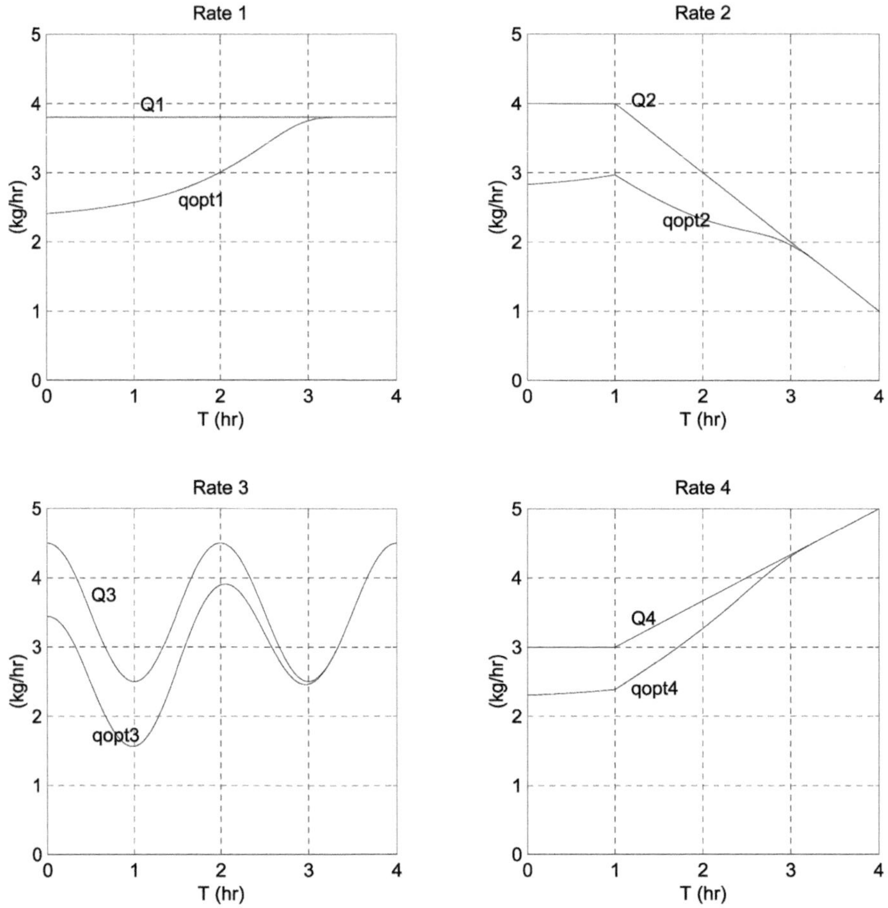

Fig. 5.5 Temporal behavior of original emission rates $Q_i(t)$ and optimal emission rates $q_{\text{opt}\,i}(t)$ in the four experiments ($i = 1,2,3,4$)

Proof Suppose that there is a number j and a subset $I \subset [0, T]$ of positive measure $|I| > 0$, such that $Q_j^*(t) > Q_j(t)$ for $t \in I$. Let

$$\tilde{q}_j(t) = \begin{cases} Q_j(t), & t \in I \\ Q_j^*(t), & t \in I^c \end{cases}$$

where $I^c = [0, T]\backslash I$. Then

$$\int_0^T g(r_j, t)\tilde{q}_j(t)dt \leq \int_0^T g(r_j, t)Q_j^*(t)dt$$

and therefore $\overrightarrow{q}(t) = \{Q_1^*(t), ..., Q_{j-1}^*(t), \tilde{q}_j(t), Q_{j+1}^*(t), ..., Q_N^*(t)\} \in \Theta$, besides,

$$F(\vec{q}) = \sum_{i=1}^{N} \int_0^T (Q_i - q_i)^2 dt = \sum_{i=1, i \neq j}^{N} \int_0^T \left(Q_i - Q_i^*\right)^2 dt$$
$$+ \int_{I^c} \left(Q_j - Q_j^*\right)^2 dt < F\left(\vec{Q}^*\right)$$

This result contradicts the fact that $\overrightarrow{Q}^*$ is the solution of optimal control problem. The lemma is proved.

Due to Lemma 5.6.1, the optimum emission rates $\overrightarrow{Q}^*$ are limited and never exceed the original emission rates $\overrightarrow{Q}$.

The next result allows reducing the feasible space Θ of optimal control problem. This fact will be important for the numerical solution of control problem.

Theorem 5.6.1 *Let $\overrightarrow{Q}^*$ be a solution of optimal control problem* (5.5.3). *Then*

$$\left(\vec{g}, \vec{Q}^*\right) \equiv \sum_{i=1}^{N} \int_0^T g(r_i, t) Q_i^*(t) dt = \alpha \tag{5.6.2}$$

Proof Let for some $j, Q_j^*(t) < Q_j(t), t \in I \subset [0, T], |I| > 0$. Such component of optimal vector $\overrightarrow{Q}^*$ must exist due to Lemma 5.6.1 and the fact that $Q_i^* = Q_i$ for each i is impossible.

Let us define a new component $\tilde{q}_j$ as$\tilde{q}_j = Q_j^* + \varepsilon(Q_j - Q_j^*), 0 < \varepsilon < 1$, and let$\overrightarrow{q}(t) = \{Q_1^*(t), ..., Q_{j-1}^*(t), \tilde{q}_j(t), Q_{j+1}^*(t), ..., Q_N^*(t)\}$. Then

$$(\vec{g}, \vec{q}) \equiv \left(\vec{g}, \vec{Q}^*\right) + \varepsilon \int_0^T g\left(r_j, t\right)\left(Q_j - Q_j^*\right) dt$$

Assume that $\left(\overrightarrow{g}, \overrightarrow{Q}^*\right) < \alpha$. Then one can define a positive number $\rho = 0.5\left[\alpha - \left(\overrightarrow{g}, \overrightarrow{Q}^*\right)\right]$ and choose a small number $\varepsilon > 0$ such that

$$\varepsilon \int_0^T g\left(r_j, t\right)\left(Q_j - Q_j^*\right) dt \leq \rho$$

It follows from the last inequality that $(\overrightarrow{g}, \overrightarrow{q}) \leq \left(\overrightarrow{g}, \overrightarrow{Q}^*\right) + \rho < \alpha$, and therefore $\overrightarrow{q} \in \Theta$. On the other hand, $Q_j - \tilde{q}_j = (1 - \varepsilon)(Q_j - Q_j^*)$, and hence,

$Q_j - \tilde{q}_j \leq Q_j - Q_j^*$. The last inequality is strict in the set $I \subset [0, T]$. Since $|I| > 0$, we obtain

$$\int_0^T (Q_j - \tilde{q}_j)^2 dt < \int_0^T \left(Q_j - Q_j^*\right)^2 dt,$$

and therefore $F(\vec{q}) < F(\vec{Q}^*)$. This contradicts the fact that $\vec{Q}^*$ is the solution of optimal control problem (5.5.3), and hence, $\left(\vec{g}, \vec{Q}^*\right) = \alpha$. The theorem is proved.

Due to Theorem 5.6.1, the feasible space (5.5.5) is reduced to

$$\Theta = \{q_i(t);\ q_i(t) \geq 0,\ (i = 1, .., N) | \sum_{i=1}^{N} \int_0^T g(r_i, t) q_i(t) dt = \alpha\} \tag{5.6.3}$$

(for simplicity, we will denote the reduced set (5.6.3) by the same letter Θ). From the point of view of searching an approximate (numerical) solution of the optimal control problem, the new feasible space (5.6.3) has a considerable advantage over the set (5.5.5). In order to obtain a discrete optimal control problem, we divide the interval $[0, T]$ into K equal subintervals $[t_k, t_{k+1}]$ of length $\Delta t > 0$ such that $t_k = k\Delta t$, $(k = 0,1, ..., K)$ and $T = K\Delta t$. Denote by $t_{k+1/2}$ the midpoint of interval $[t_k, t_{k+1}]$. Let $q_{ik} = q_i(t_{k-1/2})$, $Q_{ik} = Q_i(t_{k-1/2})$, $g_{ik} = g(r_i, t_{k-1/2})$, $(k = 0,1, ..., K)$. Then the use of the midpoint quadrature for the approximation of the integrals in control problem (5.5.3) and feasible space (5.6.3) leads to the following quadratic minimization problem:

$$\begin{gathered} \min_{q_{ik}} F \equiv \sum_{i=1}^{N} \sum_{k=1}^{K} (q_{ik} - Q_{ik})^2, \\ \text{subject to } \Delta t \sum_{i=1}^{N} \sum_{k=1}^{K} g_{ik} q_{ik} = \alpha,\ q_{ik} \geq 0 \end{gathered} \tag{5.6.4}$$

The optimization problem can equivalently be written as

$$\min_{x_i} F = \sum_{i=1}^{L} \left(x_i - x_i^0\right)^2,$$

$$\text{subject to} \begin{cases} \Delta t \sum_{i=1}^{L} b_i x_i = \tilde{\alpha}\,, \\ x_i \geq 0\,,\ i = 1, \ldots, L \end{cases} \tag{5.6.5}$$

where $x_i = q_{ik}$, $x_i^0 = Q_{ik}$, $b_i = g_{ik}$, $\tilde{\alpha} = \alpha/\Delta t$ and $L = N \cdot K$. Besides, $\sum_{i=1}^{L} b_i x_i^0 > \tilde{\alpha}$, and the discrete feasible space Θ_d is defined by the condition (5.6.5).

The optimization problem (5.6.5) has a unique solution. The existence of the solution immediately follows from the fact that the objective function F is continuous and feasible space Θ_d is compact [37]. The uniqueness of the solution is the consequence of the fact that F is strictly convex function and set Θ_d is also convex [37]. On the other hand, due to convexity, $x_i^* \le x_i^0, i = 1, 2, ..., L$ [26], and therefore, as in the continuous case (Lemma 5.6.1), the industrial sources must decrease its intensity.

The solution $x^* = (x_i^*)_{i=1}^L$ to optimization problem (5.6.5) is found by using the characterization theorem [26, 37], which states that x^* is a solution if

$$-\nabla F\left(x^*\right) \cdot \left(x - x^*\right) \le 0 \text{ for all } x \in Q_d, x \ne x^* \tag{5.6.6}$$

The first approximation of x^* is obtained as the orthogonal projection in the minimization process using Lagrange multipliers [37]:

$$x_i^* = x_i^0 - 0.5\lambda\, b_i, \quad i = 1, \ldots, L, \lambda = 2\left(\sum_{i=1}^{L} b_i x_i^0 - \tilde{\alpha}\right)\left(\sum_{i=1}^{L} b_i^2\right)^{-1} \tag{5.6.7}$$

This approximation satisfies the condition $\sum_{i=1}^{L} b_i x_i^* = \tilde{\alpha}$. To improve the approximation, successive orthogonal projections are used.

Algorithm of successive orthogonal projections. The algorithm consists of two steps.

Step I. If $x_i^* \ge 0$ for $i = 1, \ldots, L$ then $x^* \in Q_d$ and

$$-\nabla F\left(x^*\right) \cdot \left(x - x^*\right) = \lambda \sum_{i=1}^{L} b_i x_i - \lambda \sum_{i=1}^{L} b_i x_i^* = \lambda\, \tilde{\alpha} - \lambda\, \tilde{\alpha} = 0 \tag{5.6.8}$$

for all $x \in Q_d, x \ne x^*$. By the characterization theorem [37], x^* is a minimum.

Step II. If at least one of the components of vector x^* is negative, then it is necessary to redefine the projection. Without loss of generality, assume that $x_i^* \le 0$ for $i = l + 1, \ldots, L$ (since $\tilde{\alpha} > 0$, at least one of the components is always positive). Then we define $x_i^{**} = 0$ for $i = l + 1, \ldots, L$ and take a new projection with the functional F and set Θ_d restricted to the first l variables:

$$x_i^{**} = x_i^0 - 0.5\, \eta\, b_i\,, \quad i = 1, \ldots, l,\ \eta = 2\left(\sum_{i=1}^{l} b_i x_i^0 - \tilde{\alpha}\right)\left(\sum_{i=1}^{l} b_i^2\right)^{-1} \tag{5.6.9}$$

This approximation also leads to the condition $\sum_{i=1}^{L} b_i x_i^{**} = \tilde{\alpha}$. If $x_i^{**} \ge 0$ for $i = 1, \ldots, l$, then $x^{**} \in Q_d$ and

$$-\nabla F\left(x^{**}\right) \cdot \left(x - x^{**}\right) = 2 \sum_{i=l+1}^{L} x_i\left(x_i^0 - 0.5\eta b_i\right), x \in Q_d, x \ne x^{**} \tag{5.6.10}$$

Since $\sum_{i=1}^{l} b_i x_i^* > \tilde{\alpha}$, (5.6.7) implies $\eta > \lambda$ and leads to

$$-\nabla F\left(x^{**}\right)\cdot\left(x-x^{**}\right) \le 2\sum_{i=l+1}^{L} x_i x_i^* \le 0 \quad \text{for all}\, x \in Q_d,\ x \ne x^{**} \tag{5.6.11}$$

Due to characterization theorem [37], x^{**} is a minimum.

This algorithm converges to the exact solution at no more than L iterations (projections), because $\tilde{\alpha} > 0$. Moreover, it is economic (does not consume a lot of computer time), since the number of arithmetic operations does not exceed $3(L^2 + L)$ [26]. For example, if a computer performs 100,000 op/s, then all the optimal constants x_i^* can be calculated for 0.5 min if $L = 1000$, and for 12.5 min if $L = 5000$. Therefore, whenever a short-term model forecast is not favorable, the optimal control emission rates can be performed quickly. This property is especially important when the number of emission sources is large.

5.7 Control Using Convex Linear Combinations

As we know, the existence of the solution to control problem depends on α defined by (5.3.4), namely,

(a) if $\alpha < 0$ then there is no solution to control problem
(b) if $\alpha = 0$ then there is only one solution to control problem
(c) if $\alpha > 0$ then there is an infinite number of solutions to control problem.

The last case is obvious, since there is infinite number of all possible emission rates that satisfy the conditions of Theorem 5.3.1 and its generalization (see (5.3.15)). However, this is not the only way to generate an infinite number of sufficient control strategies. In fact, new control strategies can always be obtained using convex linear combinations of existing control strategies. To this end, one must choose nonnegative constants $\xi_1, \xi_2, \ldots \xi_M$ such that

$$\xi_1 + \xi_2 + \ldots + \xi_M = 1 \tag{5.7.1}$$

where M is the number of previously defined control strategies, each of which ensures compliance with health standards in the zone: $J(\phi) \le J_0$. Let us denote by $Q_{i,k}^*$ the emission rate of the i-th source defined by the k-th control strategy. Then the new emission rates

$$q_i^*(t) = \sum_{k=1}^{M} \xi_k Q_{i,k}^*(t), \quad i = 1, \ldots, N \tag{5.7.2}$$

also satisfy the air quality standard. Indeed,

$$\sum_{i=1}^{N}\int_{0}^{T} q_i^*(t)g(r_i,\, t)dt = \sum_{i=1}^{N}\sum_{k=1}^{M}\xi_k \int_{0}^{\mathrm{T}} Q_{i,k}^* g(r_i,\, t)dt$$
$$= \sum_{k=1}^{M}\xi_k\left(\sum_{i=1}^{N}\int_{0}^{\mathrm{T}} Q_{i,k}^* g(r_i,\, t)dt\right)$$

and each term in the parenthesis is less than α. Therefore,

$$\sum_{i=1}^{N}\int_{0}^{T} q_i^*(t)g(r_i,\, t)dt \le \alpha\sum_{k=1}^{M}\xi_k \le \alpha$$

Finally, considering (5.3.4) we get that $J(\phi) \le J_0$.

The new control strategy (5.7.2) is not only mathematically correct, but also represents an option to adapt the M control strategies to the needs of the industrial plants, as the choice of constants ξ_k $(k = 1 \ldots M)$ gives a wide range of possibilities. However, it should be noted that the optimal control strategy can still have certain advantages over new sufficient control strategies.

Example 5.7.1 Let us consider in an open boundary domain $D = (0{,}3\,\text{km}) \times (0{,}3\,\text{km})$ two point sources located at points $r_1 = (1.5{,}0.2)$ and $r_2 = (0.6{,}2.0)$, which emit sulfur dioxide with non-stationary rates Q_1 and Q_2 (kg/h):

$$Q_1(t) = \begin{cases} 10000 \cdot t, & 0 \le t < 0.5, \\ 5000, & 0.5 \le t \le 7, \\ 5000 - 5000 \cdot (t-7), & 7 < t \le 8. \end{cases},$$

$$Q_2(t) = \begin{cases} 5000 \cdot t, & 0 \le t < 0.5, \\ 2500, & 0.5 \le t \le 1, \\ 2500 + 2000 \cdot (t-1), & 1 < t < 1.5, \\ 3500, & 1.5 \le t \le 7.5, \\ 3500 - 7000 \cdot (t-7.5), & 7.5 < t \le 8. \end{cases}.$$

Also assume that the initial concentration of sulfur dioxide in D at $t = 0$ is

$$\phi^0(r) = 175 \cdot \exp\{-|r - r_1|^2\} + 125 \cdot \exp\{-|r - r_2|^2\},$$

i.e., the residue of SO_2, accumulated from each source during the previous period, has a Gaussian distribution. The parameters of chemical transformation and diffusion are taken as $\sigma = 0.36\,\text{h}^{-1}$ and $\mu = 1.8\,\text{km}^2/\text{h}$, respectively [39].

The velocity field $U = (u, v)$ is generated by the streamfunction $\psi = \frac{5}{3}x^2 - 5y$, whose streamlines are parabolas with direction from southwest to northeast, i.e.,

$$u = -\frac{\partial \psi}{\partial y} = 5, \qquad v = \frac{\partial \psi}{\partial x} = \frac{10}{3}x,$$

and hence $\nabla \cdot U = 0$. As in [40], the velocity components vary from 5 to 10 km/h. The process of dispersion of sulfur dioxide is observed in the zone $\Omega = [1,2] \times [1,2]$ during the eight-hour period ($T = 8$ h). The mean concentration of the pollutant in zone Ω ($\tau = T$) estimated with the model and Formula (3.5.6) is $J(\phi) = 242.43\,\mu\text{g}\,\text{m}^{-3}$. This value represents a violation of the health standard which is $J_0 = 210\,\mu\text{g}/\text{m}^3$ (according to Wark and Warner [41], this concentration (0.08 ppm) causes symptoms of bronchoconstriction in the most vulnerable populations). Figure 5.6 shows isolines of the SO_2 concentration and solution of adjoint problem at different moments.

It can be seen that due to the wind blowing from the southwest to the northeast, the SO_2 concentration in zone Ω first increases and then decreases at the end of the 8 h period. Isolines of the adjoint solution show that the original step function $p(r,t)$ (see (3.5.4) with $\omega(r,t) = 1/(\tau|\Omega|)$) moves with time in the direction opposite to wind velocity U (Fig. 5.6d). Figure 5.7 shows the respective surfaces of the SO_2 concentration and the adjoint solution.

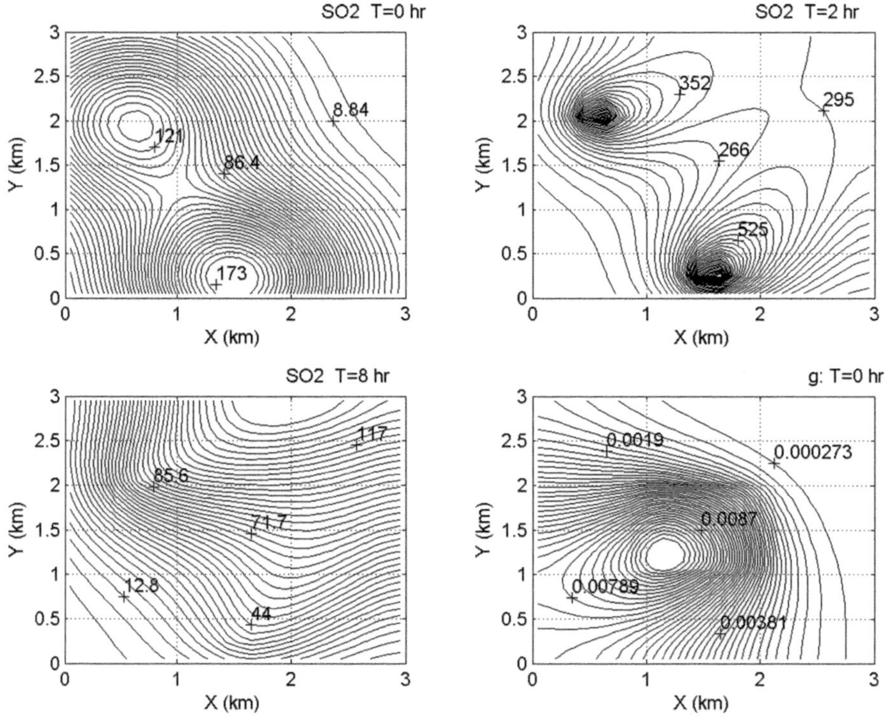

Fig. 5.6 Isolines of SO_2 concentration at $t = 0$, $t = 2h$, $t = 8h$, and isolines of adjoint problem solution g at $t = 0$

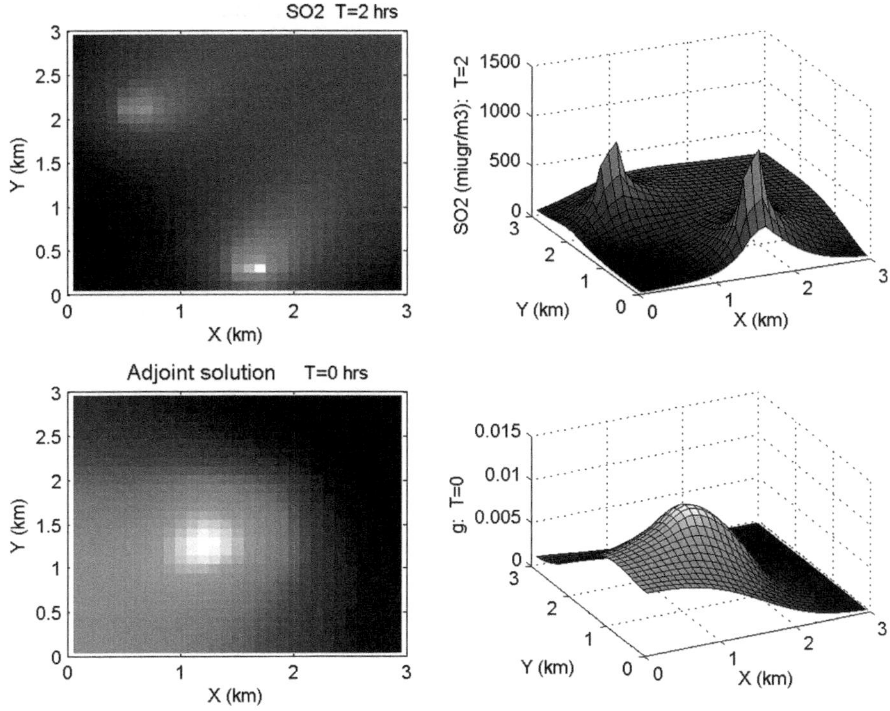

Fig. 5.7 The SO_2 concentration at $t = 2h$ (upper half), and the adjoint solution g at $t = 0$ (bottom half)

As expected, the adjoint estimate (3.5.7) gives the same result: $J(\phi) = 242.43\,\mu gm^{-3}$. Figures 5.8 and 5.9 demonstrate the original emission rates $Q_1(t)$ and $Q_2(t)$ of two sources, and the emission rates $Q_{11}^*(t)$ and $Q_{12}^*(t)$ obtained by applying the control (5.3.8), $Q_{21}^*(t)$ and $Q_{22}^*(t)$ obtained by applying the optimal control (5.4.4), and $Q_{5i}^*(t) = 0.3Q_{1i}^*(t) + 0.7Q_{2i}^*(t)$ $(i = 1{,}2)$ obtained by using the convex linear combination of two controls.

Note that in these two figures, all emission rates obtained using the control strategies (5.3.8), (5.4.4) and (5.7.2) are shown without asterisks. The results of the work of industrial enterprises with these emission rates are shown in Table 5.1.

As expected, all control strategies meet the health standard, however, only the optimal control strategies Q_{21}^* andQ_{22}^*, as well as the strategies Q_{51}^* and Q_{52}^* obtained by the convex linear combinations, generate the profiles of emission rates which are similar to the profiles of the original emission rates. In fact, these rates are the best choice in this example, since each source begins and ends the operation with zero emissions, besides, their intensities are 88% and 84% of the original rates Q_1 and Q_2, and therefore the implementation of these control strategies is rather simple.

The emission rates Q_{11}^* and Q_{12}^* obtained with the control strategy (5.3.8) can only a viable alternative only in the case that both industrial sources are able to initiate

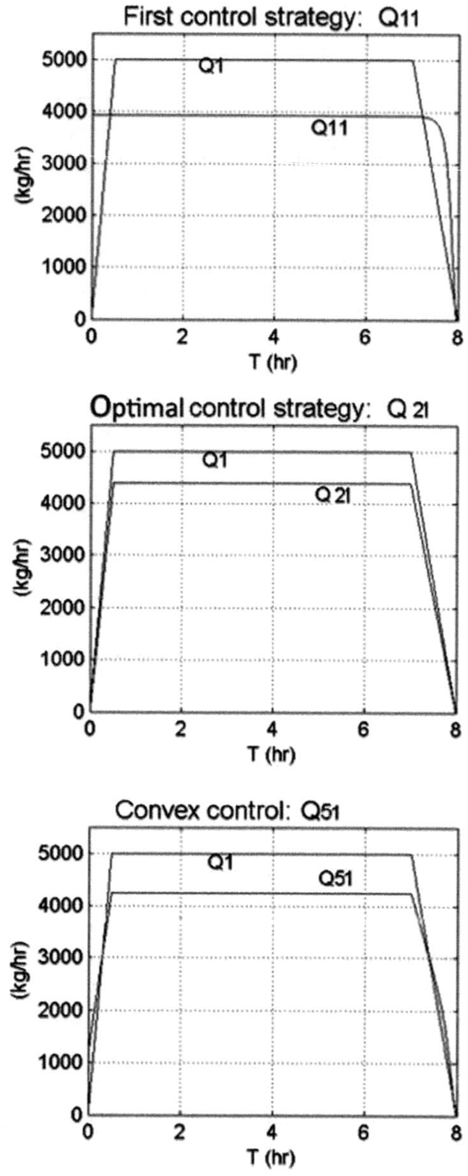

Fig. 5.8 Temporal behavior of the first source: original emission rate $Q_1(t)$, emission rate $Q_{11}(t)$ set by control (5.3.8), emission rate $Q_{21}(t)$ set by optimal control (5.4.4), emission rate $Q_{51}(t) = 0.3Q_{11} + 0.7Q_{21}$ set by a convex linear combination (5.7.2) of two controls

their activity immediately with the emission rates close to their maximum possible values.

As we know, the emission intensity control problem generally has infinite number of solutions if $\alpha > 0$. In Example 5.7.1, $\int_D g(r, 0)\phi^0(r)dr = 2.98$ and according to (5.3.4), $\alpha = 207.02$. As a result, the impact of the initial distribution of pollutant ϕ^0 on the mean concentration in zone Ω is low, even if values of ϕ^0 are not small.

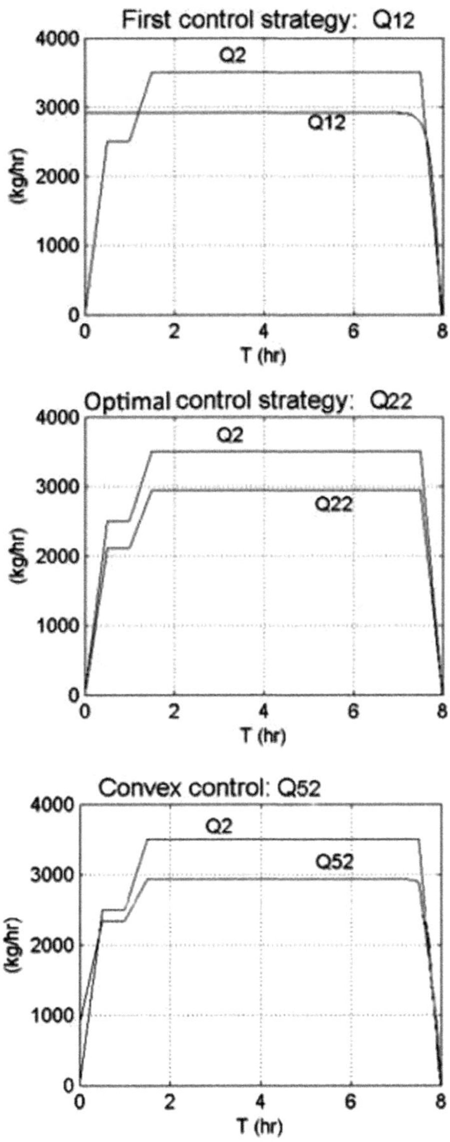

Fig. 5.9 Temporal behavior of the second source: original emission rate $Q_2(t)$, emission rate $Q_{12}(t)$ set by control (5.3.8), emission rate $Q_{22}(t)$ set by optimal control (5.4.4), emission rate $Q_{52}(t) = 0.3Q_{12} + 0.7Q_{22}$ set by a convex linear combination (5.7.2) of two controls

This is due to the dispersion caused by wind within the eight hours of operation of industrial sources, and therefore the daily operation of these sources is responsible for the violation of health standards near industrial sources.

Table 5.1 The average concentrations $J(\phi)$ in Ω

Emission rates	$J(\phi)(\mu g/m^3)$
Q_1 and Q_2	242.43
Q_{11}^* and Q_{12}^*	210.00
Q_{31}^* and Q_{32}^*	210.00
$Q_{51}^* = 0.3 \cdot Q_{11}^* + 0.7 \cdot Q_{31}^*$ $Q_{52}^* = 0.3 \cdot Q_{12}^* + 0.7 \cdot Q_{32}^*$	210.00

5.8 Optimal Control Using the Relative Contribution of Sources

In this section, we consider optimal short-term emission control, in which the emission rates of industrial sources enter the minimized function with certain weights. We would like to clarify what meaning we put into the words "mathematical methods of control". First of all, it is assumed that there are reliable meteorological data and mathematical (and numerical) models for predicting the concentrations of pollutants in an area at current emission rates from pollutant sources.

Let $\vec{Q}(t) = \{Q_1(t), ..., Q_N(t)\}$ be emission rates of N industrial sources emitting a pollutant in a domain D and time interval $(-t_0, T)$. And let $\boldsymbol{M}$ be a model for predicting the concentration $\phi_{\vec{Q}}(r, t)$ of the pollutant. Here $t_0 > 0$ is the time required to perform the forecast in the interval $(-t_0, T)$ and control in the interval $(0, T)$ (see Fig. 5.10). Let us assume that the forecast is unsatisfactory in the sense that

$$J\left(\phi_{\vec{Q}}\right) > J_0 \tag{5.8.1}$$

i.e., the average concentration of the pollutant in a monitored zone $\Omega \subset D$ and time interval $(T - \tau, T)$ exceeds the ambient air quality standard J_0. Therefore, the

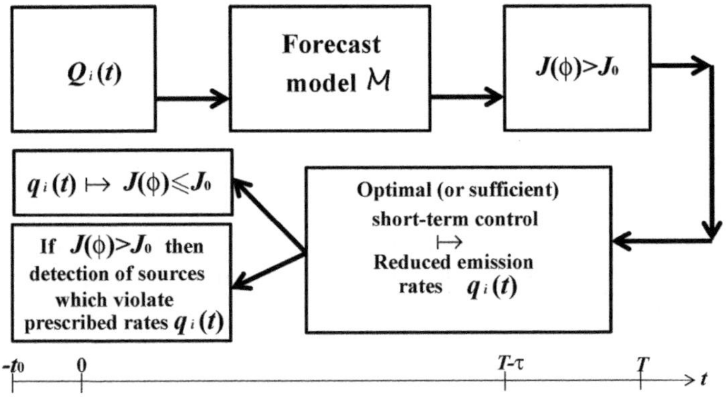

Fig. 5.10 Scheme of a short-term control within time interval $(0, T)$

purpose of control is to determine in the interval $(0, T)$ new (reduced) emission rates $\overrightarrow{q}(t) = \{q_1(t), ..., q_N(t)\}$ of the industrial sources that will guarantee the fulfillment of the sanitary standard:

$$J\left(\phi_{\vec{q}}\right) \leq J_0 \tag{5.8.2}$$

in $\Omega \times (T - \tau, T)$. Let

$$F(\vec{q}) = \frac{1}{2}\sum_{i=1}^{N} \gamma_i^2 \|Q_i - q_i\|^2 = \frac{1}{2}\sum_{i=1}^{N} \gamma_i^2 \int_0^T (Q_i - q_i)^2 dt \tag{5.8.3}$$

be defined in the domain

$$\Theta = \left\{\vec{q}(t) : q_i(t) \geq 0,\ t \in [0, T],\ i = 1, ..., N; J\left(\phi_{\vec{q}}\right) \leq J_0\right\} \tag{5.8.4}$$

In other words, Θ is the set of emission rates $\overrightarrow{q}(t) = \{q_1(t), ...q_N(t)\}$ that ensure compliance with the ambient air quality standard J_0 in Ω: $J(\phi) \leq J_0$.

The optimal control consists in finding such rates $\overrightarrow{Q}^*(t) \in \Theta$ that minimize the function $F(\overrightarrow{q})$ in the feasible set Θ:

$$F\left(\vec{Q}^*\right) = \min\{F(\vec{q}) : \vec{q} \in \Theta\} \tag{5.8.5}$$

It is clear that the control depends on the norm $\|\cdot\|$ used in (5.8.3). In this work, we use the norm in the space $L_2(0, T)$. Note that $\overrightarrow{Q}^*(t)$ is the optimal solution that represents the least constraint on the operation of industrial sources (i.e., the new emission rates $\overrightarrow{Q}^*(t)$ are as close as possible to the initial emission rates $\overrightarrow{Q}(t)$).

The weights γ_i for the objective function (5.8.3) will be chosen as follows. Let

$$c_i = \int_0^T g(r_i, t)Q_i(t)dt > 0$$

denote the mean concentration obtained in the monitored zone Ω only due to emissions from the i-th pollution source ($i = 1, ..., N$). Then $f_i = c_i/(c_1 + \cdots + c_N)$ is the portion of pollution from the i-th source, and we choose $\gamma_i = 1/f_i$. So γ_i tends to be big for small f_i.

For each monitored zone Ω, c_i and γ_i can be calculated using the adjoint model solution and Formula (3.5.7). The weights γ_i in (5.8.3) are entered in favor of low intensity sources. Indeed, a pollution source with less impact on the monitored zone Ω receives more weight. As a result, the optimal emission rate for this source, set by the control, becomes closer to the original rate. Thus, Formulas (5.8.3)–(5.8.5) represent a new control of emissions from industrial sources, which is primarily aimed at reducing the emission rates of the main sources of pollution.

Example 5.8.1 To demonstrate the possibilities of the new control strategy, we consider a numerical example of optimal short-term control using synthetic data. Six industrial sources are considered in the domain $D = (0,5) \times (0,5)$, and emission control is applied in the time interval $(0, T)$ where $T = 8$. As a monitored zone we take $\Omega = (2,3) \times (2,3)$, and $\tau = T = 8$. Figure 5.11 shows the location r_i of the point sources ($i = 1, ..., 6$), the position of zone Ω, and the nondivergent wind velocity $U = (u, v)$ where $u(x, y) = y^2/25$ and $v(x, y) = x/5$. The diffusion and transformation parameters for the dispersion model are $\mu = 0.1$ and $\sigma = 0.001$, respectively.

The emission rates of the six industrial sources during the interval $(0, T)$ are as follow:

$Q_1(t) = 3, Q_2(t) = 10e^{-(t-4)^2}, Q_3(t) = |t - 2|,$

$$Q_4(t) = 2 - 0.5\cos\left(\frac{\pi}{2}t\right),\ Q_5(t) = 2e^{-(t-1)^2}$$

and

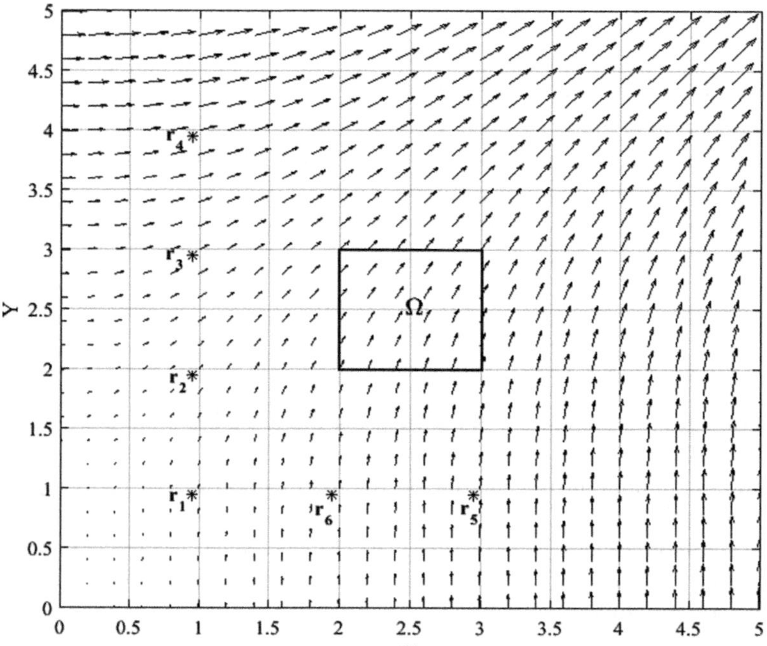

Fig. 5.11 Location of point sources r_i and monitored zone Ω in the domain D. The arrows show the direction of the wind U

$$Q_6(t) = e^{-(t-3)^2} + 2e^{-(t-5)^2}$$

In this example, the initial distribution of the pollutant at $t = 0$ is taken uniform and small: $\phi(r, 0) = 0.2$. Due to the work of sources and the initial state, the average concentration of the pollutant $J(\phi_Q)$ in the monitored zone Ω is 1.7596 and exceeds the sanitary norm $J_0 = 1.5$. The optimal control method is then used to reduce the emission rates and respect the sanitary standard J_0.

At the first stage, the adjoint problem is solved in the reverse direction of time from $t = T$ to $t = 0$. Figures 5.12 and 5.13 show the isolines of the adjoint function $g(r, t)$ in D for $t = 6$ and $t = 4$. It can be seen that the area of non-zero values of $g(r, t)$ expands in the southwest direction, opposite to the direction of the wind speed in D, as it should be. Therefore, the main attention should be paid to monitoring the intensity of emissions from sources located in the southwestern part of the domain D.

With the help of the adjoint solution $g(r, t)$, it is possible to estimate the pollution levels c_i in the controlled zone. The values c_i, f_i and γ_i for each point source are given in Table 5.2. In addition, in our case, the contribution of the initial pollution is $c_0 = \int_D g(r, 0)\phi^0(r)dr = 0.1391$ and $\alpha = J_0 - c_0 = 1.3609 > 0$.

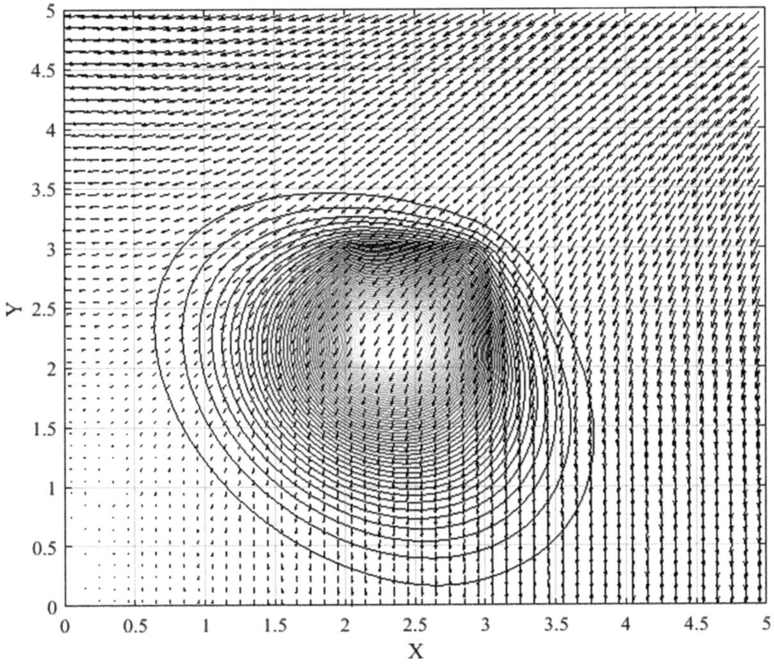

Fig. 5.12 Isolines of the adjoint function $g(r, t)$ at $t = 6$

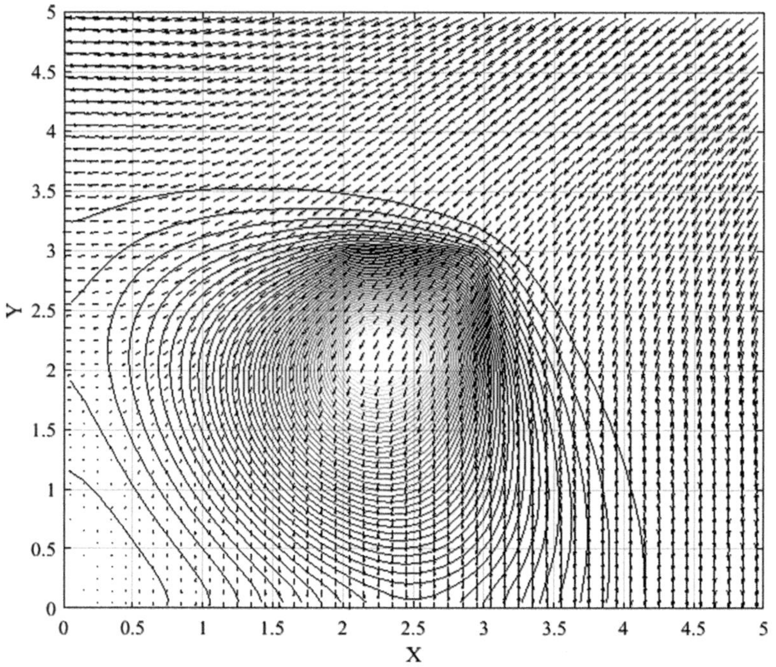

Fig. 5.13 Isolines of the adjoint function $g(r, t)$ at $t = 4$

Table 5.2 Location r_i and coefficients c_i, f_i and γ_i of each pollution source

i	r_i	c_i	f_i	γ_i
1	(0.95, 0.95)	0.4785	0.2953	3.3867
2	(0.95, 1.95)	0.5286	0.3262	3.0654
3	(0.95, 2.95)	0.1212	0.0748	13.3660
4	(0.95, 3.95)	0.0082	0.0051	196.6810
5	(2.95, 0.95)	0.1992	0.1230	8.1330
6	(1.95, 0.95)	0.2846	0.1757	5.6931

According to the fractions f_i, the sources located at points r_3 and r_4 make an insignificant contribution to the pollution of the zone Ω. Therefore, optimal control is expected to prescribe the minimum reduction in their emission rates.

Figures 5.14, 5.15, 5.16, 5.17, 5.18 and Fig. 5.19 show the initial emission rates Q_i and optimal rates Q_i^* for each source. In particular, Figs. 5.16 and 5.17 show that the initial and optimal intensity of emissions from sources, located at the points r_3 and r_4, are practically equal. Note that the CPU time used to calculate the adjoint model solution and optimal emission rates for this example is less than 2 min.

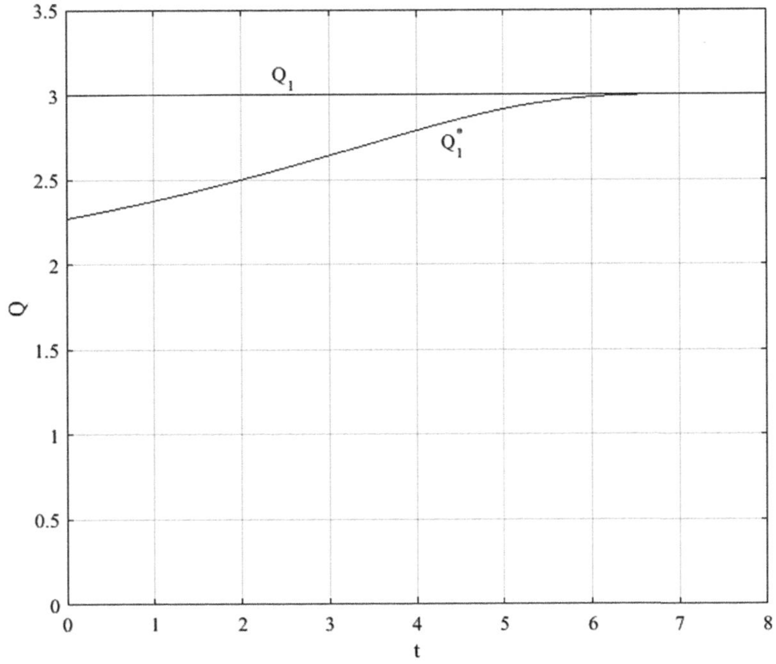

Fig. 5.14 Initial emission rate $Q_1(t)$ and optimal rate $Q_1^*(t)$ in time interval $(0, T)$

5.9 Using Linear Programming to Control Multiple Air Pollutants

In this section, we consider a three-dimensional (3D) limited-area model of dispersion of atmospheric pollution from point sources. The adjoint model and the duality principle are used to formulate a linear programming problem in order to determine the optimal emission rates of the sources and comply with air quality standards. The existence of a solution to the optimal control problem is proved. The interior point method and the simplex method are used to approximate the optimal solution. Finally, we present numerical results obtained in case of point sources that demonstrate the possibilities of the method.

We now define a short-term control problem for several pollutants. Let N be the number of pollution sources operating in a simply connected limited domain $D \subset \mathbb{R}^3$, and let K be the maximum number of quasi-passive pollutants emitted by each pollution source. Let $\boldsymbol{M}$ be a short-term dispersion model, which is used to predict the concentration ϕ_k of the k-th pollutant in the domain D and finite time interval $[0, T]$.

$$\boldsymbol{M} : \vec{q}| \rightarrow \phi_{\vec{q}}^k, \; k = 1, .., K$$

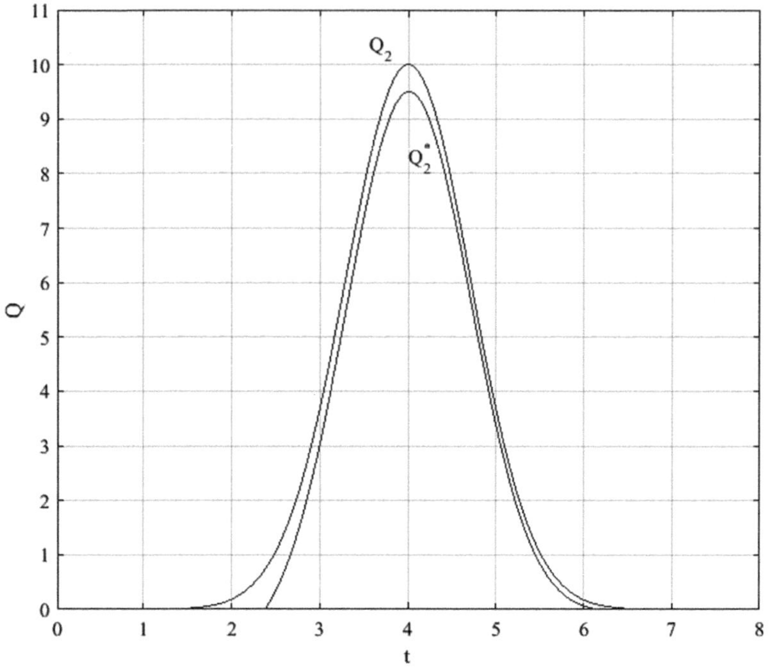

Fig. 5.15 Initial emission rate $Q_2(t)$ and optimal rate $Q_2^*(t)$ in time interval $(0, T)$

where $\overrightarrow{q} = (q_1, \ldots, q_N)$ is determined by non-stationary and non-negative pollutant emission rates q_i of the sources located at points $r_i \in D$. Note that emissions from each source may consist of different pollutants.

The average concentration of the k-th pollutant in a domain $\Omega \subset D$ and time interval $[T - \tau, T]$ of length $\tau > 0$, is determined by the functional

$$\boldsymbol{J}_{\Omega,\tau}^k = \boldsymbol{J}_{\Omega,\tau}\left(\phi_{\overrightarrow{q}}^k\right) = \frac{1}{|\Omega|\tau} \int_{T-\tau}^{T} \int_{\Omega} \phi_{\overrightarrow{q}}^k drdt, \tag{5.9.1}$$

where $|\Omega|$ is the volume of Ω. Let $\overrightarrow{Q} = (Q_1, \ldots, Q_N)$ be a vector representing the actual emission rates, and let $\boldsymbol{J}_{\Omega,\tau}(\phi_{\overrightarrow{Q}}^k) > J_k$ for some k, where $\boldsymbol{J}_k > 0$ be the air quality standard for the k-th pollutant. Then the emission rates are redundant, and the control problem is to determine such $\overrightarrow{q} = (q_1, \ldots, q_N)$ that $\boldsymbol{J}_{\Omega,\tau}(\phi_{\overrightarrow{q}}^k) \leq J_k$. Note that each subcomponent q_{ik} of q_i must be a non-negative function in the time-interval $[0, T]$.

In the general case, this problem is ill-posed, since it may have several solutions or no solutions, depending on the initial distribution of pollutant concentrations $\phi_k^0 = \phi_k(0)$. To formulate a well-posed problem, we introduce the following regularization method. Let

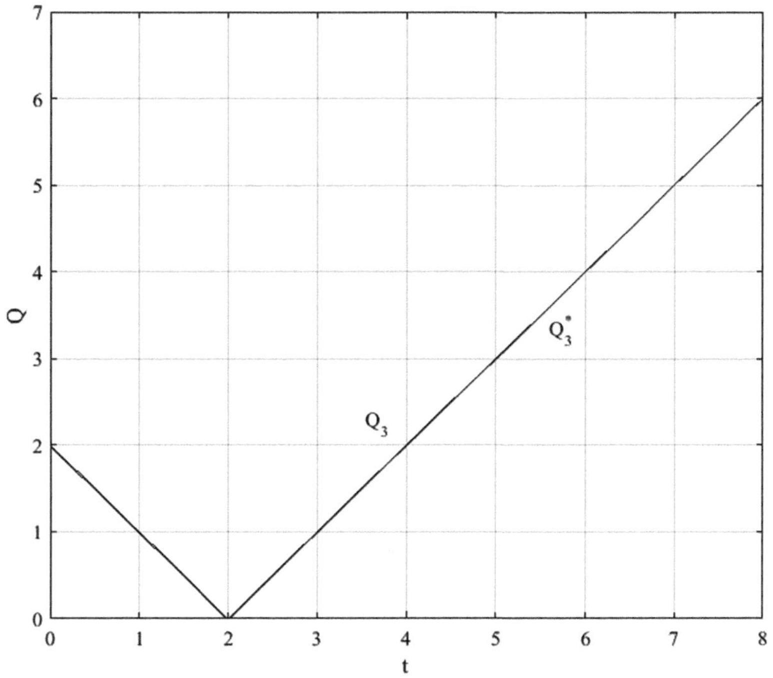

Fig. 5.16 Initial emission rate $Q_3(t)$ and optimal rate $Q_3^*(t)$ in time interval $(0, T)$

$$F(\vec{q}) = \left\| \vec{q} - \vec{Q} \right\| \tag{5.9.2}$$

be a functional defined in the domain

$$\mathbb{Q} = \{\vec{q} \mid q_i = \sum_{k=1}^{K} q_{ik}(t),\ 0 \le q_{ik}(t);\ J_{\Omega,\tau}\left(\phi_{\vec{q}}^k\right) \le J_k,\ k = 1, .., K\} \tag{5.9.3}$$

The optimal control problem is to determine such $\vec{q}^{\,*} \in \mathbb{F} \subseteq \mathbb{Q}$ that

$$F\left(\vec{q}^*\right) = \inf_{\vec{q} \in \mathbb{F}} F(\vec{q}) \tag{5.9.4}$$

where $\mathbb{F}$ is a suitable feasibility space.

Obviously $\vec{q}^{\,*}$ represents the least constraint imposed by control on point sources and depends on the norm $\|\cdot\|$ and the feasibility space $\mathbb{F}$ used. Thus, solutions to this inverse problem can be found using various mathematical programming problems or variational problems [2, 42]. It is important to note that the solution of each problem is usually sought using an iterative optimization method that applies successive estimates of the dynamic model $\boldsymbol{M}$ [43]. In general, this process is not as efficient, as it requires a lot of calculations due to complexity of model $\boldsymbol{M}$. In this section, we

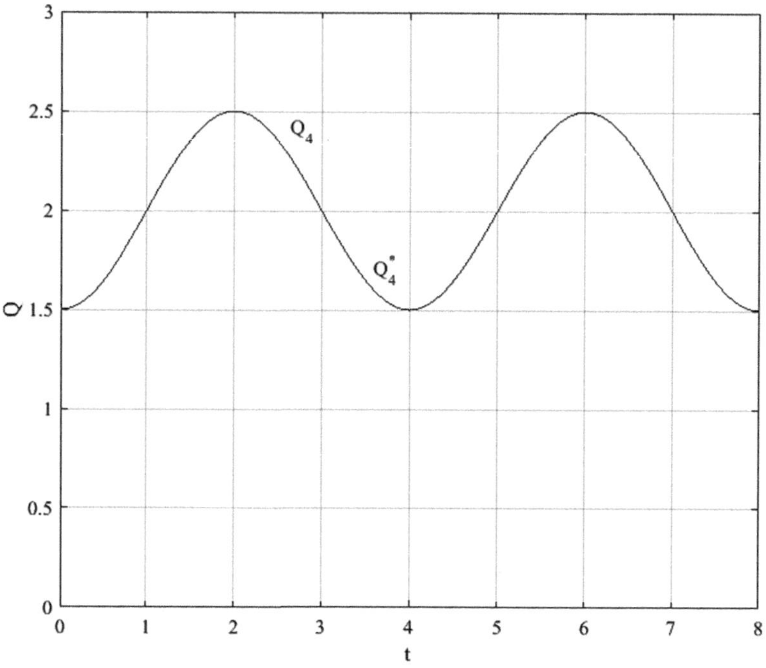

Fig. 5.17 Initial emission rate $Q_4(t)$ and optimal rate $Q_4^*(t)$ in time interval $(0, T)$

describe a method based on the use of adjoint operators and linear programming that allows us to solve the optimal control problem without using successive evaluation of model $\boldsymbol{M}$.

1. **Three-dimensional dispersion model of *k* pollutants.** Let $\mathbf{D} = \boldsymbol{D} \times (0, H)$ be a simply connected bounded domain in $\mathbb{R}^3$ with a boundary $\partial\mathbf{D} = S_0 \cup S \cup S_H$ that is the union of the cylindrical side surface S, the base S_0 at the bottom, and the top cover S_H at $z = H$ (see Fig. 5.20). The short-term dispersion model M of K passive pollutants in the domain $\mathbf{D}$ is

$$\frac{\partial \phi_k}{\partial t} + \boldsymbol{U} \cdot \nabla \phi_k + \sigma_k \phi_k - \nabla \cdot (\mu \nabla \phi_k) - \frac{\partial}{\partial z} \mu_z \frac{\partial \phi_k}{\partial z} + \nabla \cdot \boldsymbol{\phi}_k^s = f_k(\boldsymbol{r}, t) \quad (5.9.5)$$

$$\boldsymbol{\phi}_k^s = -\nu_k^s \boldsymbol{\phi}_k \boldsymbol{e}_3 \text{ in } \mathbf{D} \quad (5.9.6)$$

$$\phi_k(\boldsymbol{r}, 0) = \phi_k^0(\boldsymbol{r}) \text{ in } \mathbf{D} \quad (5.9.7)$$

$$\mu \nabla \phi_k \cdot \boldsymbol{n} - U_n \phi_k = 0 \text{ at } S^- \quad (5.9.8)$$

$$\mu \nabla \phi_k \cdot \boldsymbol{n} = 0 \text{ at } S^+ \quad (5.9.9)$$

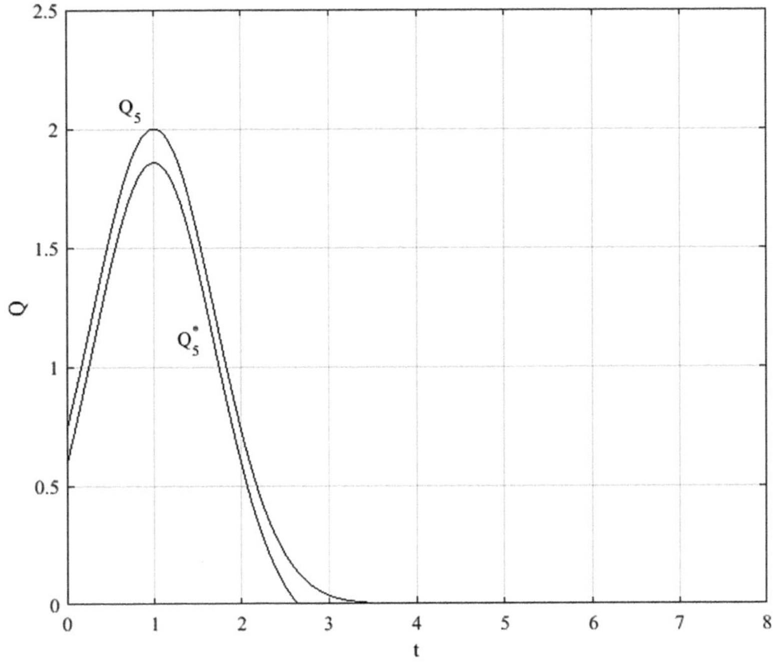

Fig. 5.18 Initial emission rate $Q_5(t)$ and optimal rate $Q_5^*(t)$ in time interval $(0, T)$

$$\hat{\mu}\nabla\phi_k \cdot \boldsymbol{n} = 0 \text{ at } S_0 \tag{5.9.10}$$

$$\mu_z \frac{\partial \phi_k}{\partial z} - U_n\phi_k = -\nu_k^s \phi_k \text{ at } S_H^- \tag{5.9.11}$$

$$\mu_z \frac{\partial \phi_k}{\partial z} = -\nu_k^s \phi_k \text{ at } S_H^+ \tag{5.9.12}$$

$$\nabla \cdot \boldsymbol{U} = \frac{\partial u}{\partial x} + \frac{\partial v}{\partial y} + \frac{\partial w}{\partial z} = 0 \text{ in } \mathbf{D} \tag{5.9.13}$$

Here $\phi_k(\boldsymbol{r}, t) \geq 0$ is the concentration of k-th primary pollutant with distribution $\phi_k^0(\boldsymbol{r})$ in $\mathbf{D}$ at the initial moment $t = 0$, $\sigma_k(\boldsymbol{r}, t) \geq 0$ is the chemical conversion coefficient, while $\mu(\boldsymbol{r}, t) > 0$ and $\widehat{\mu}(\boldsymbol{r}, t) > 0$ are the turbulent diffusion arrays,

$$\mu = \begin{pmatrix} \mu_x(\boldsymbol{r}, t) & 0 \\ 0 & \mu_y(\boldsymbol{r}, t) \end{pmatrix} \text{ and } \hat{\mu} = \begin{pmatrix} \mu_x(\boldsymbol{r}, t) & 0 & 0 \\ 0 & \mu_y(\boldsymbol{r}, t) & 0 \\ 0 & 0 & \mu_z(\boldsymbol{r}, t) \end{pmatrix} \tag{5.9.14}$$

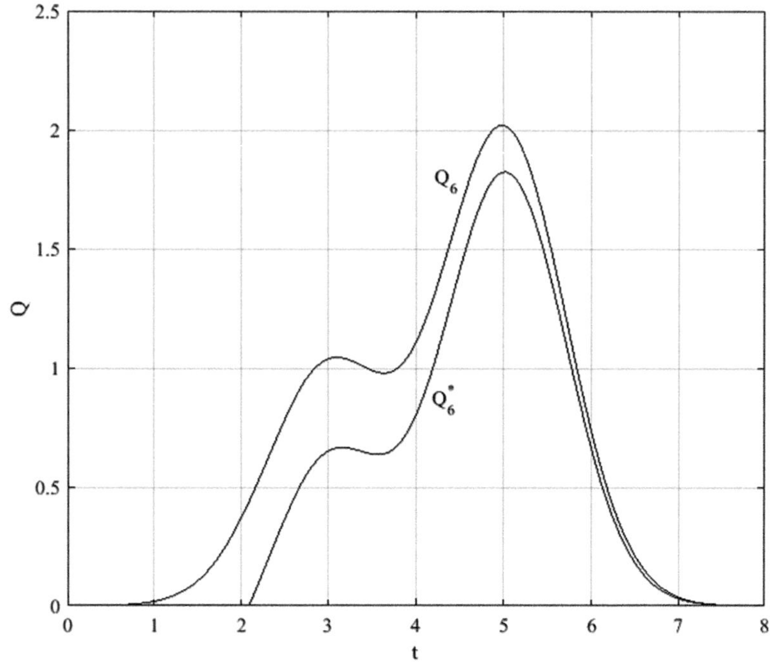

Fig. 5.19 Initial emission rate $Q_6(t)$ and optimal rate $Q_6^*(t)$ in time interval $(0, T)$

Fig. 5.20 Cross section of the **D** region

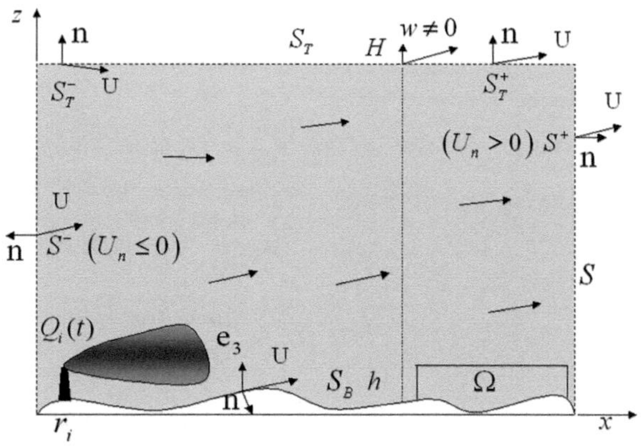

respectively. In Eq. (5.9.5), the term $\nabla \cdot \boldsymbol{\phi}_k^s$ describes the change in pollutant concentrations over time due to deposition at a constant rate $\nu_k^s > 0$. The wind velocity $\boldsymbol{U}(\boldsymbol{r}, t) = (u, v, w)$ is assumed to be known and satisfying the continuity Eq. (5.9.13).

Let's assume that

$$f_k(\boldsymbol{r},\ t) = \sum_{i=1}^{N} q_{ik}(t)\delta(\boldsymbol{r} - \boldsymbol{r}_i) \tag{5.9.15}$$

is the forcing formed by the point sources (industrial plants) located at points $\boldsymbol{r}_i \in \mathbf{D}$, $i = 1, ..., N$; in addition, $q_i(t)$ is the intensity of emissions of the i-th source formed by the emission rates of K different pollutants,

$$q_i(t) = \sum_{k=1}^{K} q_{ik}(t) \tag{5.9.16}$$

and $\delta(\boldsymbol{r} - \boldsymbol{r}_i)$ is the Dirac delta centered at the source position $\boldsymbol{r}_i$, $i = 1, ..., N$.

The conditions given on the open boundary $\partial\mathbf{D}$ of the limited domain $\mathbf{D}$ lead to a well-posed problem according to Hadamard [44]. Let us denote by $U_n = \boldsymbol{U} \cdot \boldsymbol{n}$ the projection of the velocity $\boldsymbol{U}$ on the unit outward normal $\boldsymbol{n}$ to the boundary S, which is divided into the output part S^+ where $U_n \geq 0$ (the advective pollution flow is directed out of $\mathbf{D}$) and the input part S^- where $U_n < 0$ (the advective pollution flow is directed into $\mathbf{D}$). The region $\mathbf{D}$ is assumed to be large enough to include all important pollution sources. Thus, we will assume that there are no sources outside $\mathbf{D}$, and according to (5.9.8), the total (diffusion plus advective) pollution flux is equal to zero at the input part S^-. The pollution flux is nonzero only at S^+, in addition, according to (5.9.9), the diffusion pollution flux at S^+ is considered negligible compared to the corresponding advective pollution flux. Conditions (5.9.11) and (5.9.12) have a similar meaning at S_H, where the particle sedimentation was considered. Equation (5.9.10) indicates the absence of a pollutant flow through S_0, since both $\boldsymbol{U} \cdot \boldsymbol{n}$ and v_k^s are equal to zero on irregular terrain (see Fig. 5.20). In general, (5.9.11) and (5.9.12) are necessary because $w = 0$ on S_0 and (5.9.13) lead to a non-zero vertical velocity component at S_H:

$$w(x,\ y,\ z,\ t) = -\int_0^z \left(\frac{\partial u}{\partial x} + \frac{\partial v}{\partial y}\right) dz \tag{5.9.17}$$

The boundary conditions are mathematically good, since (5.9.5)–(5.9.13) is a well-posed problem (its solution exists, it is unique and continuously depends on the initial condition and forcing [18]). This follows from the fact that the operator

$$A\phi_k = \boldsymbol{U} \cdot \nabla\phi_k + \sigma_k\phi_k - \nabla \cdot (\mu\nabla\phi_k) - \frac{\partial}{\partial z}\mu_z\frac{\partial\phi_k}{\partial z} + \nabla \cdot \boldsymbol{\phi}_k^s \tag{5.9.18}$$

is positive semidefinite:

$$(A\phi_k,\ \phi_k) = \int_{\mathbf{D}} \left\|\hat{\mu}^{\frac{1}{2}}\nabla\phi_k\right\|_2^2 d\boldsymbol{r} + \int_{\mathbf{D}} \sigma_k\phi_k^2 d\boldsymbol{r} + \frac{1}{2}\int_{S_0} v_k^s\phi_k^2 |e_3 \cdot n|\, dS$$

$$+\int_{S_H} \nu_k^s \phi_k^2 dS + \frac{1}{2}\left\{\int_{S^+\cup S_H^+} U_n\phi_k^2 dS - \int_{S^-\cup S_H^-} U_n\phi_k^2 dS\right\} \geq 0 \quad (5.9.19)$$

Here $(\phi, \eta) = \int_{\mathbf{D}} \phi\eta d\boldsymbol{r}$ is the inner product. Then one can show (see [45]) that

$$\|\phi_k\|_2 \leq T \max_{0\leq t\leq T} \|f_k(\boldsymbol{r}, t)\|_2 + \left\|\phi_k^0\right\|_2 \quad (5.9.20)$$

where $\|\varphi\|_2 = \left(\int_D \varphi^2 d\boldsymbol{r}\right)^{1/2}$. The boundary conditions are also physically appropriate, since the integration of Eq. (5.9.5) over **D** leads to the mass balance equation

$$\frac{\partial}{\partial t}\int_{\mathbf{D}} \phi_k d\boldsymbol{r} = \sum_{i=1}^{N} q_{ik}(t) - \int_{S^+\cup S_H^+} U_n\phi_k dS$$
$$- \int_{\mathbf{D}} \sigma_k\phi_k d\boldsymbol{r} - \int_{S_0} \nu_k^s \phi_k |\boldsymbol{e}_3 \cdot \boldsymbol{n}|\, dS \quad (5.9.21)$$

Thus, the total mass of pollutants increases due to non-zero emissions $q_{ik}(t)$, and decreases due to advective outflow through $S^+ \cup S_H^+$, chemical transformations and settling on the ground. Finally, to obtain a numerical solution of the dispersion model (5.9.5)–(5.9.13), and its adjoint model, a balanced and absolutely stable second-order finite-difference scheme was developed based on using the splitting method and Crank-Nicolson schemes (see Chap. 8 and [46, 47]).

2. **Three-dimensional adjoint model for *k* pollutants**. As mentioned above, the main goal of this section is to develop an optimal short-term control that keeps the average concentration (5.9.1) of each pollutant below the relevant sanitary air quality standard J_k:

$$\boldsymbol{J}_{\Omega,\tau}^k \leq J_k, \quad k = 1, \ldots, K \quad (5.9.22)$$

Functional (5.9.1) does not give an explicit dependence of $\boldsymbol{J}_{\Omega,\tau}^k$ on pollution emissions, which is necessary for the development of effective emission control and implementation of (5.9.22). Such an explicit relation can be obtained using the adjoint dispersion model [5, 48] and the duality principle (Lagrange identity) [48, 49]. The adjoint approach not only provides an efficient and economical technique for studying the sensitivity of a model solution with respect to model parameters [9], but also allows solving such important problems as the optimal allocation of new production facilities [5], control of pollution emissions [8, 14, 25, 30], identification of industrial plants that violate the established emission rates [10], remediation of polluted water systems [50, 51], etc.

To this end, we now consider in the domain $\mathbf{D} \times (0, T)$ the adjoint model related to the dispersion model (5.9.5)–(5.9.13) by means of the Lagrange identity $(A\phi, g) = (\phi, A^* g)$:

$$-\frac{\partial g_k}{\partial t} - \boldsymbol{U} \cdot \nabla g_k + \sigma_k g_k - \nabla \cdot (\mu \nabla g_k) - \frac{\partial}{\partial z} \mu_z \frac{\partial g_k}{\partial z} - \nabla \cdot \boldsymbol{g}_k^s = p(r,\ t) \quad (5.9.23)$$

$$\boldsymbol{g}_k^s = -v_k^s g_k \boldsymbol{e}_3 \quad \text{in} \quad \mathbf{D} \quad (5.9.24)$$

$$g_k(\boldsymbol{r},\ T) = 0 \text{ in } \mathbf{D} \quad (5.9.25)$$

$$\mu \nabla g_k \cdot \boldsymbol{n} + U_n g_k = 0 \text{ at } S^+ \quad (5.9.26)$$

$$\mu \nabla g_k \cdot \boldsymbol{n} = 0 \text{ at } S^- \quad (5.9.27)$$

$$\hat{\mu} \nabla g_k \cdot \boldsymbol{n} + \boldsymbol{g}_k^s \cdot \boldsymbol{n} = 0 \text{ at } S_0 \quad (5.9.28)$$

$$\mu_z \frac{\partial g_k}{\partial z} + U_n g_k = 0 \text{ at } S_H^+ \quad (5.9.29)$$

$$\mu_z \frac{\partial g_k}{\partial z} = 0 \text{ at } S_H^- \quad (5.9.30)$$

The forcing $p(\boldsymbol{r}, t)$ in (5.9.23) is defined as:

$$p(\boldsymbol{r},\ t) = \begin{cases} \frac{1}{\tau\,|\Omega|}, & \text{if } (\boldsymbol{r},\ t) \in \Omega \times (T - \tau,\ T) \\ 0, & \text{otherwise} \end{cases} \quad (5.9.31)$$

The adjoint model (5.9.23)–(5.9.30) solved backwards in time (from $t = T$ to $t = 0$) also has a unique solution that continuously depends on the forcing $p(\boldsymbol{r}, t)$. This result can be shown immediately using the variable substitution $t\prime = T - t$ [52]. By combining the solutions of the dispersion model and its adjoint [48, 53], one can obtain an alternative (5.9.1) formula for the average pollutant concentration $\boldsymbol{J}_{\Omega,\tau}^k$ in the zone Ω:

$$\boldsymbol{J}_{\Omega,\tau}\left(\phi_{\vec{q}}^k\right) = \sum_{i=1}^{N} \int_0^T g_k(\boldsymbol{r}_i,\ t) q_{ik}(t) dt + \int_{\mathbf{D}} g_k(\boldsymbol{r},\ 0) \phi_k^0(\boldsymbol{r}) d\boldsymbol{r},\ k = 1, \ldots,\ K \quad (5.9.32)$$

Estimate (5.9.32) is the desired formula, explicitly relating $\boldsymbol{J}_{\Omega,\tau}^k$ to $\phi_k^0(\boldsymbol{r})$ and $q_{ik}(t)$. Although the solution g_k of the adjoint model depends on weather conditions, pollutant characteristics (coefficients σ_k and ν_k^s), location of a monitored zone Ω, and τ and $|\Omega|$, it does not depend on the intensity of emissions $q_{ik}(t)$ and initial pollutant distributions $\phi_k^0(\boldsymbol{r})$. This solution is nonnegative and serves in (5.9.32) as a weight function for $\phi_k^0(\boldsymbol{r})$ and $q_{ik}(t)$. The last integral in (5.9.32) determines the contribution of ϕ_k^0 to $\boldsymbol{J}_{\Omega,\tau}^k$. If this term is large enough, the air quality standard can be exceeded ($\boldsymbol{J}_{\Omega,\tau}^k > J_k$), even if all emission rates are equal to zero. This integral is the key parameter that determines whether there is a solution to the control problem.

3. **Linear optimal control**. Let the dispersion model $\boldsymbol{M}$ be combined with a weather forecast model to predict both meteorological conditions and air quality conditions within a time interval $(0, T)$. Assume that the air quality forecast based on the emission rates $\overrightarrow{Q} = (Q_1, \ldots, Q_N)$ is unfavorable: $\boldsymbol{J}_{\Omega,\tau}\left(\phi_{\overrightarrow{Q}}^k\right) > J_k$, $k = 1, \ldots, K$. Then, in order to prevent excessive concentrations of pollutants in Ω, short-term optimal control can be applied with the establishment of reduced emission rates $\overrightarrow{q}$ in the interval $(0, T)$, which will guarantee compliance with the sanitary standard: $\boldsymbol{J}_{\Omega,\tau}\left(\phi_{\overrightarrow{q}}^k\right) \leq J_k$ for all k.

Remark 5.9.1 It is assumed that the concentrations of pollutants are obtained by a short-term forecast of atmospheric air quality, after which only those passive pollutants whose concentrations violate the established sanitary standards are controlled by the method. Here we assume that the number of such pollutants is K.

Remark 5.9.2 The emission intensity control is determined by introducing nonnegative damping constants $\lambda_i \leq 1$, $i = 1, \ldots, N$. Such constants, which will be obtained by an adequate optimization process, determine the optimal emission rates q_i^* as

$$q_i^* = \lambda_i Q_i, \quad i = 1, \ldots, N \tag{5.9.33}$$

where $(1 - \lambda_i) \times 100$ is the percentage of short-term reduction in emission rate of the i-th pollution source (in other words, a decrease in production activity).

Remark 5.9.3 The *control parameters* v_k are determined from (5.9.32):

$$v_k = J_k - \int\limits_{\mathbf{D}} g_k(\boldsymbol{r}, 0)\phi_k^0(\boldsymbol{r})d\boldsymbol{r}, \quad k = 1, \ldots, K \tag{5.9.34}$$

They allow us to distinguish the following control cases:

(1) If $v_j < 0$, for fixed j, then $\lambda_i = 0$ should be taken, $i = 1, .., N$, i.e., it is necessary to stop all production activities in order to avoid dangerous concentrations of pollutants.
(2) For each j such that $v_j = 0$, it is necessary to take $\lambda_m = 0$ for all sources where $b_{jm} = \int\limits_0^T g_j(\boldsymbol{r}_m, t)Q_{mj}(t)dt > 0$, namely at this stage, the industrial sources

emitting the j-th pollutant and affecting the zone Ω should stop their activity. This criterion can completely stop industrial activity in the region **D**. However, when only a few sources have interrupted their activity, then the remaining emission rates must be controlled by the following criterion.

(3) Without loss of generality, we assume that there are N pollution sources and K pollutants in **D**. Let $v_k > 0$ for all k. Due to $\boldsymbol{J}_{\Omega,\tau}\left(\phi^k_{\vec{Q}}\right) > J_k$, (5.9.32) and (5.9.34), we have

$$\sum_{i=1}^{N} \int_0^T g_k(\boldsymbol{r}_i, t) Q_{ik}(t) dt > v_k, \ \ k = 1, .., K \tag{5.9.35}$$

Now, using the L_1-norm and Formulas (5.9.16) and (5.9.33) in (5.9.2), the optimal control problem (5.9.4) is reduced to determining the damping constants λ_i which are the solution of the following mathematical programming problem:

$$\text{Minimize } F(\lambda_1, \ldots, \lambda_N) = \sum_{i=1}^{N} c_i \sum_{k=1}^{K} \int_0^T (Q_{ik} - \lambda_i Q_{ik})\, dt \tag{5.9.36}$$

$$\text{subject to : } \begin{cases} \sum\limits_{i=1}^{N} \lambda_i \int\limits_0^T g_k(\boldsymbol{r}_i, t) Q_{ik}(t) dt \le v_k, k = 1, \ldots, K, \\ \text{and } \ 0 \le \lambda_i \le 1, \ \ i = 1, \ldots, N. \end{cases} \tag{5.9.37}$$

where the coefficient c_i is the cost of a unit of mass due to a decrease in the production activity of the i-th source, and the constraints (5.9.37) are due to $\boldsymbol{J}_{\Omega,\tau}\left(\phi^k_{\vec{q}}\right) \le J_k$ and (5.9.32).

The linear programming problem (5.9.36) and (5.9.37) can be written as:

$$\text{Minimize } F(\lambda_1, \ldots, \lambda_N) = \sum_{i=1}^{N} a_i \lambda_i \tag{5.9.38}$$

$$\text{subject to : } \begin{cases} \sum\limits_{i=1}^{N} b_{ki} \lambda_i \le v_k, \ \ k = 1, \ldots, K, \\ \text{and } 0 \le \lambda_i \le 1, \ \ i = 1, \ldots, N. \end{cases} \tag{5.9.39}$$

where $a_i = -c_i \int_0^T Q_i dt$ and $b_{ki} = \int_0^T g_k(\boldsymbol{r}_i, t) Q_{ik}(t) dt$. Note that the term $-\sum_{i=1}^{N} a_i$ in the objective function (5.9.38) is omitted because it does not affect the required minimum.

If new vectors $\boldsymbol{a} = (a_1, .., a_N)$, $\boldsymbol{\lambda} = (\lambda_1, .., \lambda_N)^t$ and $\boldsymbol{v} = (v_1, .., v_K)^t$ are introduced, then the matrix form of the linear programming problem has the form

$$\text{Minimize } F = \boldsymbol{a} \boldsymbol{\lambda} \tag{5.9.40}$$

$$\text{subject to}: \begin{cases} \boldsymbol{B} \quad \boldsymbol{\lambda} \leq \boldsymbol{v}, \\ \text{and} \quad 0 \leq \boldsymbol{\lambda} \leq 1 \end{cases} \tag{5.9.41}$$

where $\boldsymbol{B}$ is the $K \times N$ matrix of non-negative transition coefficients:

$$\boldsymbol{B} = (b_{k\,i})_{K \times N} \tag{5.9.42}$$

Note that, due to (5.9.33), the optimal control problem (5.9.4) reduces to an optimization problem in the space $\mathbb{R}^N$. In this case, the corresponding feasibility space $\mathbb{F}$ is determined by the constraints (5.9.41),

$$\mathbb{F} = \{\boldsymbol{\lambda} \in \boldsymbol{R}^N \,|\, 0 \leq \boldsymbol{\lambda} \leq 1,\ \boldsymbol{B}\boldsymbol{\lambda} \leq \boldsymbol{v}\} \tag{5.9.43}$$

4. **Existence of the solutions.** The feasibility space is nonempty since $0 \in \mathbb{F}$, and is a bounded set since $0 \leq \boldsymbol{\lambda} \leq 1$. To show that $\mathbb{F}$ is a closed set, note that if $x \in \overline{\mathbb{F}}$ then there is a sequence $\{\boldsymbol{\lambda}_n\}_{n=1}^{\infty}$ in $\mathbb{F}$ such that $\boldsymbol{\lambda_n} \to \boldsymbol{x}$, namely, $\boldsymbol{\lambda}_{\boldsymbol{n}}^{\boldsymbol{i}} \to x_i$ $(1 \leq i \leq N)$. For fixed i, one can estimate that

$$x_i < \varepsilon + \boldsymbol{\lambda}_n^i \leq \varepsilon + 1 \text{ and } 0 \leq \boldsymbol{\lambda}_n^i < x_i + \varepsilon,\ \forall \varepsilon > 0 \text{ and } n > n_0$$

i.e., $0 \leq x_i \leq 1$, $i = 1, \ldots, N$. On the other hand, note that

$$\|\boldsymbol{B}\boldsymbol{x} - \boldsymbol{B}\boldsymbol{\lambda}_n\|_2 \leq \|\boldsymbol{B}\|_2 \|\boldsymbol{x} - \boldsymbol{\lambda}_n\|_2 \to 0 \text{ when } n \to \infty,$$

and $\boldsymbol{B}^k \boldsymbol{\lambda_n} \to \boldsymbol{B}^k \boldsymbol{x}$ when $n \to \infty$, $1 \leq k \leq K$. Therefore,

$$\boldsymbol{B}^k \boldsymbol{x} = \left(\boldsymbol{B}^k \boldsymbol{x} - \boldsymbol{B}^k \boldsymbol{\lambda}_n\right) + \boldsymbol{B}^k \boldsymbol{\lambda}_n < \varepsilon + v_k \forall \varepsilon > 0 \text{ and } n > n_0$$

i.e., $\boldsymbol{B}^k \boldsymbol{x} \leq v_k$, $k = 1, \ldots, K$. This means that $\boldsymbol{x} \in \mathbb{F}$, and hence $\overline{\mathbb{F}} \subset \mathbb{F}$. Since the feasibility space $\mathbb{F}$ is a non-empty, bounded and closed set in $\mathbb{R}^N$, it is a compact set.

On the other hand, to show that the feasibility space $\mathbb{F}$ is a convex set, we notice that if $t \in (0,1)$ and $\boldsymbol{\lambda}_1, \boldsymbol{\lambda}_2 \in \mathbb{F}$ then $0 \leq t\boldsymbol{\lambda}_1 \leq t$ and $0 \leq (1-t)\boldsymbol{\lambda}_2 \leq 1 - t$. Summing up the last two estimates, we obtain $0 \leq t\boldsymbol{\lambda}_1 + (1-t)\boldsymbol{\lambda}_2 \leq 1$. Since

$$\boldsymbol{B}(t\boldsymbol{\lambda}_1 + (1-t)\,\boldsymbol{\lambda}_2) = t\boldsymbol{B}\boldsymbol{\lambda}_1 + (1-t)\boldsymbol{B}\boldsymbol{\lambda}_2 \leq t\boldsymbol{v} + (1-t)\,\boldsymbol{v} \leq \boldsymbol{v},$$

$$\boldsymbol{t}\boldsymbol{\lambda}_1 + (1-\boldsymbol{t})\,\boldsymbol{\lambda}_2 \in \mathbb{F}.$$

Note that the solution of the linear programming problem (5.9.40)–(5.9.41) exists because the feasibility space $\mathbb{F}$ is a non-empty compact set in $\mathbb{R}^N$, and the objective function F is a continuous function (Weierstrass's theorem) [37]. Such a solution belongs to the boundary $\partial\mathbb{F}$ of the feasibility space [42], and due to (5.9.35) at least one optimal damping constant λ_j is strictly less than one, i.e. it is necessary to reduce emissions from at least one pollution source in order to meet the air quality standards. Note that in this air quality control strategy, some optimal damping constants may be equal to one, which means that the emission rates of respective sources do not change.

On the other hand, we note that the linear programming problem (5.9.40)–(5.9.41) can have many solutions with the same value of the objective function (5.9.40) [42], and due to the convexity of $\mathbb{F}$ and linearity of F, all these solutions form a convex set C in $\mathbb{F}$. This characteristic is an advantage, as it allows you to choose the optimal solution with the same cost among all solutions: $t\boldsymbol{\lambda}_1 + (1-t)\boldsymbol{\lambda}_2$, $t \in (0,1)$ and $\boldsymbol{\lambda}_1, \boldsymbol{\lambda}_2 \in C$. The criterion for choosing such a solution may be its technical feasibility in a real situation.

To determine the solution to the linear programming problem (5.9.40)–(5.9.41), you can use a large-scale optimization method based on LIPSOL (Linear Solver [54]), which is a variant of Mehrotra's predictor–corrector algorithm [55], the primal–dual interior point method. Also, for medium-scale optimization, the projection method can be used, which is a variation of the well-known simplex method for linear programming [56]. Because interior point methods reduce the time to solve large problems [54], they are a good alternative in air pollution problems with many point sources. In this work, we use the LINPROG function from MATLAB, which implements both of the above methods.

5. **Two-dimensional example: linear optimal control for point sources**. The dispersion of pollutants emitted into the atmosphere by stationary or mobile sources is a non-stationary three-dimensional phenomenon. However, during short-term extreme pollution events (several hours), when the vertical movement of the air mass is weak, the pollution concentration increases near the earth's surface and emission sources. In case of high atmospheric stability (thermal inversion), when pollutants accumulate in a thin surface layer with a height H, in the first approximation, the vertical structure of the process can be neglected ($w = 0$), and the three-dimensional dispersion equation can be integrated over the surface layer $[0, H]$. Dividing the resulting equation by the height H, we obtain a simplified two-dimensional dispersion model in the limited area D. Its solution $\varphi_k(r, t)$ denotes the average concentration (in a column of air of height H) of the k-th pollutant at the point $r = (x, y)$ and moment t.

Within the framework of this approximation, the average concentrations (5.9.1), over the time interval $(0, T)$ for the k-th passive pollutant is

$$\boldsymbol{J}_{\Omega,\tau}\left(\phi^k_{\vec{q}}\right) = \boldsymbol{J}_{\Omega_2,\tau}\left(\varphi^k_{\frac{1}{H}\vec{q}}\right), \ \forall\vec{q} \tag{5.9.44}$$

where $\Omega = \Omega_2 \times [0, H]$ is the control zone. Thus, by applying the control technique described in the previous subsection and using the two-dimensional model, we can find the optimal emission rates for the three-dimensional problem.

Let us now assume that in a limited domain $D = (0{,}8000\,\text{m}) \times (0{,}6000\,\text{m})$ of height $H = 200\,\text{m}$, there are four point sources located at the points $r_1 = (1000, 3000)$,

$r_2 = (3000, 2000)$, $r_3 = (3000, 4000)$ and $r_4 = (5000, 3000)$, and emitting sulfur dioxide (SO_2) with the same non-stationary emission rates $Q_i(t) = Q(t)(gs^{-1})$, $i = 1, \ldots, 4$, where

$$Q(t) = \begin{cases} \frac{20}{1296}t, & 0 \le t < 1800 \\ \frac{1000}{36}, & 1800 \le t \le 25200. \\ \frac{1000}{36} - \frac{10}{1296}(t - 25200), & 25200 < t \le 28800 \end{cases}$$

At the initial moment $t = 0$, the SO_2 concentration (in gm^{-3}) has in the domain D a Gaussian distribution

$$\varphi_1^0(r) = 10^{-2}\exp\left(-\left(\frac{x - 4000}{1000}\right)^2 - \left(\frac{y - 3000}{1000}\right)^2\right)$$

(note that $K = 1$). The coefficients of chemical decay σ and diffusion μ are assumed to be equal to $0.0002\,s^{-1}$ and $55.55\,m^2s^{-1}$, respectively. The non-divergent wind velocity $U = (u, v)$ is generated by the stream function $\psi(x, y) = 1.0y$. Thus, $u = \frac{\partial\psi}{\partial y} = 1.0\,ms^{-1}$, $v = -\frac{\partial\psi}{\partial x} = 0\,ms^{-1}$, and streamlines have the form of straight lines with a direction from east to west. The two-dimensional dispersion model and its adjoint are considered in the interval $(0, T)$ with $T = 28800$ s. We control the average concentration of the pollutant $J^1_{\Omega_2,\tau}$ (defined in (5.9.1)) in the two-dimensional zone $\Omega_2 = [6000\,m, 8000\,m] \times [2000\,m, 4000\,m]$ for the last four hours ($\tau = 14400$ s). Sanitary norm $\boldsymbol{J}_1$ for SO_2 is $80\,\mu g\,m^{-3}$ [57]. The isolines of the solution $\varphi_1(\boldsymbol{r}, t)$ of the dispersion model at the time $t = 28800$ are shown in Fig. 5.21.

There is an increase in the concentration of the pollutant in Ω_2 due to the wind. The average pollution concentration $\boldsymbol{J}^1_{\Omega_2,\tau}$ calculated by (5.9.1) is $97.58\,\mu g\,m^{-3}$, and since the result is unsatisfactory (exceeds the sanitary norm), we apply the optimal control method described in the previous subsection, namely, we solve the linear programming problem (5.9.40)–(5.9.41). For this, adjoint model solutions $g_i = g_1(r_i, t)$ are calculated (see Fig. 5.22), and various cost coefficients are shown in Table 5.3.

The table shows the optimal damping parameters for each source in five different experiments. If all sources have the same cost coefficient (see the second column of the table), then only the source located at r_4 must reduce its emission by 39% (third column) to meet the sanitary standard: $\boldsymbol{J}^1_{\Omega,\tau} = 80\,\mu g\,m^{-3}$, which is natural, since it is a closer source to the control zone Ω_2. However, if this source has a cost coefficient three times larger (forth column), then the source located at r_1 must completely stop its activity (fifth column). This result is obtained because the objective function minimizes the global cost. On the other hand, when $c_1 = c_4 = 3$ and $c_2 = c_3 = 1$ (sixth column), then the linear programming problem has many solutions (see columns 7 and 9 of the table). In this case, they can be averaged to find a more balanced solution (eleventh column). Such a solution requires that only sources located at r_2 and r_3 reduce their emissions by 52% in order to comply with sanitary standard: $\boldsymbol{J}^1_{\Omega,\tau} = 80\mu gm^{-3}$. Note that in all the above examples it is used $v_1 = J_1$

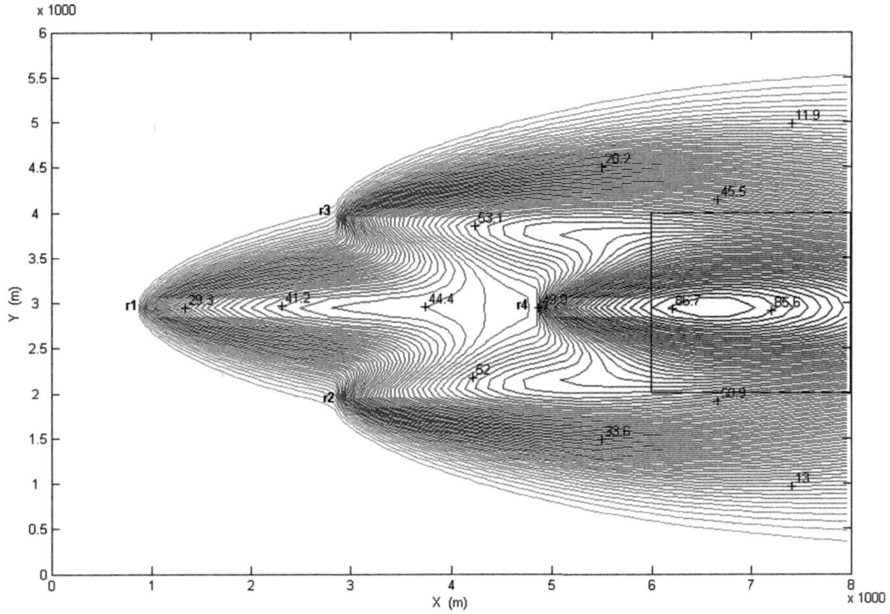

Fig. 5.21 Isolines of the dispersion model solution $\varphi_1(r, t)$ in the domain D at $t = 28800$ s. The zone Ω_2 is highlighted

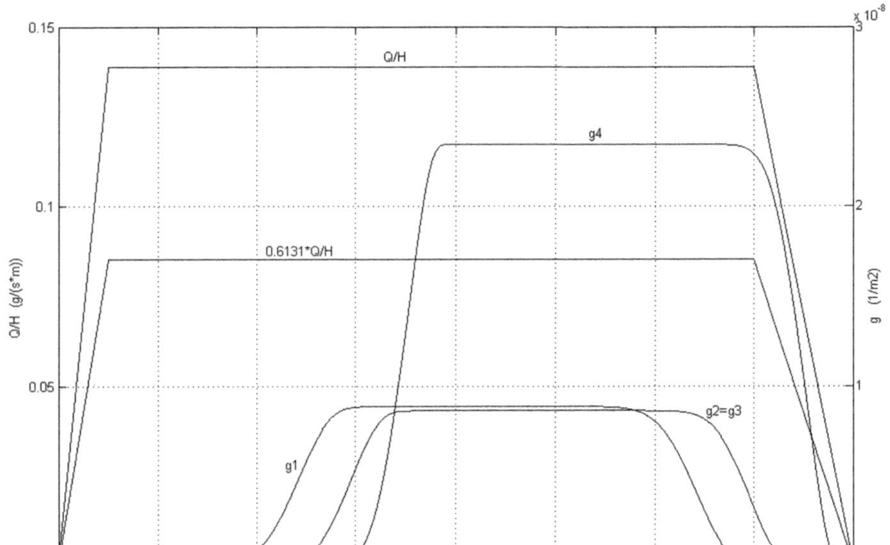

Fig. 5.22 Temporal behavior of adjoint solutions, and original and optimal emission rates

Table 5.3 Results of numerical experiments

i	c_i	λ_i	c_i	λ_i	c_i	λ_i	c_i	λ_i	c_i	λ_i
1	1	1	1	0.0072	3	1	3	1	3	1
2	1	1	1	1	1	0	1	0.9785	1	0.4825
3	1	1	1	1	1	0.9785	1	0	1	0.4825
4	1	0.6131	3	1	3	1	3	1	3	1

(see (5.9.34)), since the initial distribution of SO_2 does not affect the control zone Ω_2 during the time interval $(T - \tau, T)$.

It is interesting to note that the behavior of each optimal emission rate $q_i(t) = \lambda_i Q_i$ over time is similar to that of the corresponding initial rate Q_i (see Fig. 5.22). This fact is useful for industries, since they do not have to radically change their mode of operation.

Finally, note that the coefficient b_{ki} is a conversion factor that estimates the contribution of the emission rate Q_{ik} to the average concentration of the k-th pollutant in the zone Ω over a time interval $[0, T]$. Since such coefficients can be calculated sequentially with any dispersion model $\boldsymbol{M}$ (linear or non-linear), the linear programming model for determining the optimal emission rates can be formulated and solved as an independent second stage of the control problem. Due to the increased computation time, the sequential process for calculating the coefficients b_{ki} is more suitable for long-term air pollution studies where an immediate decision is not required. Thus, for short-term air pollution control, where an immediate decision is required, the coefficients b_{ki} should be calculated with a fast strategy, either using the adjoint model described in this section, or by introducing a new approach using parallel computing. In the next section, we'll look at this last option.

References

1. Skiba, Y.N., Parra-Guevara, D.: Pollution control methods. Int. J. Appl. Math. **25**(5), 675–710 (2012)
2. Cheng, S.Y., Li, J.B., Feng, B., Jin, Y.Q., Hao, R.X.: A Gaussian-box modeling approach for urban air quality management in a Northern Chinese City—II. Pollutant emission abatement. Water Air Soil Poll. **178**, 15–36 (2007)
3. Liu, F., Zhang, Y.H., Hu, F.: Adjoint assessment and reduction of chemical risk in open spaces. Enviorn. Model. Assess. **10**(4), 331–339 (2005)
4. Liu, F., Zhu, J., Hu, F., Zhang, Y.H.: An optimal weather condition dependent approach for emission planning in urban areas. Enviorn. Model. Softw. **22**(4), 548–557 (2007)
5. Marchuk, G.I.: Mathematical Models in Environmental Problems. Elsevier, New York (1986)
6. Moreira, D.M., Tirabassi, T., Vilhena, M.T., Carvalho, J.C.: A semi-analytical model for the tritium dispersion simulation in the PBL from the Angra I nuclear power plant. Ecol. Model. **189**, 413–424 (2005)

7. Parra-Guevara, D., Skiba, Y.N.: Industrial pollution transport. Part II: control of industrial emissions. Environ. Model. Assess. **5**(3), 177–184 (2000)
8. Penenko, V.V., Raputa, V.F.: Some models for optimizing the operation of the atmospheric-pollution sources. Sov. Meteorol. Hidrol. **2**, 46–54 (1983)
9. Skiba, Y.N.: Air pollution estimates. World Resour. Rev. **9**(4), 542–556 (1997)
10. Skiba, Y.N.: On a method of detecting the industrial plants which violate prescribed emission rates. Ecol. Model. **159**(2–3), 125–132 (2003)
11. Skiba, Y.N., Davydova-Belitskaya, V.: Air pollution estimates in Guadalajara City. Environ. Model. Assess. **7**, 153–162 (2002)
12. Skiba, Y.N., Parra-Guevara, D.: Air quality short-term control in an industrial region under adverse weather conditions. Control Theory Technol. **18**(3), 257–268 (2020)
13. Skiba, Y.N., Davydova-Belitskaya, V.: On the estimation of impact of vehicular emissions. Ecol. Model. **166**(1–2), 169–184 (2003)
14. Skiba, Y.N., Parra Guevara, D., Belitskaya, D.V.: Air quality assessment and control of emission rates. Environ. Monitor. Assess. **111**(1–3), 89–112 (2005)
15. Skiba, Y.N.: Dual oil concentration estimates in ecologically sensitive zones. Environ. Monitor. Assess. **43**(2), 139–151 (1996)
16. Skiba, Y.N.: Direct and adjoint oil spill estimates. Environ. Monit. Assess. **59**(1), 95–109 (1999)
17. Skiba, Y.N., Parra-Guevara, D.: Application of adjoint approach to oil spill problems. Environ. Model. Assess. **22**(4), 379–395 (2017)
18. Skiba, Y.N., Parra-Guevara, D.: Industrial pollution transport. Part I: formulation of the problem and air pollution estimates. Environ. Model. Assess. **5**(3), 169–175 (2000)
19. Issartel, J.-P., Baverel, J.: Inverse transport for the verification of the comprehensive nuclear test ban treaty. Atmos. Chem. Phys. **3**, 475–486 (2003)
20. Keats, A., Yee, E., Lien, F.S.: Bayesian inference for source determination with applications to a complex urban environment. Atmos. Environ. **41**, 465–479 (2007)
21. Kovgar, V.: The verification of air pollution episodes in industrial regions, Technical Report, WP-96-121. In: International Institue for Applied Systems Analysis (IIASA), 24 pp, Laxenburg, Austria (1996)
22. Penenko, V.V., Baklanov, A.: Methods of sensitivity theory and inverse modelling for estimation of source term and risk/vulnerability areas. In: Lecture notes in computer science, vol. 2074, pp. 57–66. Springer (2001)
23. Pudykiewicz, J.: Application of adjoint tracer transport equations for evaluating source parameters. Atmos. Environ. **32**, 3039–3050 (1998)
24. Parra-Guevara, D., Skiba, Yu.N.: Optimización de emisiones industriales para la protección de zonas ecológicas. Atmósfera **13**(1), 27–38 (2000)
25. Parra-Guevara, D., Skiba, Yu.N.: Elements of the mathematical modelling in the control of pollutants emissions. Ecol. Model. **167**(3), 263–275 (2003)
26. Parra-Guevara, D.: Modelación Matemática y Simulación Numérica en el Control de Emisiones Industriales, Ph. D. Thesis, Centro de Ciencias de la Atmósfera, UNAM, 130 pp., México (2001)
27. Seinfeld, J.H.: Atmospheric Chemistry and Physics of Air Pollution. Wiley, New York (1992)
28. US EPA.: Integrated Science Assessment for Particulate Matter, First External Review Draft, U. S. Environmental Protection Agency, Research Triangle Park, NC: Office of Research and Development; report no. EPA-600/R-08/139 (2008)
29. Flagan, R.C., Seinfeld, J.H.: Fundamentals of Air Pollution Engineering. Prentice-Hall, Englewood Cliffs (1988)
30. Parra-Guevara, D., Skiba, Yu.N.: On optimal solution of an inverse air pollution problem: theory and numerical approach. Math. Comp. Modell. **43**(7–8), 766–778 (2006)
31. Cheney, E.W.: Introduction to Approximation Theory. Chelsea, New York (1966)
32. Watson, G.A.: Approximation Theory and Numerical Methods. Wiley, Chichester (1980)
33. Reed, M., Simon, B.: Functional Analysis I. Academic Press, London (1980)

34. Marchuk, G.I.: Methods of Numerical Mathematics. Springer, New York (1975)
35. Marchuk, G.I., Kuzin, V.I., Skiba, Y.N.: Projection-difference method for calculating the adjoint functions for a model of heat transfer in the atmosphere-ocean-soil system. In: Actual Problems of Numerical and Applied Mathematics, pp. 149–154, Nauka, Novosibirsk (1983)
36. Marchuk, G.I., Skiba, Yu.N.: Role of the adjoint equation solution in estimating the mean temperature anomalies. Atmósfera **5**(3), 119–133 (1992)
37. Luenberger, D.G.: Linear and Nonlinear Programming. Kluwer Academic Publishers, Boston (2003)
38. Cacuci, D.G., Wever, C.F., Oblow, E.M., Marable, J.H.: Sensitivity theory for general systems of nonlinear equations. Nucl. Sci. Eng. **75**(1), 88–110 (1980)
39. Shutyaev, V.P.: Solvability of the data assimilation problem in the scale of Hilbert spaces for quasilinear singularly perturbed evolutionary problems. Russ. J. Numer. Anal. Math. Modell. **12**(1), 53–66 (1997)
40. Prince, R.C., Bare, R.E., Garrett, R.M., Grossman, M.J., Haith, C.E., Keim, L.G., Lee, K., Holtom, G.J., Lambert, P., Sergy, G.A., Owens, E.H., Guénette, C.C.: Bioremediation of a marine oil spill in the Arctic. In: Alleman, B.C., Leeson, A. (eds.) In Situ Bioremediation of Petroleum Hydrocarbon and Other Organic Compounds, pp. 227–232. Battle Press, Columbus, OH (1999)
41. Wark, K., Warner, C.F.: Air Pollution, its Origin and Control, Harper & Row (1999)
42. Bazaraa, M.S., Jarris, J.J., Sherali, H.D.: Linear Programming and Network Flows. Wiley, Hoboken, New Jersey (2009)
43. Robertson, L., Persson, C.: Attempts to apply four dimensional data assimilation of radiological data using the adjoint technique. In: Radiation Protection Dosimetry, vol. 50, pp. 333–337. Nuclear Technology Publishing (1993)
44. Hadamard, J.: Lectures on Cauchy's Problem in Linear Partial Differential Equations. Yale University Press, New Haven (1923)
45. Rudin, L.I., Osher, S., Fatemi, E.: Nonlinear total variation based noise removal algorithms. Phys. D **60**, 259–268 (1992)
46. Crank, J., Nicolson, P.: A practical method for numerical evaluation of solutions of partial differential equations of the heat conduction type. Proc. Cambridge Philos. Soc. **43**, 50–67 (1947)
47. Skiba, Y.N.: Balanced and absolutely stable implicit schemes for the main and the adjoint pollutant transport equations in the limited area. Rev. Int. Contam. Ambient. **9**(2), 39–51 (1993)
48. Marchuk, G.I.: Adjoint Equations and Analysis of Complex Systems. Kluwer, Dordrecht (1995)
49. Vladimirov, V.S., Marchuk, G.I.: A definition of an adjoint operator for nonlinear problems. Dokl. Math. **61**(3), 438–441 (2000)
50. Parra-Guevara, D., Skiba, Y.N.: A variational model for the remediation of aquatic systems polluted by biofilms. Int. J. Appl. Math. **20**(7), 1005–1026 (2007)
51. Parra-Guevara, D., Skiba, Y.N.: A strategy for bioremediation of marine shorelines polluted with oil by using several nutrient release points. In: Ehrhardt, M. (ed.) Mathematical Modelling and Numerical Simulation of Oil Pollution Problems. Series "The Reacting Atmosphere", vol. 2, pp. 25–56. Springer, Heidelberg (2015)
52. Davydova-Belitskaya, V., Skiba, Yu.N.: Climate of Guadalajara city (Mexico); its variation and change within latest 120 years. World Resour. Rev. **11**(2), 258–270 (1999)
53. Marchuk, G.I., Agoshkov, V.I., Shutyaev, V.P.: Adjoint Equations and Perturbation Algorithms in Nonlinear Problems. CRC Press, Boca Raton (1996)
54. Zhang, Y.: Solving large-scale linear programs by interior-point methods under the MATLAB environment. Technical Report TR96-01, Department of Mathematics and statistics, University of Maryland (1995)

55. Mehrotra, S.: On the implementation of a primal-dual interior point method. SIAM J. Optim. **2**, 575–601 (1992)
56. Dantzig, G.B., Orden, A., Wolfe, P.: Generalized simplex method for minimizing a linear form under linear inequality restraints. Pacific J. Math. **5**, 183–195 (1955)
57. Seinfeld, J.H., Pandis, S.N.: Atmospheric Chemistry and Physics: from Air Pollution to Climate Change. Wiley, New York (1998)

Chapter 6
Bioremediation Problem

This chapter proposes a strategy for remediation of an oiled marine environment that uses hydrodynamics in a limited water area to distribute nutrients (nitrogen or phosphorus) and stimulate biodegradation in several oiled zones. For example, some recreational or aquaculture areas may be selected as such zones. In accordance with the strategy, the nutrient released at several points of the domain with certain discharge rates is distributed by currents and turbulent diffusion and reaches all contaminated zones. In addition, a critical average nutrient concentration (above natural concentration) must be reached and maintained in each oiled zone for a certain period of time in order to properly stimulate the growth of the oil-degrading microorganisms. To solve the control problem, both variational and linear programming methods are used. A linear objective function estimates the mass of the nutrient introduced into the system, and limits are set to protect the marine environment. In the variational method, a quadratic functional and equality constraints are considered, and an analytical solution is obtained using the methods of the calculus of variations. An alternative to the quadratic-variational method is the linear programming method, which allows solving control problems in a more general and strictly numerical form. Linear programming is a mathematical and operations research technique used in administrative and economic planning to achieve the best results (for example, maximum profit or minimum cost) in a mathematical model, the requirements of which are represented by linear equalities or linear inequalities.

Section 6.1 discusses the issue of bioremediation of oiled shorelines. The general 3D model of dispersion of a cleaner in the marine environment and the corresponding adjoint model are described in Sect. 6.2. Adjoint functions and the duality principle are considered in Sect. 6.3. Section 6.4 proves the existence and uniqueness of a solution to the variational formulation of the remediation problem. Optimal discharge parameters and the adjoint functions are discussed in Sect. 6.5. Section 6.6 considers the modulation and point control of basic discharge rates. Numerical examples of remediation in a channel are discussed in Sect. 6.7.

Y. N. Skiba and D. Parra Guevara, *Mathematical Methods for the Assessment and Control of Industrial Emissions*, Mathematical Engineering,
https://doi.org/10.1007/978-3-031-58109-0_6

6.1 Bioremediation of Marine Oiled Shorelines

Crude oil is one of the most important organic pollutants in the marine environments. It is estimated that around the world about 1.3×10^6 tons of oil annually enter sea waters and estuaries [1]. Mass emissions from pipelines, wells, and tankers receive the most public attention, but they actually represent only a relatively small fraction of the total amount of oil released into the environment. Almost 50% is due to natural seepage and less than 9% to catastrophic emissions. Consumption and urban runoff account for almost 40% of the intake [1]. Regardless of the source of pollution, significant amounts of small releases of oil regularly occur in coastal waters [2], making the release of oil ashore a common problem requiring attention.

It is well known that oil is made up of many different toxic compounds that pose a risk to the marine environment due to spills, but there are many naturally occurring indigenous microorganisms that are not only capable of but thrive on the breakdown of these toxic compounds. This process of using microorganisms for such shoreline cleanup efforts is known as bioremediation and has proven to be a successful method for cleaning up marine areas affected by oil spills [3]. Two different types of bioremediation are used in oil spill response: bioaugmentation and biostimulation. Bioaugmentation is the addition of microorganisms capable of breaking down toxic hydrocarbons in order to reduce the content of pollutants. Biostimulation is the addition of nutrients needed by local hydrocarbon degrading microorganisms to achieve maximum degradation of toxic compounds present in oil. The degradation of hydrocarbons (biodegradation) begins with the conversion of an alkane chain or a polycyclic aromatic hydrocarbon (PAH) into an alcohol. Oxidation then converts the compound to an aldehyde, then to an acid, and ultimately to water, carbon dioxide, and biomass. In case of PAHs, fission occurs, which ultimately leads to mineralization [4]. More than 170 genera of microorganisms have been identified in the environment that are capable of degrading hydrocarbons, due to such diversity and different conditions at the site of a spill, not all hydrocarbons biodegrade at the same rate, and not all hydrocarbons degrade, but estimates for the biodegradability of various types of crude oil range from 70 to 97%. What remains are mostly asphaltenes and resin compounds, which are mostly biologically inert [5].

Although biodegradation is a particularly important mechanism for the removal of non-volatile oil components from the environment, it is a relatively slow natural process and it can take months or years before a significant portion of oil is degraded by microorganisms in sediments of marine and/or freshwater environment [6]. The simplest way to stimulate biodegradation, and the only one that has been experimentally tested in the field, is the careful addition of nitrogen and phosphorus nutrients. This was first used on a large scale in Alaska, after the 1989 spill from the tanker Exxon Valdez [7–9]. Two fertilizers have been used in large scale applications: an oleophilic liquid product designed to stick to oil, called Inipol EAP22 [10]; and an agricultural slow release granular product called Customblen [5]. Bioremediation was very successful, as shown in a joint monitoring program conducted by Exxon, USEPA and the Alaska Department of Environmental Conservation [8]. In addition, this was achieved without any noticeable negative impact on the environment [7–9]. Since then, bioremediation has been used in a limited area as part of the Sea Empress spill response [11] and has been demonstrated on experimental spills in marine or brackish environments in the Delaware Bay [12], Texas wetland [13], a fine-sand beach in England [14], mangroves in Australia [15], and an Arctic shoreline in Spitsbergen [16].

Because of these successes, it is desirable to include bioremediation in response to future spills when oil strands on rocky or inaccessible shorelines. In this situation, currents can be used to transport nutrients to contaminated areas rather than throwing them directly onto the site. At the same time, an important factor in achieving the success of biostimulation is obtaining the ideal (critical) concentration of nutrients necessary for maximum growth of organisms, and maintaining this concentration for as long as possible. This can be challenging given that no suitable point for nutrient release is not known, and also due to physical influences such as density differences, wave motions, and tidal effects. Trace studies are often used to study the effect of water and nutrient movement in various situations [17, 18].

This chapter proposes a strategy for remediation of an oiled marine environment that uses hydrodynamics in a limited water region D to distribute a nutrient (nitrogen or phosphorus) and promote biodegradation in several oiled zones Ω_i of $(1 \leq i \leq N)$. For example, some recreational or aquaculture areas may be selected as such zones. In accordance with the strategy, the nutrient emitted at the points $r_1, r_2, \dots, r_N$ of the domain D with discharge rates $Q_1(t), Q_2(t), \dots, Q_N(t)$ is distributed by currents and turbulent diffusion and reaches all polluted zones. In addition, a critical average nutrient concentration c_i (above the natural concentration) must be reached and maintained in each oil-polluted zone Ω_i for a certain time in order to properly stimulate the growth of oil-degrading microorganisms [17]. This time interval is referred to as $(T - \tau, T)$. It should be noted that there is not always an adequate set of release rates $\{Q_i(t)\}_{i=1}^N$, that is, at times this strategy fails. In particular, this can happen when the release points $\{r_i\}_{i=1}^N$ are chosen incorrectly in relation to the flow and location of the zones Ω_i, or when T is not enough time for the nutrient to reach all the zones.

To prevent such situations, the problem is solved in two stages. At the first stage, each zone Ω_i is considered separately from the others and contains only one source. A variational problem is posed and solved to find both the optimal position of the release point r_i in the zone and the optimal release rate $Q_i(t)$ to achieve the concentration c_i in Ω_i, $1 \leq i \leq N$. We prove that this problem has a unique solution. At the second stage, the process of dispersion of the nutrient in all zones together is considered. Due to advection by currents, the nutrient released in one zone can reach other polluted zones. Therefore, we need to specify (modulate) the power of all emission rates $Q_i(t)$ in order to fulfill the critical average concentrations c_i in all polluted zones during the time interval $(T - \tau, T)$. To this end, a positive coefficient γ_i in introduced to modulate each release rate $Q_i(t)$; such factors are chosen as a solution to a quadratic programming problem.

We also prove the existence and uniqueness of this optimization problem. Note that a strategy is considered optimal if it solves the problem and at the same time minimizes the total mass of the nutrient entering the water system to mitigate nutrient impacts on the marine environment and reduce restoration costs. Thus, by applying the least amount of nutrients, optimal control not only cleans the zones, but also protects the entire ecosystem.

The new strategy provides for multiple release points, located in such a way that there is only one release point per each oiled zone. It generalizes the previous strategy, when a single source was used to distribute the nutrient to all oiled areas. Analytical and numerical results for the case of one source were obtained by considering variational formulations, quadratic programming problems and linear programming problems [19, 20].

Considering all the above remarks, the variational problem of the optimal two-stage remediation strategy is formulated as follows:

$$\text{minimize} \quad m(Q_1, \ldots, Q_N) = \frac{1}{2} \sum_{i=1}^{N} \int_0^T Q_i^2(t)dt \tag{6.1.1}$$

$$\text{subject to:} \quad c_i - \alpha_i \leq J_i(\varphi) = \frac{1}{\tau|\Omega_i|} \int_{T-\tau}^{T} \int_{\Omega_i} \phi(r,t)drdt \leq c_i + \beta_i, \; 1 \leq i \leq N \tag{6.1.2}$$

$$0 \leq Q_i(t), \; 0 \leq t \leq T, \; 1 \leq i \leq N \tag{6.1.3}$$

where m is the functional representing the total mass of a nutrient released into the water system over a time interval $(0, T)$ and $\phi = \phi(r, t)$ is the concentration of this substance at the point r in D at time $t > 0$. This concentration will be determined using the dispersion model described in Sect. 6.2. Besides, the functional $J_i(\varphi)$ is the average concentration of the nutrient in the i-th zone Ω_i over the time interval $(T - \tau, T)$ $(1 \leq i \leq N)$. In what follows, we will call this functional a direct estimate of nutrient concentration. Without loss of generality, all zones Ω_i considered in this chapter are disjoint. The constraints in Eq. (6.2) are imposed to keep the concentration $J_i(\varphi)$ close to the critical concentration c_i required for optimal biodegradation $(1 \leq i \leq N)$. Thus, $c_i - \alpha_i$ is the minimum concentration of the nutrient in the oil-polluted zone Ω_i acceptable for efficient stimulation of the biodegradation process, and $c_i + \beta_i$ is the maximum allowable concentration of the nutrient in the oil-polluted zone Ω_i set to protect the water system. Note that the introduction of small positive parameters α_i and β_i increases the number of feasible solutions to problem (6.1.1)–(6.1.3), and therefore this problem is less restrictive than that described by Parra-Guevara and Skiba in [21]. Finally, we note that problem (6.1.1)–(6.1.3) can be used to determine the optimal discharge parameters in a fairly common case when it is necessary to reintroduce nutrients in oil-contaminated areas due to slow oil degradation in the marine environment. Sufficient conditions for such a methodology are given in Sect. 6.3.

6.2 A 3D Dispersion Problem in a Limited Water Area

The concentration of a nutrient $\phi(r, t)$ in a limited domain $D \subset \mathbf{R}^3$ and time interval $(0, T)$ is estimated by the following dispersion model

$$\frac{\partial \phi}{\partial t} + \vec{U} \cdot \nabla \phi - \nabla \cdot \mu \nabla \phi + \sigma \phi + \nabla \cdot \vec{\phi}_s = \sum_{i=1}^{N} Q_i(t)\delta(r - r_i) \tag{6.2.1}$$

$$\vec{\phi}_s = -v_s \phi \vec{k} \text{ in } D \tag{6.2.2}$$

$$\mu \frac{\partial \phi}{\partial n} = \vec{\phi}_s \cdot \vec{n} - \zeta \phi \vec{k} \cdot \vec{n} \text{ on } S_T \tag{6.2.3}$$

$$\mu \frac{\partial \phi}{\partial n} = 0 \text{ on } S^+ \tag{6.2.4}$$

$$\mu \frac{\partial \phi}{\partial n} - U_n \phi = 0 \text{ on } S^- \tag{6.2.5}$$

$$\mu \frac{\partial \phi}{\partial n} = 0 \text{ on } S_B \tag{6.2.6}$$

$$\phi(r, 0) = \phi^0(r) \text{ in } D \tag{6.2.7}$$

$$\nabla \cdot \vec{U} = 0 \text{ in } D \tag{6.2.8}$$

Here (6.2.1) is the advection–diffusion equation, $\vec{U}(r, t)$ is the known flow velocity that satisfies the incompressibility condition (6.2.8), $\mu(r, t)$ is the turbulent diffusion coefficient, $\sigma(r, t)$ is the chemical conversion coefficient characterizing the rate of nutrient decomposition in water. Note that the first-order (linear) kinetics $\sigma\phi$ describing the process of chemical transformation is a reasonable approximation for such nutrients in water as the nitrogen and phosphorus. The term $\nabla \cdot \vec{\phi}_s$ in (6.2.1), describes the change in nutrient concentration per unit time due to sedimentation at a constant rate $v_s > 0$, and $\delta(r - r_i)$ represents the Dirac delta centered on the discharge point r_i. Equation (6.2.3) is the boundary condition on the free surface S_T of the domain D, where $\zeta(r, t)$ is the coefficient characterizing the process of nutrient evaporation, and (6.2.6) is the boundary condition at the bottom S_B of the domain D. Equations (6.2.7) and (6.2.8) are the corresponding conditions on the lateral boundary of D, besides, S^+ is the rigid part or output part of the boundary where $U_n = \vec{U} \cdot \vec{n} \geq 0$, and S^- is its input part where $U_n < 0$ (Fig. 6.1). Finally, Eq. (6.2.7) represents the initial nutrient distribution at $t = 0$. In all equations, $\vec{n}$ is the unit vector of the outer normal to the boundary $\partial D = S_T \cup S^+ \cup S^- \cup S_B$ of the domain D, $\partial/\partial n$ is the derivative in the direction of the normal, and $\vec{k} = (0, 0, 1)^T$ is the unit column vector directed upwards in the Cartesian coordinate system (Fig. 6.2). We observe that

$$\vec{k} \cdot \vec{n} = 0 \text{ on } S^{+} \cup S^{-} \text{ and } \vec{U} \cdot \vec{n} = 0 \text{ on } S_T \cup S_B \tag{6.2.9}$$

Also note that the boundary conditions (6.2.3)–(6.2.5) are general (i.e., not only for horizontal free and bottom surfaces S_T and S_B), and hence, the dispersion model can take into account free surface wave motion and marine topography.

First of all we will show that the solution of dispersion model (6.2.1)–(6.2.8) satisfies the mass balance equation. Indeed, integrating Eq. (6.2.1) over domain D we obtain

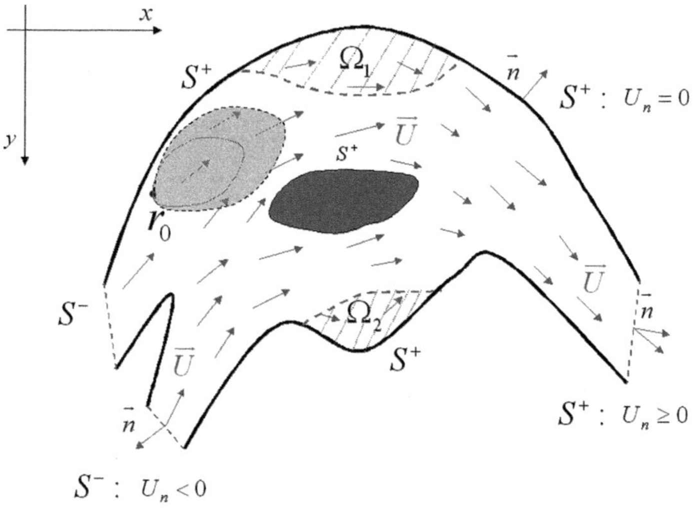

Fig. 6.1 View of domain D from above

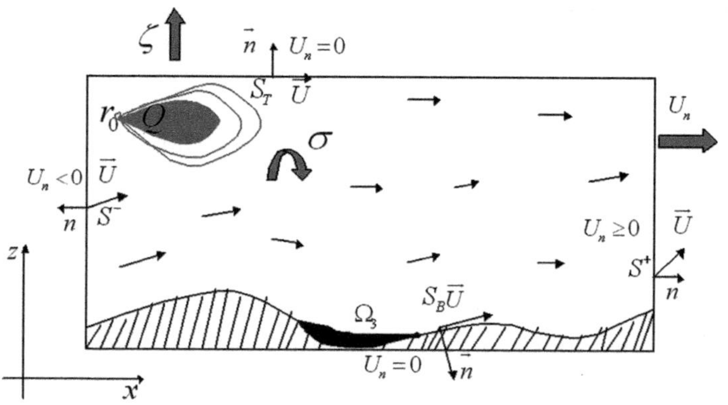

Fig. 6.2 Cross-sectional area of domain D

$$\frac{\partial}{\partial t}\int_D \phi dr + \int_D \vec{U}\cdot\nabla\phi dr - \int_D \nabla\cdot\mu\nabla\phi dr + \int_D \sigma\phi dr + \int_D \nabla\cdot\vec{\phi}_s dr = \sum_{i=1}^{N}\int_D Q_i(t)\delta(r-r_i)dr$$

Applying the divergence theorem [22], some integrals can be rewritten as follows:

$$\int_D \vec{U}\cdot\nabla\phi dr = \int_D \nabla\cdot\left(\vec{U}\phi\right)dr = \int_{\partial D}\vec{U}\cdot\vec{n}\phi dS$$

$$\int_D \nabla\cdot\mu\nabla\phi dr = \int_{\partial D}\mu\nabla\phi\cdot\vec{n}dS = \int_{\partial D}\mu\frac{\partial\phi}{\partial n}dS$$

$$\int_D \nabla\cdot\vec{\phi}_s dr = \int_{\partial D}\vec{\phi}_s\cdot\vec{n}dS$$

Finally, dividing each integral over boundary ∂D into four integrals over S_T, S^+, S^- and S_B, and applying Eqs. (6.2.3) to (6.2.6) and observation (6.2.9), we obtain the mass balance equation:

$$\begin{aligned}\frac{\partial}{\partial t}\int_D \phi dr = &\sum_{i=1}^{N} Q_i(t) - \int_D \sigma\phi dr - \int_{S^+} U_n\phi dS\\ &- \int_{S_T}\zeta\phi\vec{k}\cdot\vec{n}dS + \int_{S_B} v_s\phi\vec{k}\cdot\vec{n}dS\end{aligned} \tag{6.2.10}$$

Since $\vec{k}\cdot\vec{n} > 0$ at S_T and $\vec{k}\cdot\vec{n} < 0$ at S_B, the total mass of the nutrient increases due to the discharge processes ($Q_i(t) > 0$), and decreases because of the chemical transformations ($\sigma > 0$), advective outflow through S^+ ($U_n > 0$), superficial evaporation ($\zeta > 0$) and sedimentation ($v_s > 0$).

We now show that the dispersion problem (6.2.1)–(6.2.8) is well posed. Indeed, the model operator is:

$$A\phi = \vec{U}\cdot\nabla\phi - \nabla\cdot\mu\nabla\phi + \sigma\phi + \nabla\cdot\vec{\phi}_s \tag{6.2.11}$$

Defining the inner product in $L_2(D)$ as $(A\phi,\varphi) = \int_D \varphi A\phi dr$ we get the expression

$$(A\phi,\phi) = \int_D \phi\vec{U}\cdot\nabla\phi dr + \int_D \sigma\phi^2 dr - \int_D \phi\nabla\cdot\mu\nabla\phi dr + \int_D \phi\nabla\cdot\vec{\phi}_s dr$$

The divergence theorem allows us to change some of the integrals in the last equation:

$$\int_D \phi\vec{U}\cdot\nabla\phi dr = \frac{1}{2}\int_{\partial D}\phi^2\vec{U}\cdot\vec{n}dS$$

$$\int_D \phi\nabla\cdot(\mu\nabla\phi)dr = \int_{\partial D}\phi\mu\frac{\partial\phi}{\partial n}dS - \int_D \mu|\nabla\phi|^2 dr$$

$$\int_D \phi\nabla\cdot\vec{\phi}_s dr = \frac{1}{2}\int_{\partial D}\phi\vec{\phi}_s\cdot\vec{n}dS$$

Finally, dividing each integral over ∂D into four integrals over S_T, S^+, S^- and S_B, and applying conditions (6.2.3) to (6.2.6) and observation (6.2.9), we obtain

$$(A\phi,\phi) = \int_D \sigma\phi^2 dr + \int_D \mu|\nabla\phi|^2 dr + \int_{S_T}\zeta\phi^2\vec{k}\cdot\vec{n}dS$$
$$+\frac{1}{2}\left\{\int_{S^+}U_n\phi^2 dS - \int_{S^-}U_n\phi^2 dS + \int_{S_T}v_s\phi^2\vec{k}\cdot\vec{n}dS - \int_{S_B}v_s\phi^2\vec{k}\cdot\vec{n}dS\right\} \tag{6.2.12}$$

Since $U_n < 0$ at S^-, $\vec{k}\cdot\vec{n} > 0$ at S_T and $\vec{k}\cdot\vec{n} < 0$ at S_B, Eq. (6.2.12) can be rewritten as

$$(A\phi,\phi) = \int_D \sigma\phi^2 dr + \int_D \mu|\nabla\phi|^2 dr + \int_{S_T}\zeta\phi^2\vec{k}\cdot\vec{n}dS$$
$$+\frac{1}{2}\left\{\int_{S^+\cup S^-}|U_n|\phi^2 dS + \int_{S_T\cup S_B}v_s\phi^2\left|\vec{k}\cdot\vec{n}\right|dS\right\}$$

Thus, operator A is positive semidefinite: $(A\phi,\phi)\geq 0$.

Taking the inner product of every term of Eq. (6.2.1) with ϕ we obtain

$$\left(\frac{\partial\phi}{\partial t},\phi\right) = (f,\phi) - (A\phi,\phi),\quad f(r,t) = \sum_{i=1}^{N} Q_i(t)\delta(r-r_i)$$

Using the condition $(A\phi,\phi)\geq 0$ and the Schwarz inequality, the last equation implies the inequality

$$\left(\phi,\frac{\partial\phi}{\partial t}\right)\leq\|\phi\|\|f\|,\quad \|\phi\| = \sqrt{(\phi,\phi)}$$

Further,

$$\left(\phi,\frac{\partial\phi}{\partial t}\right) = \frac{1}{2}\frac{\partial}{\partial t}\|\phi\|^2 = \|\phi\|\frac{\partial}{\partial t}\|\phi\|$$

and hence

$$\frac{\partial}{\partial t}\|\phi\| \leq \|f\|$$

Finally, integration over time interval $(0, T)$ leads to

$$\|\phi\| \leq T \max_{0\leq t\leq T} \|f(r,t)\| + \left\|\phi^0(r)\right\| \tag{6.2.13}$$

Since the dispersion model (6.2.1)–(6.2.8) is linear, estimate (6.2.13) ensures the uniqueness of the solution to problem (6.2.1)–(6.2.8) depends continuously on the initial conditions and forcing. Also, using the method described in [23], one can prove the existence of a generalized solution to problem (6.2.1)–(6.2.8), i.e., this model is well posed according to Hadamard [24]. We also note that the positive semi-definiteness of the operator A makes it possible to split the problem into coordinate directions and use the numerical schemes of Marchuk [25] and Crank–Nicolson [26] to construct a direct (iterationless), balanced and unconditionally stable numerical algorithm of the second order of accuracy in space and time for the effective solution of problem (6.2.1)–(6.2.8) [27].

6.3 Adjoint Functions and the Duality Principle

It is quite difficult to analyze and solve the variational problem (6.1.1)–(6.1.3), since the constraints in (6.1.2) are implicitly related to the solutions Q_i of the control problem through the solution ϕ of the dispersion model (6.2.1)–(6.2.8). To establish the explicit dependence of constraints on the control functions Q_i, we now introduce an adjoint model whose operator A^* is adjoint to the operator A of the model (6.2.1)–(6.2.8) in the sense of the Lagrange identity

$$(A\varphi, g) = \left(\varphi, A^* g\right) \tag{6.3.1}$$

where $(\cdot,\cdot)$ is the inner product in the Hilbert space $L_2(D)$. The adjoint model solutions will be used to establish the principle of duality for average concentrations of a nutrient entering the marine environment. The operator A^* can be constructed by analogy with the construction of the adjoint operator in Sect. 3.4. We get

$$\begin{aligned}(A\phi, g) &= \int_D g\vec{U}\cdot\nabla\phi dr + \int_D \sigma g\phi dr - \int_D g\nabla\cdot\mu\nabla\phi dr + \int_D g\nabla\cdot\vec{\phi}_s dr\\ &= \int_D \phi\left(-\vec{U}\cdot\nabla g - \nabla\cdot\mu\nabla g + \sigma g - \nabla\cdot\vec{g}_s\right)dr\\ &\quad + \int_{\partial D} g\phi\vec{U}\cdot\vec{n}dS + \int_{\partial D}\phi\mu\frac{\partial g}{\partial n}dS - \int_{\partial D} g\mu\frac{\partial\phi}{\partial n}dS + \int_{\partial D} g\vec{\phi}_s\cdot\vec{n}dS\end{aligned}$$

where $\vec{g}_s = -v_s g\vec{k}$.

Consider the following adjoint dispersion model:

$$-\frac{\partial g}{\partial t} - \vec{U}\cdot\nabla g - \nabla\cdot\mu\nabla g + \sigma g - \nabla\cdot\vec{g}_s = p(r,t) \tag{6.3.2}$$

$$\vec{g}_s = -v_s g\vec{k} \text{ in } D \tag{6.3.3}$$

$$\mu\frac{\partial g}{\partial n} + \zeta g\vec{k}\cdot\vec{n} = 0 \text{ on } S_T \tag{6.3.4}$$

$$\mu\frac{\partial g}{\partial n} + U_n g = 0 \text{ on } S^+ \tag{6.3.5}$$

$$\mu\frac{\partial g}{\partial n} = 0 \text{ on } S^- \tag{6.3.6}$$

$$\mu\frac{\partial g}{\partial n} = -\vec{g}_s\cdot\vec{n} \text{ on } S_B \tag{6.3.7}$$

$$g(r, T) = 0 \text{ in } D \tag{6.3.8}$$

$$\nabla\cdot\vec{U} = 0 \text{ in } D \tag{6.3.9}$$

where

$$A^*g = -\vec{U}\cdot\nabla g - \nabla\cdot\mu\nabla g + \sigma g - \nabla\cdot\vec{g}_s \tag{6.3.10}$$

Is the adjoint operator with boundary conditions (6.3.4)–(6.3.7) guaranteeing the fulfilment of the Lagrange identity.

It can be seen that the operators $-\vec{U}\cdot\nabla g$ and $-\nabla\cdot\vec{g}_s$ are skew symmetric, and a comparison of the equations and boundary conditions of models (6.2.1)–(6.2.8) and (6.3.2)–(6.3.10) leads to an important result: if the adjoint model (6.3.2)–(6.3.10) is solved in the opposite direction in time (from $t = T$ to $t = 0$) then it also has a unique solution, that depends continuously on the forcing $p(r, t)$. This result can be shown immediately by the substitution $t' = T - t$ [23].

In addition, the forcing $p(r, t)$ of Eq. (6.3.2) can be determined in such a way that the average concentration of the nutrient

$$J_i(\phi) = \frac{1}{\tau|\Omega_i|}\int_{T-\tau}^{T}\int_{\Omega_i}\phi(r, t)drdt \tag{6.3.11}$$

in the oiled zone $\Omega_i \subset D$ will be explicitly related to all discharge rates $Q_j(t), j = 1, \ldots, N$ and initial nutrient concentration $\phi^0(r)$ through the adjoint solution g. Indeed, let's take

$$p(r, t) = \begin{cases} \frac{1}{\tau|\Omega_i|}, & r \in \Omega_i \text{ and } t \in (T-\tau, T) \\ 0, & \text{otherwise} \end{cases} \tag{6.3.12}$$

where $|\Omega_i|$ is the volume of oil-contaminated zone, and τ is the time required for the nutrient to reach the critical concentration in the zone. Then the use of this formula (see Sect. 3.5) leads to the adjoint estimate

$$J_i(\phi) = \sum_{j=1}^{N} \int_0^T g_i(r_j, t) Q_j(t) dt + \int_D g_i(r, 0) \phi^0(r) dr \tag{6.3.13}$$

Formulas (6.3.11) and (6.3.13) are known as dual estimates. For the first discharge of nutrients, $\phi^0(r) = 0$ can be taken, so the last formula reduces to

$$J_i(\phi) = \sum_{j=1}^{N} \int_0^T g_i(r_j, t) Q_j(t) dt \tag{6.3.14}$$

The use of (6.3.14) in (6.1.2) for each zone Ω_i ($i = 1, \ldots, N$), transforms the variational problem (6.1.1)–(6.1.3) to a form more convenient for analysis:

$$\text{minimize} \quad m(Q_1, \ldots, Q_N) = \frac{1}{2} \sum_{j=1}^{N} \int_0^T Q_j^2(t) dt \tag{6.3.15}$$

$$\text{subject to:} \quad c_i - \alpha_i \le \sum_{j=1}^{N} \int_0^T g_i(r_j, t) Q_j(t) dt \le c_i + \beta_i, \quad 1 \le i \le N \tag{6.3.16}$$

$$0 \le Q_j(t), \; 0 \le t \le T, \; 1 \le j \le N \tag{6.3.17}$$

Note that problem (6.3.15)–(6.3.17) uses N adjoint functions $g_i(r, t)$, whose values at the discharge points $r_j, j = 1, \ldots, N$, generate N^2 time-dependent influence functions $g_i(r_j, t)$. They are included in the constraints (6.3.16) and do not depend on the discharge rates $Q_j(t)$. In accordance with the duality principle (6.3.14), each function $g_i(r_j, t)$ contains the dynamical information necessary to quantify how the release of a nutrient at the point r_j affects the zone Ω_i ($j = 1, \ldots, N$) (see Fig. 6.1). However, when re-discharging the nutrient, a non-zero initial nutrient concentration must be taken into account (see (6.3.13)). It should be noted that over time, for a number of natural reasons, the concentration of the nutrient in the region D decreases to its natural value. Therefore, the following conditions for the mean nutrient concentration must be satisfied since the moment $t_0 > T$:

$$\frac{1}{|\Omega_i|} \int_{\Omega_i} \phi(r, t) dr < c_i + \beta_i, \quad i = 1, \ldots, N, \quad t \ge t_0 \tag{6.3.18}$$

The moment t_0 can be determined by observing the average nutrient concentration in the region D or by using the solution ϕ predicted by the model (6.2.1)–(6.2.8) with the forcing $Q_j(t)$ equal to zero for $t > T$ and $j = 1, \ldots, N$. After conditions (6.3.18) are met, the initial concentration for modeling the next application of nutrient is selected as $\phi^0(r) = \phi(r, t_0)$ and the next simulation time interval is $[t_0, t_0 + T]$. Due to conditions (6.3.18), the contribution of the new initial condition $\phi^0(r)$ to the average nutrient concentrations in time interval $[t_0 + T - \tau, t_0 + T]$ is less than the upper bounds $c_i + \beta_i$ in Ω_i, $(i = 1, ..., N)$. Note that without such conditions the feasibility space of the problem (6.1.1)–(6.1.3) is empty and there is no solution to the control problem. Thus, if conditions (6.3.18) are met, we can again take $t_0 = 0$ and consider problem (6.3.15)–(6.3.17) for modeling the second discharge of nutrient with the following positive parameters:

$$c_i' = c_i - \int_D g_i(r, 0)\phi^0(r)dr, \quad i = 1, \ldots, N \tag{6.3.19}$$

In addition, the adjoint functions in (6.3.19) must be calculated on time interval $[t_0, t_0 + T]$. Also note that if negative values appear on the left side of the constraints (6.3.16), they are replaced by zero without loss of generality. With these remarks in mind, the variational problem (6.3.15)–(6.3.17) is a general remediation strategy that can be applied repeatedly.

6.4 Existence and Uniqueness of a Solution

To find the optimal discharge points r_i in D, and the basic discharge rates $Q_i(t)$ at these points, we now consider the variational problem (6.3.15)–(6.3.17) for the case when $N = 1$ and $\alpha_1 = \beta_1 = 0$. That is, at the first stage of the strategy we consider exactly the local problem of remediation, in which the critical concentration c_1 is reached exactly. Then the adjoint function $g_1(r, t)$ for the oil-polluted zone Ω_1 is calculated and the variational problem takes the form

$$\text{minimize} \quad m(Q) = \frac{1}{2}\int_0^T Q^2(t)dt \tag{6.4.1}$$

$$\text{subject to:} \quad \int_0^T g_1(r_1, t)Q(t)dt = c_1 \tag{6.4.2}$$

$$Q(t) \geq 0, \quad 0 \leq t \leq T \tag{6.4.3}$$

where, for simplicity, we have omitted the sub-index in the release rate, i.e., $Q(t) = Q_1(t)$. The site $r_1 \in D$ is any point such that

$$P(r_1) = \int_0^T g_1(r_1, t)dt > 0 \tag{6.4.4}$$

The set of points at which condition (6.4.4) holds is called the support of the function P. Condition (6.4.4) is necessary for constraint (6.4.2) to hold. This condition is satisfied for any point $r_1 \in \Omega_1$, as well as for some points sufficiently close to Ω_1. To prove the existence and uniqueness of a solution to the variational problem (6.4.1)–(6.4.3), we recall some properties of the Hilbert space $H = L_2(0, T)$ together with a strong result of approximation theory (the minimum distance theorem).

Theorem 6.4.1 [28]. *A non-empty closed convex set in a uniformly convex Banach space has a unique point closest to a given point.*

Lemma 6.4.1 [28]. *The space $H = L_2(0, T)$ is a uniformly convex Banach space.*

Note that the meaning of condition (6.4.1) is to minimize the norm (distance) in the space H. It is for this reason that Theorem 6.4.1 is useful in proving existence and uniqueness. Let us now consider a specific set and a point in space H for which Theorem 6.4.1 applies.

Definition 6.4.1 The feasible space F for the variation problem (6.4.1)–(6.4.3) is given as follows

$$F = \left\{ Q \in H;\, Q(t) \geq 0, 0 \leq t \leq T, \text{ and } \int_0^T Q(t) g_1(r_1, t) dt = c_1 \right\} \tag{6.4.5}$$

Lemma 6.4.2 *The feasible space F is a non-empty set in space H.*

Proof Since the adjoint solution $g_1(r_1, t)$ is a non-negative square-integrable function, we have that

$$Q^*(t) = \frac{c_1 g_1(r_1, t)}{\int_0^T g_1^2(r_1, t) dt} \tag{6.4.6}$$

is a function from H that satisfies the constraints (6.4.2) and (6.4.3), and hence belongs to the feasible space F. The lemma is proved.

The meaning and usefulness of the function (6.4.6) are established in the next section.

Lemma 6.4.3 *The feasible space F is a convex set in H.*

Proof In fact, let $Q_1, Q_2 \in F$ and $\lambda \in (0, 1)$. Then, evidently, $\lambda Q_1 + (1 - \lambda) Q_2 \geq 0$, $0 \leq t \leq T$. Besides,

$$\int_0^T (\lambda Q_1 + (1 - \lambda) Q_2) g_1(r_1, t) dt = \lambda c_1 + (1 - \lambda) c_1 = c_1$$

and hence, F is a convex set in H. The lemma is proved.

Lemma 6.4.4 *The feasible space F is a closed set in H.*

Proof To show this, we must prove that $F = \overline{F}$ [29]. Let Q_0 be an element of $\overline{F}$. Then there is a sequence $\{Q_k\}_{k=1}^{\infty}$ in F such that $\|Q_k - Q_0\| \to 0$ as $k \to \infty$. Suppose that $Q_0(t) < 0$ in some interval $I \subset (0, T)$ of positive measure $|I| > 0$. Then

$$\|Q_k - Q_0\|^2 = \int_0^T (Q_k - Q_0)^2 dt \geq \int_I (Q_k - Q_0)^2 dt \geq \int_I Q_0^2 dt = l > 0$$

The last inequality contradicts the convergence of sequence $\{Q_k\}_{k=1}^{\infty}$ in H, and hence, $Q_0(t)$ is non-negative in $(0, T)$. On the other hand, applying the Schwarz inequality we obtain

$$\left| c_1 - \int_0^T Q_0 g_1(r_1, t) dt \right| = \left| \int_0^T (Q_k - Q_0) g_1(r_1, t) dt \right| \leq \|Q_k - Q_0\| \|g_1(r_1, t)\| \to 0$$

$$\text{as } k \to \infty$$

and therefore $\int_0^T Q_0 g_1(r_1, t) dt = c_1$, that is, $Q_0 \in F$. The lemma is proved.

Note that the null function $q(t) \equiv 0$, $0 \leq t \leq T$, does not belong to the feasible set F. Indeed, constraint (6.4.2) is not satisfied for such a function since $c_1 > 0$. This remark allows us to establish the most important result of this section.

Theorem 6.4.2 *The variational problem (6.4.1)–(6.4.3) has a unique non-trivial solution in the space H.*

Proof By Lemma 6.4.1, the space H is a uniformly convex Banach space. Moreover, by Lemmas 6.4.1–6.4.3, the feasibility space F is a non-empty closed convex set in H. Therefore, by Theorem 6.4.1, there is a unique function $Q^* \in F$ that minimizes the distance between the set F and the point $q \equiv 0$. That is, according to (6.4.1), the function Q^* minimizes the objective functional $m(Q)$. Finally, we note that $Q^* \neq 0$, since $q \notin F$, and therefore the unique solution of problem (6.4.1)–(6.4.3) is non-trivial. The theorem is proved. The next section shows that the function Q^*, mentioned in Theorem 6.4.2, is the function (6.4.6).

6.5 First Stage: Optimal Discharge Parameters

Analytical expression for the optimal discharge rate Q^*, i.e., for the solution of variational problem (6.4.1)–(6.4.3), can be obtained using the Lagrange multiplier method [32]. Let

$$L(Q) = \frac{1}{2} \int_0^T Q^2(t) dt - \lambda \left\{ \int_0^T g_1(r_1, t) Q(t) dt - c_1 \right\} \tag{6.5.1}$$

be the Lagrange functional corresponding to the problem (6.4.1)–(6.4.3), where λ is the corresponding Lagrange multiplier. The first variation of $L(Q)$ according to Gateaux [32] is calculated as

$$\delta L(Q; \delta Q) = \frac{\partial}{\partial \varepsilon} L(Q + \varepsilon \delta Q)_{\varepsilon=0} = \int_0^T \{Q(t) - \lambda g_1(r_1, t)\} \delta Q dt \tag{6.5.2}$$

where δQ is the variation of Q. A necessary condition for Q^* to be the minimum of $L(Q)$ is $\delta L(Q^*; \delta Q) = 0$ for any δQ [32]. Therefore, from Eq. (6.5.2) we obtain

$$Q^*(t) = \lambda g_1(r_1, t) \tag{6.5.3}$$

where the Lagrange multiplier λ is defined using the constraint (6.4.2) as follows

$$\lambda = \frac{c_1}{\int_0^T g_1^2(r_1, t) dt} \tag{6.5.4}$$

The final result is obtained by substituting Eq. (6.5.4) into (6.5.3).

Note that due to the Schwarz inequality [30],

$$0 < \int_0^T g_1(r_1, t) dt \le T^{\frac{1}{2}} \left\{ \int_0^T g_1^2(r_1, t) dt \right\}^{\frac{1}{2}}$$

and therefore $\int_0^T g_1^2(r_1, t) dt > 0$, i.e., the function Q^* is well-defined by Eqs. (6.5.3) and (6.5.4). Moreover, since $g_1(r_1, t) \ge 0$, we conclude that $Q^*(t) \ge 0$, $0 \le t \le T$.

Let us now show that Q^* defined by formulas (6.5.3) and (6.5.4), also satisfies the sufficient minimum condition. Indeed, let $Q_0 = Q^* + \delta Q$ be a feasible discharge rate. From constraint (6.4.2) we have

$$\int_0^T g_1(r_1, t) \delta Q dt = 0 \tag{6.5.5}$$

where $\delta Q \ne 0$ is an arbitrary variation of Q^*. Then,

$$m(Q_0) - m\big(Q^*\big) = \int_0^T Q^*(t) \delta Q dt + \frac{1}{2} \int_0^T \delta^2 Q dt \tag{6.5.6}$$

By virtue of Eq. (6.5.3) and condition (6.5.5), Eq. (6.5.6) can be written as

$$m(Q_0) - m\big(Q^*\big) = \lambda \int_0^T g_1(r_1, t) \delta Q dt + \frac{1}{2} \int_0^T \delta^2 Q dt = \frac{1}{2} \int_0^T \delta^2 Q dt > 0$$

where λ is given by (6.5.4). Thus, $m(Q_0) > m(Q^*)$, and hence, Q^* defined by (6.5.3) and (6.5.4) is the global minimum of the variational problem (6.4.1)–(6.4.3). Note that Theorem 6.4.2 guarantees that this minimum is unique.

On the other hand, the mass of nutrient introduced into the water system by means of the discharge rate Q^* is estimated as

$$m(Q^*) = c_1^2 \Big/ \left\{ 2 \int_0^T g_1^2(r_1, t)dt \right\} \tag{6.5.7}$$

so that, to minimize the amount of mass, the integral

$$I(r_1) = \int_0^T g_1^2(r_1, t)dt \tag{6.5.8}$$

should take the maximum value. Thus, the optimal discharge point r_1 is chosen in such a way as to maximize the area under the graph of the function $g_1^2(r_1, t)$, $t \in (0, T)$. Note that $I(r_1)$ defined by (6.5.8) is a continuous non-linear function of three real variables $r_1 = (x, y, z)$, which has a global maximum in the closed set $\overline{\Omega_1}$. Indeed, according to the definition of the adjoint model forcing $p(r, t)$, the largest values of the adjoint function are always reached at the points of the domain Ω_1.

Since all these results can be consistently applied to each oil-polluted zone, we conclude that at the first stage of the remediation strategy, the method allows us to determine the discharge points r_i^*, one in each oil-polluted zone Ω_i, and also determine from Eqs. (6.5.3) and (6.5.4) the corresponding basic discharge rates of the nutrient:

$$Q_i^*(t) = \lambda_i g_i\left(r_i^*, t\right) = \frac{c_i}{\int_0^T g_i^2\left(r_i^*, t\right)dt} g_i\left(r_i^*, t\right), \quad (i = 1, \ldots, N) \tag{6.5.9}$$

Note that all discharge parameters are calculated using the adjoint model solutions.

6.6 Second Stage: Modulation of the Basic Discharge Rates

At the second stage of the remediation strategy, we determine positive parameters $\gamma_1, \gamma_2 \ldots, \gamma_N$ so that the new discharge rates

$$Q_i(t) = \gamma_i Q_i^*(t), \quad 1 \leq i \leq N \tag{6.6.1}$$

of the nutrient would satisfy the (global) variational problem (6.3.15)–(6.3.17). These parameters modulate the intensity of the basic release rates to meet the nutrient concentration requirements in all oil-contaminated zones Ω_i. Such a correction for the basic discharge rates is necessary, since the nutrient discharged in one zone could reach other zones during the time interval $(0, T)$ due to advection and diffusion processes.

Substituting Eq. (6.6.1) into the variational problem (6.3.15)–(6.3.17) we obtain a quadratic programming problem, the solution of which determines the optimal parameters γ_i^*, and hence, the optimal discharge rates at the points r_i, $i = 1, \ldots, N$:

$$\text{minimize} \quad m(\gamma_1, \ldots, \gamma_N) = \frac{1}{2} \sum_{j=1}^{N} p_j^2 \gamma_j^2 \tag{6.6.2}$$

$$\text{subject to:} \quad c_i - \alpha_i \leq \sum_{j=1}^{N} a_{ij} \gamma_j \leq c_i + \beta_i, \quad i = 1, \ldots, N \tag{6.6.3}$$

$$\gamma_j \geq 0, \quad j = 1, \ldots, N \tag{6.6.4}$$

where $p_j^2 = \int_0^T \left[Q_j^*(t) \right]^2 dt$ and $a_{ij} = \int_0^T Q_j^*(t) g_i\big(r_j, t\big) dt, \quad i, j = 1, \ldots, N.$

The solution of the quadratic programming problem (6.6.2)–(6.6.4) exists because the corresponding feasible space is a compact set in $\mathbf{R}^N$ and the objective function m is a continuous function of several real variables [30]. Moreover, such a solution is unique, since m is also a strictly convex function and the feasibility space is a convex set in $\mathbf{R}^N$ [22]. It is assumed that the feasibility space is a non-empty set due to the introduction of suitable (sufficiently large) parameters α_i and β_i.

The quadratic programming problem (6.6.2)–(6.6.4) can be solved using the MATLAB *quadprog* program after defining the adjoint functions. Regarding this procedure, note that when the only constraints of the problem are the upper and lower bounds of the variables, i.e. no linear inequalities or equalities are specified, the default *quadprog* algorithm is the large-scale method. Moreover, if the problem has only linear equalities, i.e., no upper and lower bounds or linear inequalities are specified, the default *quadprog* algorithm is also a large-scale method. This method is a subspace trust-region method based on Newton's internal reflection method described by Coleman and Li [31]. Each iteration includes an approximate solution of a large linear system using the preconditioned conjugate gradient method (PCG). Otherwise, medium-scale optimization is used, and *quadprog* uses an active set method, which is also a projection method similar to that described in Gill et al. [33]. It finds an initial feasible solution by solving a linear programming problem [34, 35]. Due to the structure of the quadratic programming problem (6.6.2)–(6.6.4), the examples use the second method of the *quadprog* routine.

6.7 Numerical Examples of Remediation in a Channel

To illustrate the developed method, let us now consider a two-dimensional example of remediation in a one hundred and twenty meter long channel [0, 120] and ten meters wide [0, 10]. The channel contains three oil-polluted zones Ω_i ($N = 3$):

$$\Omega_1 = [20, 30] \times [9, 10],\ \Omega_2 = [40, 60] \times [9, 10] \text{ and } \Omega_3 = [96, 100]\times[0, 2].$$

Critical nutrient concentrations c_i $\left(\text{gm}^{-3}\right)$ in the zones vary from experiment to experiment in accordance with Table 6.1. The parameters of the adjoint model (6.3.3)–(6.3.10) are as follows: the velocity vector $\vec{U}$ is directed along the channel and is equal to 30 m h^{-1}, the diffusion coefficient μ is 6 m^2 h^{-1}, the chemical decay coefficient σ is 1 h^{-1}, and $\zeta = 0$, $v_s = 0$. The release of the nutrient is carried out from the optimal points within four hours, $(0, T) \equiv (0, 4)$, and the average concentration is monitored during the last hour interval $(3, 4)$, i.e., $\tau = 1$ h.

For each oil-contaminated zone, the adjoint model (6.3.3)–(6.3.10) was solved using a two-dimensional version of the splitting-up method described in Sect. 8.9. Discretization parameters are the same in all numerical experiments. Regular grids are used in space ($\Delta x = \Delta y = 0.4$) and time ($\Delta t = 0.005$). The function I, given by Eq. (6.5.8), was built for each oil-polluted zone through the corresponding adjoint solution. In each case, by maximizing the function I, the following optimal discharge points were determined: $r_1^* = (20.2, 9.8)$, $r_2^* = (40.2, 9.8)$ and $r_3^* = (96.2, 0.2)$. For this grid (as well as for finer grids), we found that the upper left corners of the zones Ω_1 and Ω_2, and the lower left corner of the zone Ω_3 tend to be the optimal discharge points, as they should be for maximum nutrient exposure in each oil-contaminated zone.

The adjoint solutions $g_{ij} = g_i\left(r_j^*, t\right)$, for the i-th oil-polluted zone and the j-th optimal discharge point, are presented in Figs. 6.3, 6.4 and 6.5. According to Eq. (6.5.9), the basic discharge rate for each polluted zone Ω_i is a multiple of the adjoint function $g_{ii} = g_i\left(r_i^*, t\right)$. From the form of these functions, presented in Figs. 6.3, 6.4 and 6.5, we can conclude that the basic discharge rates are zero in the time interval $[0, 2.25]$. According to Eq. (6.3.13), this means that the basic discharge rate affects the concentration of the nutrient in the polluted zone only if the adjoint function of the zone is non-zero in the time interval $[2.25, 4.0]$. Figure 6.3 shows that g_{12} and g_{13} do not satisfy this condition, which means that the discharge of nutrients at the points r_2^* and r_3^* does not affect the nutrient concentration in the zone Ω_1, as expected due to the flow direction and location of zones in the channel.

A similar result follows from Fig. 6.5, since the adjoint functions g_{31} and g_{32} are practically equal to zero within time interval $[2.25, 4.0]$, which means that the discharge of nutrient at points r_1^* and r_2^* practically does not affect its concentration in the zone Ω_3. However, from Fig. 6.4 it follows that function g_{21} is positive on the time interval $[2.25, 4.0]$, which means that the discharge of nutrient at point r_1^* affects its concentration in the zone Ω_2, as expected. Finally, the temporal behavior

Table 6.1 Concentrations c_i $\left(\text{gm}^{-3}\right)$ in the three oil-polluted zones

Concentration	Experiment				
	1	2	3	4	5
c_1	0.8	1.0	0.5	1.2	0.6
c_2	0.8	0.8	1.0	0.5	1.2
c_3	0.8	0.5	1.5	1.2	0.6

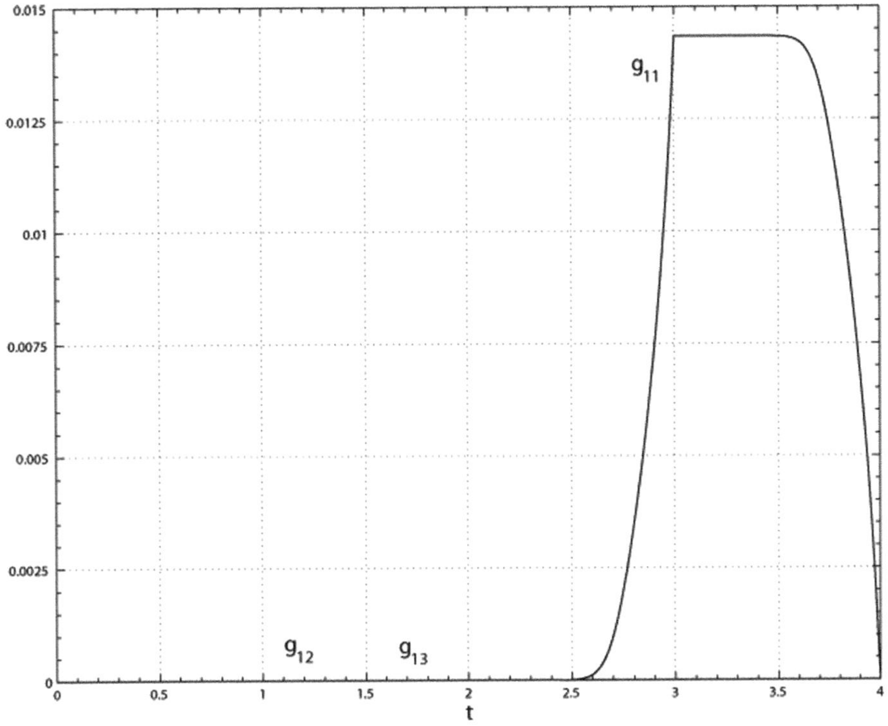

Fig. 6.3 Adjoint functions $g_{1j} = g_1\left(r_j^*, t\right)$ corresponding to the zone Ω_1 and calculated at the optimal discharge points $r_j^*(j = 1, 2, 3)$

of the adjoint function g_{23} allows us to conclude that the discharge of nutrient at point r_3^* does not affect the concentration of the nutrient in the zone Ω_2.

Thus, the oil-polluted zones are not independent of the dispersion process, since the release of the nutrient in one zone can affect concentrations in other zones. If it is the case, then the application of the second stage of the remediation strategy is necessary to adjust the intensity of the basic discharge rates. For this, the problem of quadratic programming (6.6.2)–(6.6.4) was solved using the adjoint functions $g_{ij} = g_i\left(r_j^*, t\right)$, critical concentrations c_i given in Table 6.1 and corresponding basic discharge rates. Table 6.2 shows the optimal modulation parameters γ_i^* obtained for each experiment.

For all experiments, the slack variables of the quadratic programming problem (6.6.2)–(6.6.4) are taken equal to zero: $\alpha_i = \beta_i = 0$ $(i = 1, 2, 3)$, so each critical concentration c_i is reached exactly in the corresponding oil-polluted zone.

Table 6.2 shows that the only discharge rate that needs to be corrected is in the zone Ω_2 ($\gamma_2^* < 1$ in the five experiments). This is a consequence of the impact of the discharge of nutrient at the point r_1^* on the zone Ω_2. The optimal discharge rates for experiments 1 and 4 are shown in Figs. 6.6 and 6.7, respectively. Compared to

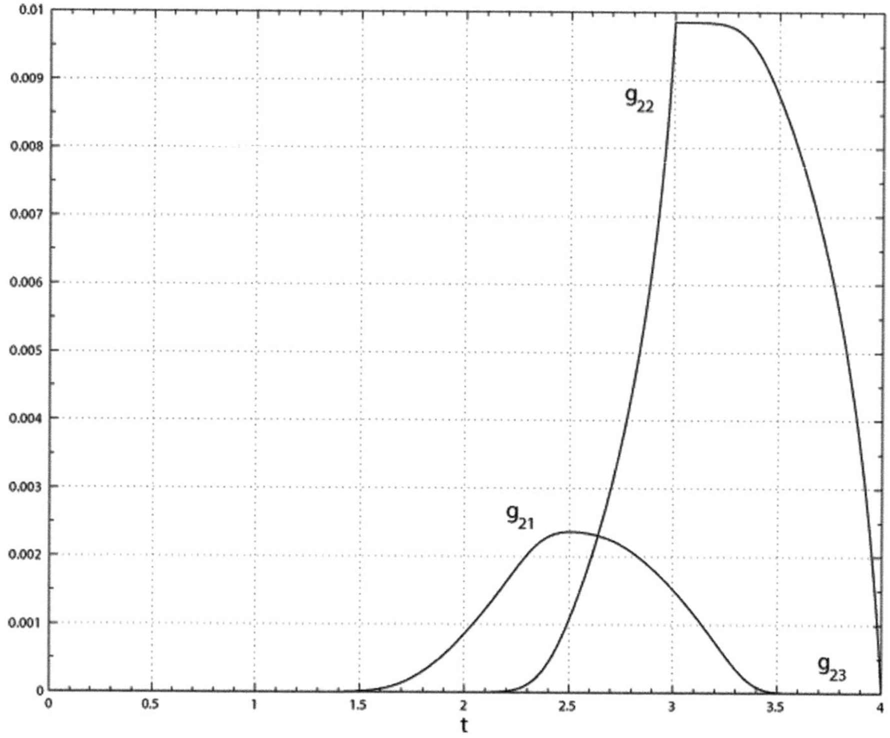

Fig. 6.4 Adjoint functions $g_{2j} = g_2\left(r_j^*, t\right)$ corresponding to the zone Ω_2 and calculated at the optimal discharge points $r_j^* (j = 1, 2, 3)$

Fig. 6.6, the intensity of functions Q_1 and Q_3 in Fig. 6.7 has increased. This is the result of increasing critical concentrations from 0.8 to 1.2 (see Table 6.1). At the same time, the decrease in the intensity of the function Q_2 in Fig. 6.7 compared to Fig. 6.6 is explained by the drop in the critical concentration of the nutrient from 0.8 to 0.5, as well as by the correction of Q_2 through the parameter γ_2^* (see Table 6.2).

It should be noted that in all experiments, slack variables are optional, since the feasible space of problem (6.6.2)–(6.6.4) is non-empty for $\alpha_i = \beta_i = 0$ $(i = 1, 2, 3)$, and the existence of an optimal solution is guaranteed. However, such variables are necessary for the overall formulation of the strategy. For example, when the critical concentrations for three polluted zones are $c_1 = 19.0$, $c_2 = 0.8$ and $c_3 = 0.8$, then the feasible problem space is empty. Indeed, in this case, the basic discharge rate at the point r_1^* is so intense that the concentration of the nutrient in the zone Ω_2 cannot be maintained at the level of 0.8.

On the other hand, if we introduce nonzero slack variables as $\alpha_1 = \beta_1 = 0.1$ and $\alpha_i = \beta_i = 0$ $(i = 2, 3)$, then the feasible space of problem (6.6.2)–(6.6.4) is non-empty and we have the optimal solution: $\gamma_1^* = 0.9947$, $\gamma_2^* = 0.0048$ and $\gamma_3^* = 1.0000$. Figure 6.8 shows the optimal discharge rates obtained for the three

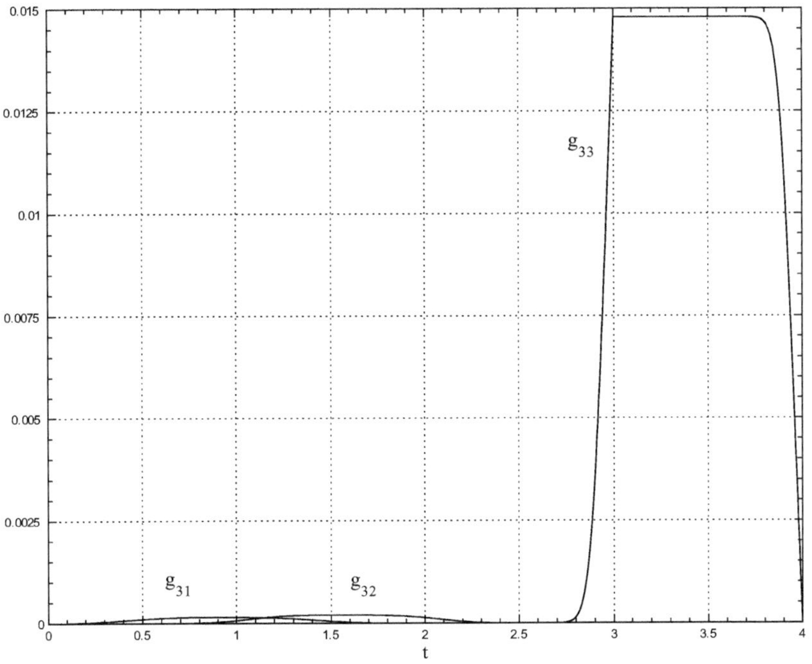

Fig. 6.5 Adjoint functions $g_{3j} = g_3\left(r_j^*, t\right)$ corresponding to the zone Ω_3 and calculated at the optimal discharge points $r_j^*(j = 1, 2, 3)$

Table 6.2 Modulation parameters γ_i^* for the basic discharge rates

Parameter	Experiment				
	1	2	3	4	5
γ_1^*	1.0000	1.0000	1.0000	1.0000	1.0000
γ_2^*	0.9579	0.9473	0.9789	0.8989	0.9789
γ_3^*	1.0000	1.0000	1.0000	1.0000	1.0000

oil-polluted zones. Note that Q_2 is practically zero, which means that the discharge rate Q_1 is responsible for the concentration achieved in the zone Ω_2.

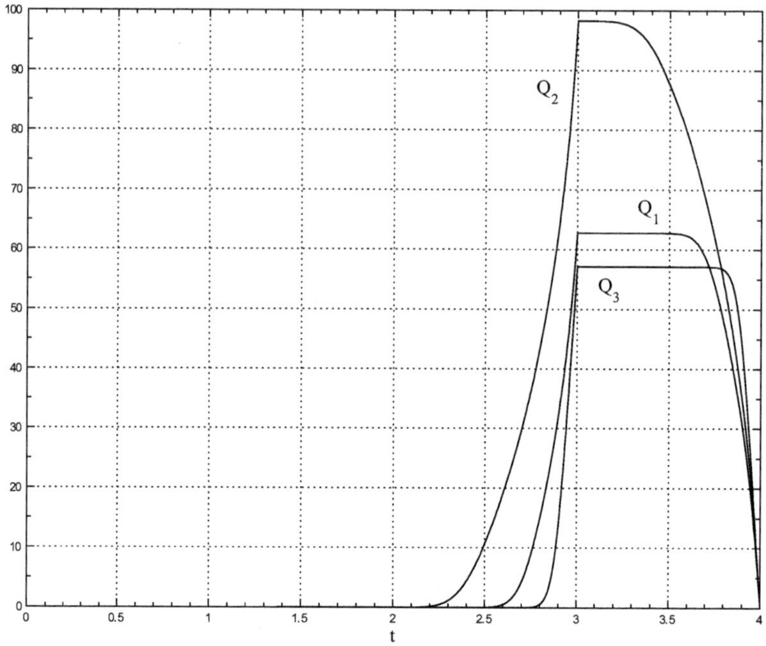

Fig. 6.6 Optimal discharge rates $Q_i(t) = \gamma_i^* Q_i^*(t)$, $i = 1, 2, 3$ for experiment 1

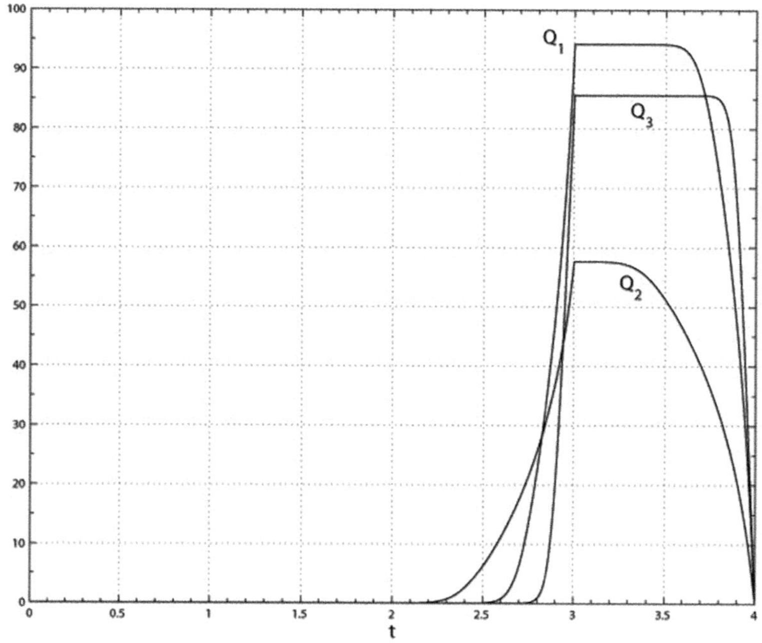

Fig. 6.7 Optimal discharge rates $Q_i(t) = \gamma_i^* Q_i^*(t)$, $i = 1, 2, 3$ for experiment 4

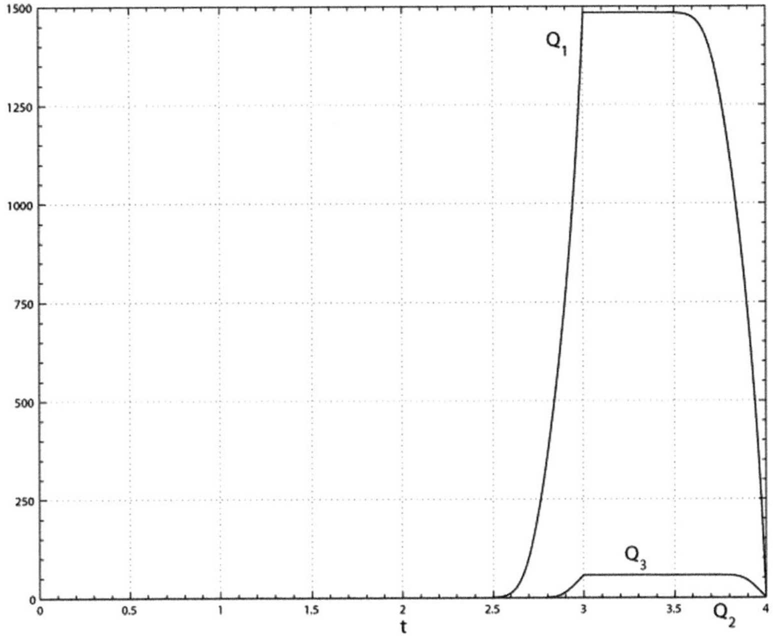

Fig. 6.8 Optimal discharge rates $Q_i(t) = \gamma_i^* Q_i^*(t)$, $i = 1, 2, 3$ for the critical concentrations $c_1 = 19.0, c_2 = 0.8$ and $c_3 = 0.8$, and slack variables $\alpha_1 = \beta_1 = 0.1$ and $\alpha_i = \beta_i = 0$ $(i = 2, 3)$

References

1. National Research Council, Oil in the Sea III: Inputs, Fates and Effects, National Academy of Sciences, Washington DC, 2002.
2. Head, M., Swannell, R.P.J.: Bioremediation of petroleum hydrocarbon contaminants in marine habitats. Curr. Opin. Biotechnol. **10**(3), 234–239 (1999)
3. Coulon, F., McKew, B.A., Osborn, A.M., McGenity, T.J., Timmis, K.N.: Effects of temperature and biostimulation on oil-degrading microbial communities in temperate estuarine waters. Environ. Microbiol. **9**(1), 177–186 (2006)
4. Venosa, A.D.: Oil spill bioremediation on coastal shoreline: a critique. In: Sikdar, S.K., Irvine, R.I. (eds.) Bioremediation: Principles and Practice, vol. III, pp. 259–301. Bioremediation Technologies. Technomic, Lancaster, PA (1998).
5. Prince, R.C., Lessard, R.R., Clark, J.R.: Bioremediation of marine oil spills. Oil Gas Sci. Technol. **58**, 463–468 (2003)
6. Zhu, X., Venosa, A.D., Suidan, M.T., Le, K.: Guidelines for the Bioremediation of Marine Shorelines and Freshwater Wetlands. U.S. Environmental Protection Agency, Cincinnati, OH (2001)
7. Bragg, J.R., Prince, R.C., Harner, E.J., Atlas, R.M.: Effectiveness of bioremediation for the Exxon Valdez oil spill. Nature **368**, 413–418 (1994)
8. Prince, R.C., Bragg, J.R.: Shoreline bioremediation following the Exxon Valdez oil spill in Alaska. Bioremediat. J. **1**, 97–104 (1997)
9. Prince, R.C., Clark, J.R., Lindstrom, J.E., Butler, E.L., Brown, E.J., Winter, G., Grossman, M.J., Parrish, R.R., Bare, R.E., Braddock, J.F., Steinhauer, W.G., Douglas, G.S., Kennedy, J.M., Barter, P.J., Bragg, J.R., Harner, E.J., Atlas, R.M.: Bioremediation of the Exxon Valdez

oil spill: monitoring safety and efficacy. In: Hinchee, R.E., Alleman, B.C., Hoeppel, R.E., Miller, R.N. (eds.) Hydrocarbon Remediation, pp. 107–124. Lewis Publishers, Boca Raton, FL (1994)
10. Ladousse, A., Tramier, B.: Results of 12 years of research in spilled oil bioremediation: Inipol EAP22. In: Hildrew, J.C., Ludwigson, J. (eds.) Proceedings of the 1991 International Oil Spill Conference, vol. 1991, no. 1, pp. 577–582. American Petroleum Institute, Washington DC (1991)
11. Swannell, R.P.J., Mitchell, D., Lethbridge, G., Jones, D., Heath, D., Hagley, M., Jones, M., Petch, S., Milne, R., Croxford, R., Lee, K.: A field demonstration of the efficacy of bioremediation to treat oiled shorelines following the Sea Empress incident. Environ. Technol. **20**(8), 863–873 (1999)
12. Venosa, A.D., Suidan, M.T., Wrenn, B.A., Strohmeier, K.L., Haines, J.R., Eberhart, B.L., King, D., Holder, E.: Bioremediation of an experimental oil spill on the shoreline of Delaware Bay. Environ. Sci. Technol. **30**, 1764–1775 (1996)
13. Mills, M.A., Bonner, J.S., Simon, M.A., McDonald, T.J., Autenrieth, R.L.: Bioremediation of a controlled oil release in a wetland. In: Proceedings of Environment Canada's 20th Arctic and Marine Oil Spill Program Technical Seminar, pp. 609–616. Environment Canada, Ottawa (1997)
14. Swannell, R.P.J., Mitchell, D., Jones, D.M., Petch, S., Head, I.M., Wilis, A., Lee, K., Lepo, J.E.: Bioremediation of oil-contaminated fine sediment. In: Marshall, S. (ed.) Proceedings of the 1999 International Oil Spill Conference, vol. 1999, no. 1, pp. 751–756. American Petroleum Institute, Washington DC (1999)
15. Ramsay, M.A., Swannell, R.P.J., Shipton, W.A., Duke, N.C., Hill, R.T.: Effect of bioremediation on the microbial community in oiled mangrove sediments. Mar. Pollut. Bull. **41**, 413–419 (2000)
16. Prince, R.C., Bare, R.E., Garrett, R.M., Grossman, M.J., Haith, C.E., Keim, L.G., Lee, K., Holtom, G.J., Lambert, P., Sergy, G.A., Owens, E.H., Guénette, C.C.: Bioremediation of a marine oil spill in the Arctic. In: Alleman, B.C., Leeson, A. (eds.) In Situ Bioremediation of Petroleum Hydrocarbon and Other Organic Compounds, pp. 227–232. Battle Press, Columbus, OH (1999)
17. Boufadel, M.C., Suidan, M.T., Venosa, A.D.: Tracer studies in laboratory beach simulating tidal influences. J. Environ. Eng. **132**(6), 616–623 (2006)
18. Boufadel, M.C., Suidan, M.T., Venosa, A.D.: Tracer studies in a laboratory beach subjected to waves. J. Environ. Eng. **133**(7), 722–732 (2007)
19. Parra-Guevara, D., Skiba, Y.N.: A linear-programming based strategy for bioremediation of oil-polluted marine environments. Environ. Model. Assess. **18**(2), 135–146 (2013)
20. Parra-Guevara, D., Skiba, Y.N.: A strategy for bioremediation of marine shorelines polluted with oil by using several nutrient release points. In: Ehrhardt, M. (ed.) Mathematical Modelling and Numerical Simulation of Oil Pollution Problems. Series "The Reacting Atmosphere", vol. 2, pp. 25–56. Springer Verlag, Heidelberg (2015)
21. Parra-Guevara, D., Skiba, Yu.N.: A variational model for the remediation of aquatic systems polluted by biofilms. Int. J. Appl. Math. **20**(7), 1005–1026 (2007)
22. Cheney, E.W.: Introduction to Approximation Theory. Chelsea, New York (1966)
23. Skiba, Y.N., Parra-Guevara, D.: Industrial pollution transport. Part I: formulation of the problem and air pollution estimates. Environ. Model. Assess. **5**(3), 169–175 (2000)
24. Hadamard, J.: Lectures on Cauchy's Problem in Linear Partial Differential Equations. Yale University Press, New Haven (1923)
25. Marchuk, G.I.: Mathematical Models in Environmental Problems. Elsevier, New York (1986)
26. Crank, J., Nicolson, P.: A practical method for numerical evaluation of solutions of partial differential equations of the heat conduction type. Proc. Camb. Philos. Soc. **43**, 50–67 (1947)
27. Skiba, Y.N.: Balanced and absolutely stable implicit schemes for the main and the adjoint pollutant transport equations in the limited area. Rev. Int. Contam. Ambient. **9**(2), 39–51 (1993)
28. Cacuci, D.G., Wever, C.F., Oblow, E.M., Marable, J.H.: Sensitivity theory for general systems of nonlinear equations. Nucl. Sci. Eng. **75**(1), 88–110 (1980)
29. Dieudonné, J.: Foundations of Modern Analysis. Academic Press, New York (1969)

30. Kreyszig, E.: Introductory Functional Analysis with Applications. John Wiley & Sons (1989)
31. Coleman, T.F., Li, Y.: An interior trust region approach for nonlinear minimization subject to bounds. SIAM J. Optim. **6**(2), 418–445 (1996)
32. Smith, D.R.: Variational Methods in Optimization. Dover Publ, New York (1998)
33. Gill, P., Murray, W., Wright, M.: Practical Optimization. Academic Press, London (1981)
34. Mehrotra, S.: On the implementation of a primal-dual interior point method. SIAM J. Optim. **2**, 575–601 (1992)
35. Zhang, Y.: Solving Large-Scale Linear Programs by Interior-Point Methods Under the MATLAB Environment, Technical Report TR96-01. Department of Mathematics and Statistics, University of Maryland (1995)

Chapter 7
Identification of the Intensity and Location of Unknown Point Sources

In this chapter, we consider inverse problems associated with estimating the location and intensity of unknown air pollution sources, using a given set of observed concentrations of pollutants. The cases of constant, instantaneous and time-varying emission rates of the sources are considered. Dispersion models are the main tool for solving such inverse problems and assessing the impact of harmful substances emitted into the atmosphere. Inverse problems are used to identify the *causes* of some desired or actually observed *effects*, or for calibrating the parameters of a mathematical model in order to reproduce the observations with a high precision.

Inverse problems are discussed in Sect. 7.1. Measurements at monitoring sites are presented in Sect. 7.2. The motivation for identification strategies is given in Sect. 7.3. Section 7.4 describes methods for assessing the intensity of unknown constant and instantaneous point sources, if their positions are known. In the case of an unknown time-varying point source, its average intensity is determined and its lower estimate is given. Methods for determining the location of a permanent unknown source, as well as the location and time of emission of an unknown instantaneous point source, are described in Sect. 7.5. Section 7.6 discusses Tikhonov regularization for identifying time-variable emissions. The problem of quadratic programming for the numerical solution of the variational inverse problem is described in Sect. 7.7. Sections 7.8 and 7.9 provide examples of identifying the parameters of various unknown sources.

7.1 Inverse Problems

The dispersion models are routinely used to assess the impact of emissions of pollutants on the air quality under different meteorological conditions. Also, they are used at nuclear power plants and industrial enterprises for emergency response and assessments of impact of hazardous substances accidentally released into the atmosphere [1]. Moreover, Chap. 6 shows the importance of using dispersion models and the

Y. N. Skiba and D. Parra Guevara, *Mathematical Methods for the Assessment and Control of Industrial Emissions*, Mathematical Engineering,
https://doi.org/10.1007/978-3-031-58109-0_7

adjoint approach in developing optimal emission control strategies. In this chapter, we will show that the same methods can be used to estimate the basic parameters of an unknown permanent, instantaneous and non-stationary point source of pollution.

As already noted, inverse problems are used to identify the *causes* of some desired or actually observed *effects*, or for calibrating the parameters of a mathematical model in order to reproduce the observations with high precision. As the *effects* we consider observed or desired values of state variables of a dynamical model, while the initial condition, forcing and internal parameters of the model are the *causes* [2, 3].

The inverse problems are usually ill-posed in the sense of Hadamard [4, 5], i.e. they may have several solutions or no solution, and the solutions can be unstable with respect to variations of external inputs and internal parameters of the problem. Therefore, the mathematical analysis of an inverse problem is not trivial. Special regularization methods should be used for suppressing the instability and choosing an appropriate solution of an inverse problem. Tikhonov regularization, named for Andrey Tikhonov [6], is the most commonly used method of regularization of ill-posed problems. It consists in the minimization of a functional that combines data errors with the norms of the control variable and its derivatives.

The inverse problems discussed in this chapter are to estimate the intensity and location of an unknown source of air pollution using a given set of observed pollutant concentrations. The solutions of such problems are relevant in the light of the efforts made to establish the liability among the parties potentially responsible for local and regional pollution, to assess the environmental impact and to protect the population in polluted zones. Besides, solutions of inverse problems may be used for evaluating and testing emissions inventories of many toxic and radioactive substances [7]. Also, these solutions permit the identification of unknown sources of atmospheric tracers [1, 7, 8].

When determining the location and intensity of an unknown source, its intensity is usually assumed to be constant or in the form of a pulse. Therefore, these methods do not cover all real events and are of limited use [7, 9, 10]. In addition, the developed methods often use analytical solutions of a simple pollution dispersion model (Gaussian plume model), which limits their application only to cases of constant scattering conditions in the atmosphere [7, 11, 12].

7.2 Measurements at Monitoring Sites

Let there be N sources in the region D, the position and intensity of which are known. The concentration of pollutants created by them will be considered known, background. Now suppose that a single event, such as an explosion or release of a hazardous substance, occurs in a time interval $(0, T)$ at a known location $r_0 \in D$, but the emission rate $Q(t)$ of the new source is unknown and must be determined.

Since the dispersion problem (2.4.1)–(2.4.6) is linear, we can, without loss of generality, exclude from consideration all known sources in the domain D and the concentrations of pollutants created by them. Therefore, we will assume that only one source is operating located at point $r_0 \in D$ with an unknown emission rate $Q(t)$, and $\phi^0(r) \equiv 0$. Then the problem of studying the distribution of a pollutant over the domain D and the time interval $(0, T)$ can be studied using the following 2D (vertically averaged) dispersion model:

$$\phi_t + \operatorname{div}(U\phi) + \sigma\phi - \operatorname{div}(\mu\nabla\phi) = Q(t)\,\delta(r - r_0) \tag{7.2.1}$$

where $Q_i(t)$ is the emission rate of the source, $\delta(r - r_0)$ is the Dirac function, $U(r, t) = \{u(r, t), v(r, t)\}$ is the nondivergent wind velocity vector,

$$\operatorname{div} U = 0, \tag{7.2.2}$$

(known from observations or some dynamical model), $\sigma(r, t) > 0$ characterizes the rate of exponential decay of $\phi(r, t)$ due to various physical and chemical processes, and $\mu(r, t) > 0$ is the turbulent diffusion coefficient. Equation (7.2.1) is solved with the initial condition

$$\phi(r, 0) = 0 \quad \text{at } t = 0 \tag{7.2.3}$$

and the following boundary conditions (see 2.4.5 and 2.4.6):

$$\mu\frac{\partial}{\partial n}\phi - U_n\phi = 0 \text{ at } S^- \tag{7.2.4}$$

$$\mu\frac{\partial}{\partial n}\phi = 0 \text{ at } S^+ \tag{7.2.5}$$

Since the background concentrations of the pollutant are assumed to be known, we will assume that time series of observations obtained at any monitoring site **R** also do not contain the values of these background concentrations.

Let $\mathbf{R} \in D$ and $t_j \in (0, T)$ be the monitoring site (place of observation) and time, respectively ($j = 1, 2, \ldots, M$). Let us denote by $\phi(\mathbf{R}, t_j)$ the concentration of the pollutant calculated by the model (7.2.1)–(7.2.5) at the place **R** and at the moment of time t_j. To calculate this concentration, we take the function

$$p(r, t) = \delta(r - \mathbf{R})\,\delta\big(t - t_j\big), (j = 1, 2, \ldots, M) \tag{7.2.6}$$

as the forcing of the adjoint problem (3.4.9)–(3.4.12). Then the adjoint estimate (3.5.7) of the concentration $\phi(\mathbf{R}, t_j)$ reduces to

$$\phi\big(\mathbf{R}, t_j\big) = \int_0^{t_j} Q(t) g_j(r_0, t) dt, \ (j = 1, 2, \ldots, M) \tag{7.2.7}$$

In (7.2.7), we considered that, by virtue of (7.2.6), the adjoint model solution $g_j(r, t)$ is equal to zero on the segment (t_j, T). Therefore, the j-th solution $g_j(r, t)$ of the adjoint problem (3.4.9)–(3.4.12) must be calculated only in the interval $(0, t_j)$ with the "initial" condition $g(r, t_j) = 0$ in D ($j = 1, 2, \ldots, M$). In addition, each solution $g_j(r, t)$ depends only on the atmospheric scattering conditions, the monitoring site $\mathbf{R}$, the moment of sampling t_j ($j = 1, 2, \ldots, M$), and does not depend on the source intensity $Q(t)$.

Formulas (7.2.7) represent the desired explicit relationships between the emission rate $Q(t)$ (the *cause*) and the pollutant concentration $\phi(\mathbf{R}, t_j)$ at the point $\mathbf{R}$ and moments t_j (the *effect*). Note that in order to use Eq. (7.2.7) it is necessary to calculate M different solutions of the adjoint problem (3.4.9)–(3.4.12).

In the particular case when the process of atmospheric dispersion is stationary, operator of the dispersion problem does not depend on time, and

$$g_j(r, t) = g_\alpha\big(r, T - t_j + t\big) \text{ in } (0, t_j),\ j = 1, 2, \ldots, M \tag{7.2.8}$$

where g_α is the solution of adjoint problem (3.4.9)–(3.4.12) with $p(r, t) = \delta(r - \mathbf{R})\delta(t - T)$. This function is known as the basic kernel. Thus in the stationary case, the calculations are much simpler, since formulas (7.2.7) accept the form

$$\phi\big(\mathbf{R}, t_j\big) = \int\limits_0^{t_j} Q(t) g_\alpha\big(r_0, T - t_j + t\big) dt, j = 1, 2, \ldots, M \tag{7.2.9}$$

and therefore only one adjoint function g_α must be calculated to establish explicit relationships between $Q(t)$ and $\phi(\mathbf{R}, t_j), j = 1, 2, \ldots, M$.

If we take $p(r, t) = \delta(r - \mathbf{R}_i)\delta(t - T)$ as the forcing of the adjoint problem (3.4.9)–(3.4.12) then the adjoint estimate (3.5.7) takes the form

$$\phi(\mathbf{R}_i, T) = \int\limits_0^{T} Q(t) g_i(r_0, t) dt, \ (i = 1, 2, \ldots, N) \tag{7.2.10}$$

The choice between formulas (7.2.7) and (7.2.10) depends on the type of available monitoring information. Formulas (7.2.10) are useful when pollutant concentrations are available simultaneously from different monitoring stations in D, while formulas (7.2.7) are applied if there is a time series of pollutant concentrations obtained from a single monitoring station in D. Hereafter, only formulas (7.2.7) will be used.

Assume that all M measurements

$$\phi_j(\mathbf{R}) = \phi\big(\mathbf{R}, t_j\big) + \delta\phi_j,\ j = 1, 2, \ldots, M, \tag{7.2.11}$$

show that the concentrations $\phi(\mathbf{R}, t_j)$ calculated by formula (7.2.7) (or model (7.2.1)–(7.2.5)) at the monitoring site $\mathbf{R}$ contain small errors $\delta\phi_j$, the mean value of which is zero:

$$\frac{1}{M}\sum_{j=1}^{M}\delta\phi_j = 0 \tag{7.2.12}$$

It should be noted that the methods described below are also applicable with minimal modifications in the general case when condition (7.2.12) is not satisfied.

7.3 Motivation of Identification Strategies: Time Series Filtering

This section presents a method for filtering time series in order to eliminate the high frequency error contained in the data. The method is introduced to justify the application of such a method to identify the intensity of unknown point sources. The filtering method is formulated in terms of a variational problem. The quadratic functional to be minimized consists of a term that approximates the data and two weighted terms that estimate the size of the first and second derivatives. Weight parameters are introduced to regulate the filtration process. It is proved that the minimum of such a functional is a solution of the fourth-order linear Euler–Lagrange equation. To discretize such a differential equation and obtain a linear system of equations, finite-difference formulas of the second order of approximation are used. The solution of this linear system is a filtered time series. Two criteria are described for choosing the parameters that regulate the filtering process, and numerical examples of time series filtering are presented, confirming the high performance of the method.

Time series are datasets that are of great importance in various fields of science and technology, as they are a tool that contains up-to-date information both for predicting phenomena and for estimating model parameters. In recent decades, applications of time series have been extended to truly dissimilar areas [13, 14], including predicting the behavior of financial markets, predicting epileptic seizures, predicting extinction events in a degraded natural environment, evaluating rising global temperatures or reconstructing the emission history of pollution sources [15]. Depending on the type of application, time series may require pre-processing that separates or filters relevant information in order to have more accurate data and thus improve forecasts or estimates. There are various types of filters for time series, some of them are defined by differential equations, in particular, the Euler–Lagrange equations.

The Euler–Lagrange equations are differential equations whose solution space contains extreme functions (maximum and minimum) of variational problems. These equations appear formally in Leonhard Euler's 1744 work: *Methodus inveniendi curvas lineas maximi minimive proprietate gaudentes sive solution problematis isoperimetrici latissimo* sensu *accepti* [16], where some problems of geometry and kinematics are systematically analyzed. However, the theory of the calculus of variations and the Euler–Lagrange equations as they are known today arose from a method introduced by Joseph-Louis Lagrange in several papers published between 1750 and 1760 [16]. The mentioned theory, which is more general than that considered by

Euler, is based on the concept of the first variation of a functional, which allows one to analyze the extrema of functionals in a large number of practical problems [17, 18], including problems of geometry, mechanics, optics, optimal design and control, data assimilation, parameter estimation and image filtering [19, 20] and others.

One of the most important applications of filters is the removal of errors in the available dataset, which can be more or less harmful depending on the type of problem. In particular, the high frequency errors contained in the time series represent a difficulty for solving inverse problems related to the identification of parameters in pollutant dispersion models. This is due to the instability that these inverse problems exhibit with respect to disturbances in the data, in particular, it is known that high-frequency errors induce large changes in the parameters to be estimated, in addition to generating problems of existence of solutions [15, 21]. Since this type of error appears as a random component in the monitoring of signals, such as, for example, the concentration of contaminants, it is important to filter this type of error contained in the time series as an a priori process in order to be able to use various types of data in the solution of inverse problems related to the dispersion of substances in the atmosphere.

The following example shows how detrimental to solving an inverse problem is the presence of a high-frequency error contained in a time-dependent signal (data). We consider a zero-dimensional dispersion model or box model. It is assumed that the concentration of a pollutant $\phi(t)$ is described by the following equations [22]:

$$\frac{d\phi}{dt} + \left(\sigma + \frac{1}{V}ua^2\right)\phi = \frac{1}{V}Q(t), \quad 0 < t < T \tag{7.3.1}$$

$$\phi(0) = 0 \tag{7.3.2}$$

Here it is assumed that, at all times, the pollutant forms a homogeneous mixture with air in the region $\mathbf{D} = [0, a] \times [0, a] \times [0, a]$, which is a cube with edge length a y volumen $V = a^3$. The wind speed u is assumed to be stationary, with the direction perpendicular to the vertical wall of the box. In addition, σ denotes the chemical transformation coefficient of the pollutant, and $Q(t)$ is the emission rate of the specified substance for a point source in the region **D**.

The purpose of the inverse problem is to use the pollutant concentration function to obtain the emission rate. Such a problem, apparently, is solved by substituting the concentration of a pollutant into the left side of Eq. (7.3.1). However, if we take into account that the concentration of the controlled pollutant is $\phi(t) + \delta\phi(t)$, where $\delta\phi(t)$ is a small perturbation, then solution (7.3.1) corresponding to these data has the form $Q(t) + \delta Q(t)$, where $\delta Q(t)$ is the corresponding perturbation or error in the emission rate. Since Eq. (7.3.1) is linear, these disturbances can be grouped by the following differential equation:

$$\delta Q(t) = V\left\{\delta'\phi(t) + r\delta\phi(t)\right\} \tag{7.3.3}$$

where the coefficient r is defined as $r = \sigma + V^{-1}ua^2$. The disturbance is now considered to be of high frequency, that is, it has the form $\delta\phi(t) = \varepsilon \sin(4\omega\pi t/T)$, where $\varepsilon > 0$ is the amplitude of the disturbance ($\varepsilon \ll 1$), and the frequency ω is a positive integer. Substituting $\delta\phi(t)$ in Eq. (7.3.3) we obtain the following:

$$\delta Q(t) = \varepsilon V \left\{ \frac{4\omega\pi}{T} \cos\left(\frac{4\omega\pi t}{T}\right) + r \sin\left(\frac{4\omega\pi t}{T}\right) \right\} \tag{7.3.4}$$

Note that when $t = T/2$ in Eq. (7.3.4) then

$$\delta Q\left(\frac{T}{2}\right) = \varepsilon V \frac{4\omega\pi}{T} \cos(2\omega\pi) = \varepsilon V \frac{4\omega\pi}{T}, \quad \omega = 1, 2, \ldots \tag{7.3.5}$$

Therefore, $\delta Q(T/2) \to \infty$ when $\omega \to \infty$. This means that even with a small amplitude data error ($\varepsilon \ll 1$), the error in the reconstructed emission rate $\delta Q(t)$ becomes arbitrarily large as the disturbance frequency ω increases. This is one of the most important reasons for using filters that remove these types of errors in time-dependent signals. The classical theory of creating filters that eliminate or reduce various types of errors uses the Fourier transform (F) and some properties of signal convolution [23]. The general argument exploits the property that such a transform has when applied to the convolution of a function $h(t)$, known as the impulse response function, with a time-dependent signal $\phi(t)$:

$$\mathrm{F}\{(h * \phi)(t)\} = \mathrm{F}\{h(t)\}\mathrm{F}\{\phi(t)\} \tag{7.3.6}$$

Since time-dependent functions are transformed into functions in the frequency domain, the right-hand side of Eq. (7.3.6) can be written as follows:

$$\mathrm{F}\{h(t)\}\mathrm{F}\{\phi(t)\} = H(\omega)\Phi(\omega) \tag{7.3.7}$$

Now, the H function can be chosen to eliminate certain undesirable frequencies contained in the Φ function, for example, if

$$H(\omega) = \begin{cases} 1, & \omega \leq \omega_0 \\ 0, & \omega > \omega_0 \end{cases} \tag{7.3.8}$$

then the function Φ reduces only to the low frequencies $\omega \leq \omega_0$ after multiplying by H. The advantage of this procedure is that when the signal $\phi(t)$ has errors whose frequencies are contained in the interval (ω_0, ∞) then the spurious frequencies disappear due to the product $H(\omega)\Phi(\omega)$, and therefore the data error disappears. In fact, the resulting function $H(\omega)\Phi(\omega)$ is the frequency spectrum of the filtered signal ϕ_f. Applying the inverse Fourier transform (F^{-1}) to both parts of Eqs. (7.3.6) and (7.3.7) we obtain the following general result:

$$\phi_f(t) = \mathrm{F}^{-1}\{H(\omega)\Phi(\omega)\} = (h * \phi)(t) \tag{7.3.9}$$

This means that it is possible to create different filters for signals by choosing different transfer functions $H(\omega)$. In all these cases, the filter acts as a convolution of the impulse response function $h(t)$ with the signal $\phi(t)$ to be filtered (see Eq. 7.3.9). Note that the low-pass filter defined by Eq. (7.3.8) is only one of many possible filters. It is also important to note that in the product $H(\omega)\Phi(\omega)$ frequencies that make up the real signal can also be suppressed, so perfect information recovery, in general, is unlikely. Signal filtering strongly depends on the parameters that define the filter, such as ω_0 in Eq. (7.3.8), which may depend on the knowledge that the user has a priori about the error to be suppressed. It is clear that the application of Eq. (7.3.9) requires the inverse Fourier transform when the filter is defined through the transfer function $H(\omega)$, which implies a computational effort that is not trivial. To avoid this type of calculation, filters can be defined by taking advantage of the convolution operation in (7.3.9) together with the following theorem.

Theorem 7.3.1 *The solution of a linear ordinary differential equation of order n, with constant coefficients and homogeneous initial conditions, is the convolution of the impulse response function h(t) with the forcing of the equation.*

Proof The following initial value problem is considered:

$$L[y] = \sum_{i=0}^{n} c_i \frac{d^i y}{dt^i} = \phi(t) \tag{7.3.10}$$

$$y^i(0) = 0, i = 0, 1, \ldots, n-1 \tag{7.3.11}$$

where the coefficients c_i are real numbers and $c_n \neq 0$. The function $h(t)$ is defined as a solution to problem (7.3.10) and (7.3.11) with the forcing $\phi(t) = \delta(t)$, where $\delta(t)$ is the Dirac delta function centered at the point $t = 0$, also known as impulse function (hence the name given to the function $h(t)$). By multiplying Eq. (7.3.10) by the function $h(t-\tau)$ and integrating over the interval $(0, t)$ we get

$$\sum_{i=0}^{n} \int_0^t h(t-\tau) c_i \frac{d^i y}{d\tau^i} d\tau = \int_0^t \phi(\tau) h(t-\tau) d\tau = (h * \phi)(t) \tag{7.3.12}$$

On the other hand, the technique of integration by parts together with condition (7.3.11) allows us to write

$$\int_0^t h(t-\tau) c_i \frac{d^i y}{d\tau^i} d\tau = \int_0^t y(\tau) c_i \frac{d^i h}{d\tau^i}(t-\tau) d\tau, i = 1, \ldots, n \tag{7.3.13}$$

Substituting Eq. (7.3.13) into the left side of Eq. (7.3.12), we obtain

$$\int_0^t y(\tau)\left\{\sum_{i=0}^{n} c_i \frac{d^i h}{d\tau^i}(t-\tau)\right\}d\tau = (h * \phi)(t) \tag{7.3.14}$$

By definition of the function $h(t)$, we have

$$\sum_{i=0}^{n} c_i \frac{d^i h}{d\tau^i}(t-\tau) = \delta(t-\tau),\ \tau \in (0, t) \tag{7.3.15}$$

and from Eq. (7.3.14) we get

$$\int_0^t y(\tau)\delta(t-\tau)d\tau = (h * \phi)(t) \tag{7.3.16}$$

The change of variable $\xi = t - \tau$ in the integral of Eq. (7.3.16) implies

$$\int_0^t y(\tau)\delta(t-\tau)d\tau = \int_0^t y(t-\xi)\delta(\xi)d\xi \tag{7.3.17}$$

Finally, using the filtering property of the Dirac delta function in (7.3.17) and substituting the result into Eq. (7.3.16) completes the proof of the theorem

$$y(t) = (h * \phi)(t) \tag{7.3.18}$$

This result implies that every initial value problem of the type (7.3.10) and (7.3.11) is a filter for the signal $\phi(t)$. Furthermore, the filtered signal $\phi_f(t)$ is the solution $y(t)$ of this problem. In this case, the way in which the signal frequencies are modified will depend on the transfer function $H(\omega)$ (defined as the Fourier transform of $h(t)$), which in turn will depend on the coefficients c_i $(i = 0, 1, \ldots, n)$ in Eq. (7.3.10). Thus, the filtering process is reduced to the search for certain coefficients c_i together with the solution of an initial value problem. The so-called RC filter is a particular case of this theory [23].

In what follows, a different approach to the filtering problem is shown in order to eliminate high-frequency errors contained in time series. No fast Fourier transform or convolution routines are used, nor is a c_i parameter lookup done for the initial value problem (7.3.10) and (7.3.11). These filters are based on boundary value problems for linear ordinary differential equations, specifically, Euler–Lagrange equations [18]. The advantage of this method is that these equations are solved efficiently through finite difference schemes, that is, time series filtering is done quickly by solving a linear system of equations.

7.3.1 Variational Formulation and Euler–Lagrange Equation

The time-dependent function $\phi_d(t)$, which is the sum of the reconstructed signal $\phi(t)$ with a high-frequency error $\varepsilon(t)$, is considered to be known:

$$\phi_d(t) = \phi(t) + \varepsilon(t), t \in (0, T) \tag{7.3.19}$$

It is assumed that the data set for filtering is the corresponding discrete time series ϕ_d^k, which has the following structure:

$$\phi_d^k = \phi_k + \varepsilon_k, k = 1, 2, \ldots, K \tag{7.3.20}$$

where $\phi_k = \phi(t_k)$ and $\varepsilon_k = \varepsilon(t_k)$ are the exact value of the signal and its error at instant $t_k \in (0, T)$, respectively. For signal filtering, the following variational problem is proposed:

$$\text{Minimize } J(\phi) = \frac{1}{2}\int_0^T (\phi_d - \phi)^2 dt + \frac{1}{2\gamma}\int_0^T (\phi')^2 dt + \frac{1}{2\eta}\int_0^T (\phi'')^2 dt \tag{7.3.21}$$

$$\text{Subject to}: \phi(0) = \phi_0, \phi'(0) = \phi_0', \phi(T) = \phi_T, \phi'(T) = \phi_T' \tag{7.3.22}$$

The first integral of the quadratic functional $J(\phi)$ allows, through the minimization process, to recover the signal ϕ approximating the data ϕ_d in the interval $(0, T)$, while the last two integrals eliminate the high frequency error ε by reducing the size of the derivatives ϕ' and ϕ''. The positive parameters γ and η have been introduced in the functional $J(\phi)$ to regulate the weight between the term that facilitates the approximation to the data and the terms that correspond to the filtering of the error. In this way, when γ and η tend to infinity then the last two terms in the functional $J(\phi)$ have a small weight, which favors the approximation of the data with smooth functions. At the other extreme, when γ and η tend to zero then the last two terms in the functional $J(\phi)$ acquire greater weight, which increases the smoothness of the solution of problem (7.3.21) and (7.3.22) but not the approximation to the data. The boundary conditions given by Eq. (7.3.22) are assumed to be known, since a simple interpolation using the first and last data of the time series allows obtaining such values. In order to find the solution ϕ of the variational problem (7.3.21) and (7.3.22) the first variation of the functional $J(\phi)$ is calculated through the Gateaux formula [18],

$$\delta J(\phi; \delta\phi) = \left\{\frac{\partial}{\partial\lambda} J(\phi + \lambda\delta\phi)\right\}_{\lambda=0} \tag{7.3.23}$$

where $\delta\phi$ is a variation of the function ϕ. In the case of problem (7.3.21) and (7.3.22), this variation must satisfy the homogeneous boundary conditions

$$\delta\phi(0) = 0,\ \delta'\phi(0) = 0,\ \delta\phi(T) = 0,\ \delta'\phi(T) = 0 \tag{7.3.24}$$

This is due to the fact that the function $\phi + \lambda\delta\phi$ satisfies the boundary conditions (7.3.22) for each real number λ.

Applying Eq. (7.3.23) to the functional $J(\phi)$ defined in (7.3.21) leads to

$$\delta J(\phi; \delta\phi) = -\int_0^T (\phi_d - \phi)\delta\phi dt + \frac{1}{\gamma}\int_0^T \phi'\delta'\phi dt + \frac{1}{\eta}\int_0^T \phi''\delta''\phi dt \tag{7.3.25}$$

Integration by parts in the last two terms of the Eq. (7.3.25) and use of boundary conditions (7.3.24) gives

$$\int_0^T \phi'\delta'\phi dt = -\int_0^T \phi''\delta\phi dt \quad and \quad \int_0^T \phi''\delta''\phi dt = \int_0^T \phi''''\delta\phi dt \tag{7.3.26}$$

Therefore, substituting equalities (7.3.26) into Eq. (7.3.25), we obtain

$$\delta J(\phi; \delta\phi) = \int_0^T \left\{\phi - \phi_d - \frac{1}{\gamma}\phi'' + \frac{1}{\eta}\phi''''\right\}\delta\phi dt \tag{7.3.27}$$

On the other hand, the necessary condition of extreme of the functional $J(\phi; \delta\phi)$ requires

$$\delta J(\phi; \delta\phi) = 0, \forall\delta\phi \tag{7.3.28}$$

Smith [18], and therefore

$$\int_0^T \left\{\phi - \phi_d - \frac{1}{\gamma}\phi'' + \frac{1}{\eta}\phi''''\right\}\delta\phi dt = 0, \forall\delta\phi \tag{7.3.29}$$

It follows that the integrand between curly brackets in Eq. (7.3.29) is equal to zero, since the variation $\delta\phi$ is arbitrary. This means that the solution ϕ of the variational problem (7.3.21) and (7.3.22) is the solution of the following linear boundary value problem:

$$\frac{1}{\gamma}\phi'' - \frac{1}{\eta}\phi'''' - \phi = -\phi_d \quad in\ (0, T) \tag{7.3.30}$$

$$\phi(0) = \phi_0, \phi'(0) = \phi_0', \phi(T) = \phi_T, \phi'(T) = \phi_T' \tag{7.3.31}$$

Equation (7.3.30), related to the variational problem and known as the Euler–Lagrange equation, is a necessary extremal condition. Furthermore, due to the convexity properties in the variational problem (7.3.21) and (7.3.22), the solution of problem (7.3.30) and (7.3.31) is the only solution of said optimization problem, that is, the boundary value problem (7.3.30) and (7.3.31) also represents a sufficient condition for the extremum of the functional $J(\phi; \delta\phi)$. The proof is similar to the one given in [24]. In this way, it is concluded that the solution of problem (7.3.30) and (7.3.31) represents the signal that has been filtered from the known signal ϕ_d.

When the signal ϕ_d is known analytically on the interval $(0, T)$ it is possible to obtain the exact solution of problem (7.3.30) and (7.3.31). For this purpose, the four basic solutions of Eq. (7.3.30) are sought in the form $\phi(t) = \exp(rt)$. This leads to the characteristic equation $p(r) = \eta^{-1}r^4 - \gamma^{-1}r^2 + 1 = 0$. The change of variable $s = r^2$ reduces this equation to the quadratic equation $\eta^{-1}s^2 - \gamma^{-1}s + 1 = 0$, whose roots are

$$s = \frac{\eta}{2}\left[\frac{1}{\gamma} \pm \sqrt{\frac{1}{\gamma^2} - \frac{4}{\eta}}\right] \tag{7.3.32}$$

When $\eta > 4\gamma^2$ Eq. (7.3.32) gives two positive real roots, and therefore all basic solutions are exponential functions. In the case $\eta = 4\gamma^2$, a positive real root is obtained from (7.3.32), which implies that the basic solutions are two exponential functions and two functions that are the product of indicated exponentials and the identity function. In the last case, $\eta < 4\gamma^2$, two complex conjugate roots are obtained from (7.3.32), so the basic functions are an exponential multiplied separately by the sine and cosine functions, and the respective functions that are the product of these and the identity function. The general solution of Eq. (7.3.30) is the linear combination of these basic functions plus a particular solution of the inhomogeneous differential equation. Such a particular function is obtained by the parameter variation method also from the basic functions and the signal ϕ_d [25]. Finally, the analytical solution of problem (7.3.30) and (7.3.31) is obtained by calculating the coefficients contained in the general solution from the numerical conditions (7.3.31). The mentioned calculation consists of solving a linear system of four equations. This analysis shows that the introduction of the third term in the functional $J(\phi)$, defined by (7.3.21), widens the space of functions from which the signal ϕ is intended to be approximated, and therefore, a better representation of said signal can be expected.

Unfortunately, knowing the signal ϕ_d in analytical form is an unusual situation, rather, the usual situation is that there is a time series $\{\phi_d^k\}$ given by Eq. (7.3.20). In this case, it is necessary to apply a numerical method to filter the data. The proposed method to discretize problem (7.3.30) and (7.3.31) is based on second-order approximation centered finite difference formulas. For this purpose, a regular partition of the interval $(0, T)$ is considered into sub-intervals of length $\Delta t = T/(K-1)$, together with the nodes $t_k = (k-1)\Delta t$, $k = 0, 1, \ldots, K+1$. Note that the nodes t_0 and t_{K+1} are artificial, that is, they are outside the interval $(0, T)$. Considering this mesh in the interval $(0, T)$, the discretization of Eq. (7.3.30) is given as

$$\frac{1}{\gamma}\frac{\phi_{k-1} - 2\phi_k + \phi_{k+1}}{\Delta t^2} - \frac{1}{\eta}\frac{\phi_{k-2} - 4\phi_{k-1} + 6\phi_k - 4\phi_{k+1} + \phi_{k+2}}{\Delta t^4} - \phi_k = -\phi_d^k$$
$$k = 2, \ldots, K - 1.$$

In addition, the discretization of boundary conditions (7.3.31) has the form

$$\phi'(0) = \phi'(t_1) = \frac{\phi_2 - \phi_0}{2\Delta t} = \phi_0' \ \textit{and} \ \ \phi'(T) = \phi'(t_K) = \frac{\phi_{K+1} - \phi_{K-1}}{2\Delta t} = \phi_T'$$

This leads to the following system of linear equations:

$$\phi_{k-2} - \left(4 + \frac{\eta}{\gamma}\Delta t^2\right)\phi_{k-1} + \left(6 + \frac{2\eta}{\gamma}\Delta t^2 + \eta\Delta t^4\right)\phi_k$$
$$- \left(4 + \frac{\eta}{\gamma}\Delta t^2\right)\phi_{k+1} + \phi_{k+2} = -\eta\Delta t^4\phi_d^k$$
$$k = 2, \ldots, K - 1 \tag{7.3.33}$$

$$\phi_0 = \phi_2 - 2\Delta t\phi_0' \ \textit{and} \ \phi_{K+1} = \phi_{K-1} + 2\Delta t\phi_T' \tag{7.3.34}$$

whose solution is the filtered signal of the time series. Note that the matrix of the linear system is symmetric and five-diagonal, moreover, it does not depend on the data $\{\phi_d^k\}$, information about which appears on the right side of the system of linear equations.

7.3.2 Numerical Examples

The difficulty of the filtration process can be understood by looking at Eq. (7.3.20). In this equation, for each data point ϕ_d^k there are two unknowns: ϕ_k and ε_k. This makes it impossible to solve the problem and obtain all the variables exactly. For this reason, the global approximation of the signal is sought through the formulation (7.3.21) and (7.3.22), which requires us to solve the system (7.3.33) and (7.3.34).

To demonstrate the effectiveness of this time series filtering method, an example with typical results is presented. For this we consider the signal

$$\phi(t) = \sin(5t) + \left\{2\cos(3t) + 5\cos\left(\sqrt{2}t\right)\right\} - t + e^{0.01t}, 0 \le t \le 4\pi \tag{7.3.35}$$

being the sum of a periodic function, a quasi-periodic function, a strictly decreasing function, and a strictly increasing function. Synthetic time series $\{\phi_d^k\}$ are generated by adding a high frequency random error to this function (see Eq. 7.3.20). In the examples, the amplitude of the error is considered to be a representative value of the signal size ϕ, that is, the errors ε_k are uniformly distributed in the interval

$(-p\max|\phi|/2,\ p\max|\phi|/2)$, where the coefficient p represents the percentage. In addition, a percentage error is used to essess the quality of the filtration process:

$$E\% = \left(\left\|\phi - \phi_f\right\|_2 / \|\phi\|_2\right) \times 100 \tag{7.3.36}$$

where ϕ_f is a filtered time series, i.e., the solution of the linear system (7.3.33) and (7.3.34).

Numerical experiments show that with an increase in γ and η, the approximation of the exact signal ϕ improves, that is, the signal ϕ_d is filtered by removing the high-frequency error. However, there is no advantage in considering such parameters independently. Therefore, without loss of generality, it is assumed that $\gamma = \eta$, and the filtering of the time series is performed when γ increases in value. Figure 7.1 shows this process for various values of this parameter, $K = 800$ and $p = 15/100$. If $\gamma = 1$ then E% $= 53$, one can observe in the first graph that the filtered signal ϕ_f (blue line) is separated from the data $\{\phi_d^k\}$ (red line) and from the exact signal ϕ (green line).

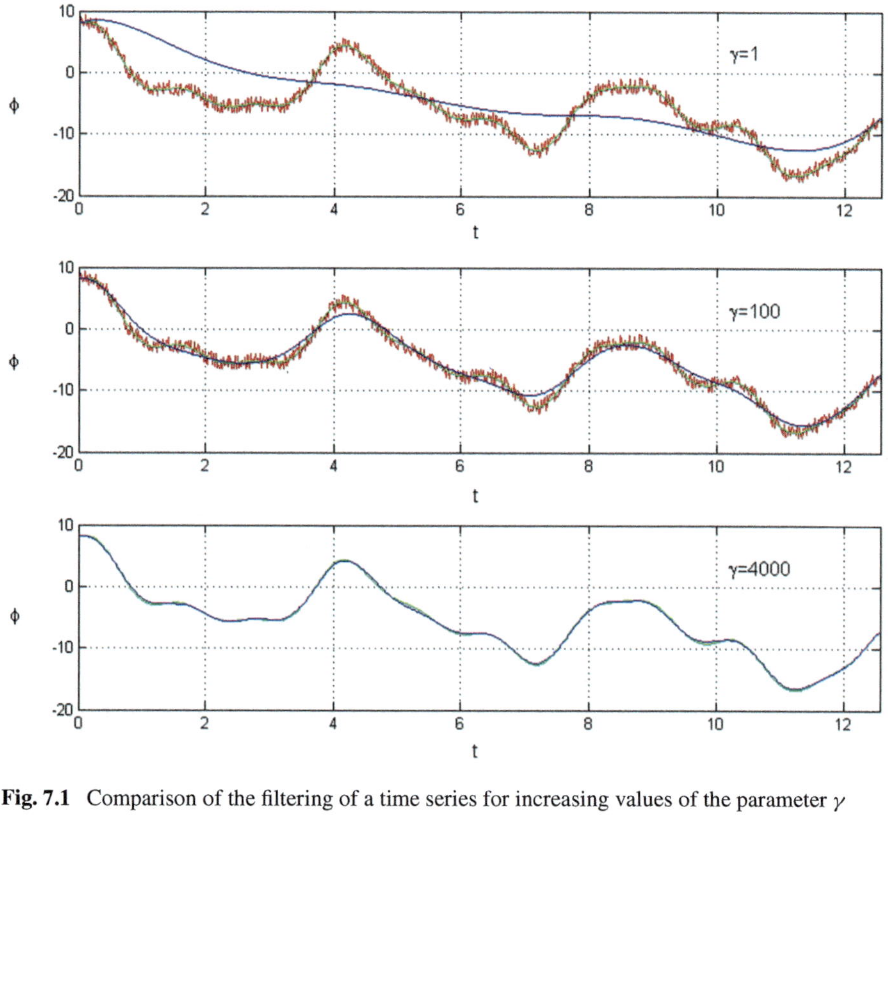

Fig. 7.1 Comparison of the filtering of a time series for increasing values of the parameter γ

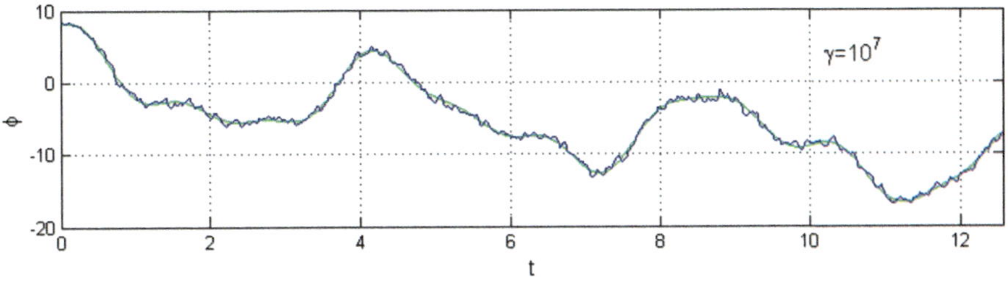

Fig. 7.2 Comparison of the exact signal ϕ (green line) with the solution of the linear system (7.3.33) and (7.3.34) for $\gamma = 10^7$

When γ increases, for example, up to $\gamma = 100$, the error decreases up to E% $= 12$ (second graph); and finally, for $\gamma = 4000$, a better result is obtained from the filtering of the time series with E% $= 2$ (third graph). However, if the value of the parameter γ continues to increase, then the solution of the linear system (7.3.33) and (7.3.34) is contaminated with the high-frequency error contained in the data, that is, spurious oscillations of small amplitude appear. Figure 7.2 shows this behavior for $\gamma = 10^7$. In this example, the approximation error is E% $= 5$, and although it does not differ much from the error that initially exists in the data (E% $= 10$), oscillations that are not part of the exact signal ϕ can be seen.

This indicates that it is necessary to introduce in this filtering method a criterion to select an appropriate value of the parameter γ. Such a parameter value should allow the approximation of the exact signal ϕ avoiding contamination with high frequency error contained in the data. A general criterion for selecting this parameter is obtained by observing that the functional defined by Eq. (7.3.21) is a regularization of the Tikhonov type with the regularization parameter γ. In this case, the regularization parameter is chosen by applying the L-curve criterion [26]. This criterion consists of establishing a balance between the approximation to the data and the size of the first derivative. The procedure begins with the trace of the trajectory (curve L) defined by

$$\gamma \mapsto \left(\ln\left\|\phi_d - \phi_f\right\|_2, \ln\left\|\phi_f'\right\|_2\right) \tag{7.3.37}$$

Subsequently, the parameter γ is selected as the value that corresponds to the point of the trajectory closest to the origin. Considering the data of Fig. 7.1, the first graph of Fig. 7.3 shows the respective curve L (blue line) for $1 < \gamma < 10^4$, and the point of this that is closest to the origin (in red). The value of the parameter that corresponds to that point is $\gamma = 794$. For this value of the parameter the filtered series ϕ_f does not contain spurious oscillations (second graph in Fig. 7.3), and the approximation error is E% $= 5$.

The value of γ that provides the L curve criterion can be improved. For this, we observe that when γ increases, the behavior of ϕ_f when approaching ϕ is divided into three stages. First there is a transition process, after a certain threshold, there is a steady state, and then there is a contamination process with high frequency error in

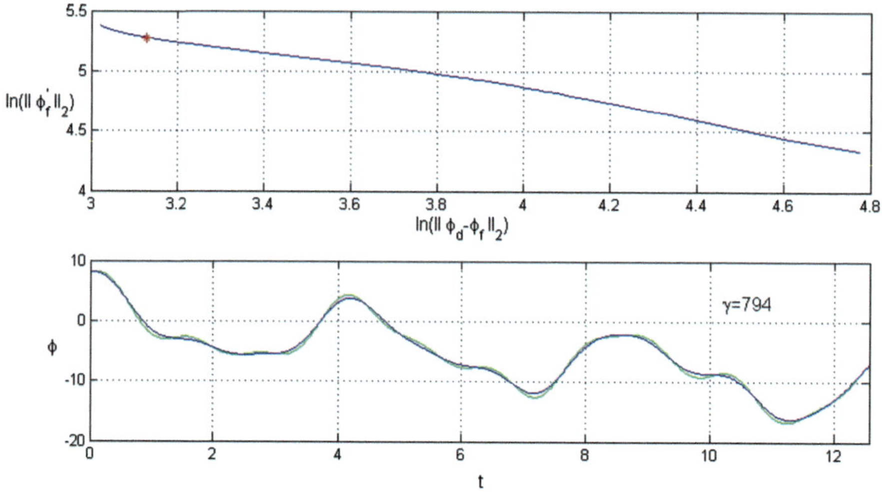

Fig. 7.3 First graph: Curve L; Second graph: Comparison of ϕ with ϕ_f when $\gamma = 794$

the data. These stages are characterized through the CSD function, which is formed by counting the total number of local changes in the sign of the derivative ϕ_f. Note that if $(\phi_f^{k+1} - \phi_f^k)(\phi_f^{k+2} - \phi_f^{k+1}) < 0$ then there is one such local change at $t = t_{k+1}$.

When the CSD function reaches equilibrium, ϕ_f correctly approximates the signal ϕ, that is, it does not yet incorporate the high-frequency errors of the data. This is due to the weight of the last two terms of the functional (7.3.21). Therefore, an appropriate value of γ for time series filtering is the beginning of such a stability interval. Figure 7.4 shows the CSD function corresponding to the data in Fig. 7.1. The stability interval is limited by red lines. The mentioned interval starts at $\gamma = 6450$, which is the value of the parameter chosen for filtering the time series. In this case, the filtered time series ϕ_f approximates the exact signal ϕ correctly, i.e. without spurious oscillations. The approximation error is E% = 2, which improves the approximation given by the L curve criterion. The profile of the ϕ_f signal is similar to that observed in the third graph of Fig. 7.1.

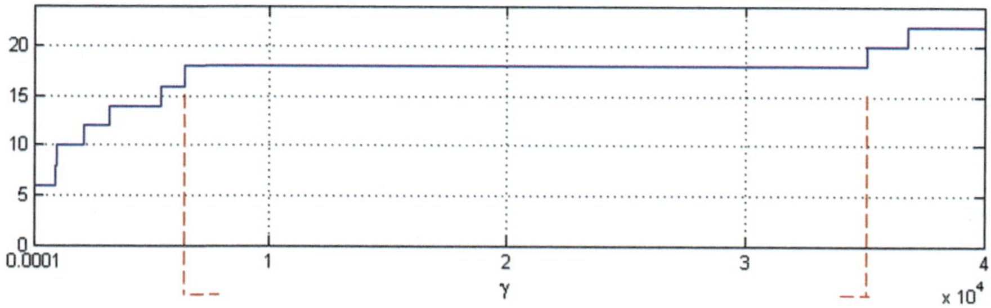

Fig. 7.4 CSD function plotted on the interval (1,40000) and stability interval

To conclude, two operational advantages of the method stand out: (i) the programming of the filtering method in the MATLAB environment is quite simple. Also, the developed programs are efficient on a desktop computer; and (ii) the filtering of very long time series can be done by filtering the signal by adjacent intervals, the boundary conditions (7.3.31) guarantee the continuity and differentiability of the filtered signal at the ends of the indicated intervals. This allows solving low order linear systems.

7.4 Identification of the Intensity of Unknown Point Sources

This section estimates the intensity of various unknown point sources, provided their location r_0 is known [27].

1. **Unknown source of constant intensity**. Assume that an unknown source of constant intensity is located at the point r_0. In this case, its intensity can be immediately identified by using formulas (7.2.7) and (7.2.11).

Let ϕ_j be the concentrations of a pollutant registered at the monitoring site **R** at time points t_j ($j = 1, 2, \ldots, M$), and $\phi(\mathbf{R}, t_j)$ be the concentration values calculated at the site **R** and at the same time points t_j using model (7.2.1)–(7.2.5).

Let us assume that all measurements $\phi_j(\mathbf{R})$ at the monitoring point **R** contain small errors $\delta\phi_j$ (see 7.2.11):

$$\phi_j(\mathbf{R}) = \phi(\mathbf{R}, t_j) + \delta\phi_j, j = 1, 2, \ldots, M \tag{7.4.1}$$

where $\phi(\mathbf{R}, t_j)$ are the concentrations calculated according to the model (7.2.1)–(7.2.5). Also assume that the mean value of the errors is zero:

$$\frac{1}{M}\sum_{j=1}^{M} \delta\phi_j = 0 \tag{7.4.2}$$

Since the intensity of the point source is constant, we have

$$Q(t) = Q_c, 0 < t < T$$

Then relations (7.2.7) accept the form

$$\phi(\mathbf{R}, t_j) = Q_c \int_0^{t_j} g_j(r_0, t)dt, (j = 1, 2, \ldots, M) \tag{7.4.3}$$

If we sum the relations (7.4.3) over j from 1 to M and use (7.4.2) we obtain

$$\sum_{j=1}^{M}\phi(\mathbf{R},t_j)=Q_c\sum_{j=1}^{M}\int_0^{t_j}g_j(r_0,t)dt \tag{7.4.4}$$

In view of (7.4.1) and (7.4.2), $\sum_{j=1}^{M}\phi_j(\mathbf{R})=\sum_{j=1}^{M}\phi(\mathbf{R},t_j)$, and (7.4.4) leads to

$$\sum_{j=1}^{M}\phi_j(\mathbf{R})=Q_c\sum_{j=1}^{M}\int_0^{t_j}g_j(r_0,t)dt$$

Therefore,

$$Q_c=\left(\sum_{j=1}^{M}\phi_j(\mathbf{R})\right)\left\{\sum_{j=1}^{M}\int_0^{t_j}g_j(r_0,t)dt\right\}^{-1} \tag{7.4.5}$$

Thus, the measurements ϕ_j of pollutant concentrations recorded at the monitoring site $\mathbf{R}$ and at time points t_j ($j=1,2,\ldots,M$), and the values $\phi(\mathbf{R},t_j)$ of pollutant concentrations calculated at the same site $\mathbf{R}$ and at the same time points t_j using model (7.2.1)–(7.2.5), make it possible to determine the intensity of unknown constant source if its location r_0 is known.

2. **Unknown instantaneous source**. Suppose now that we need to estimate the intensity Q_e of instantaneous point source at a moment t_e and at some (known) point r_0, i.e.,

$$Q(t)=Q_e\delta(t-t_e),\ 0<t_e<T \tag{7.4.6}$$

Then, due to (7.2.7),

$$\phi(\mathbf{R},t_j)=Q_e\int_0^{t_j}\delta(t-t_e)\,g_j(r_0,t)dt,\ (j=1,2,\ldots,M) \tag{7.4.7}$$

By summing again relations (7.4.7) over j again from 1 to M, and taking into account Eqs. (7.4.1) and (7.4.2), we obtain

$$Q_e=\left(\sum_{j=1}^{M}\phi_j(\mathbf{R})\right)\left\{\sum_{j=1}^{M}\int_0^{t_j}\delta(t-t_e)g_j(r_0,t)dt\right\}^{-1} \tag{7.4.8}$$

Since $g_j(r_0,t_e)=\int_0^{t_j}\delta(t-t_e)g_j(r_0,t)dt$, formula (7.4.8) leads to

$$Q_e = \left(\sum_{j=1}^{M} \phi_j(\mathbf{R})\right)\left\{\sum_{j=1}^{M} g_j(r_0, t_e)\right\}^{-1} \tag{7.4.9}$$

3. **A lower estimate of the total mass of pollutant released from a non-stationary source.**

Although the estimation of the non-steady emission rate of a point source is a more difficult problem, the zero-mean value of errors in observed data is also helpful to assess the lower bound of the total mass of pollutant, which enters the environment in the case of an accident. Such estimation can be made provided that the source location is known. A recent example of such a problem is the accidental release of radioactive substances into the ocean from the nuclear plant at Fukushima [28]. Hereinafter

$$m = \int_0^T Q(t)dt$$

denotes the total mass of a pollutant emitted at the source point r_0 during the time interval $(0, T)$.

Definition 7.4.1 A monitoring site $\mathbf{R}$ is called *useful* if the time series $\{\phi_j(\mathbf{R})\}_{j=1}^{M}$ at the point $\mathbf{R}$ satisfies the condition $\sum_{j=1}^{M}\phi_j(\mathbf{R}) > 0$.

Definition 7.4.2 Let $\mathbf{R}$ be a useful monitoring site. The parameter $\widehat{\phi}_R$ and function $\widehat{g}_R(t)$ are defined as follows:

$$\hat{\phi}_R = \sum_{j=1}^{M} \phi_j(\mathbf{R}) \text{ and } \hat{g}_R(t) = \sum_{j=1}^{M} g_j(r_0, t)$$

Theorem 7.4.1 *Let* $\mathbf{R}_1, \ldots, \mathbf{R}_N$ *be useful monitoring sites. Then, the following estimate from below for the total mass of emitted pollutant is valid*:

$$m \geq \max_{1 \leq i \leq N} \left\{ \frac{\hat{\phi}_{R_i}}{\hat{G}_{R_i}} \right\}, \text{ where } \hat{G}_{R_i} = \max_{0 \leq t \leq T} \left\{ \hat{g}_{R_i}(t) \right\} > 0, i = 1, \ldots, N \tag{7.4.10}$$

Proof As before, we assume that condition (7.4.2) holds. Then, summing relations (7.2.7) over all j, substituting Eqs. (7.4.1) and (7.4.2) in the sum, and taking into account Definition 7.4.2, we obtain

$$\hat{\phi}_{R_i} = \int_0^T Q(t)\hat{g}_{R_i}(t)dt, i = 1, \ldots, N$$

Due to definition of $\widehat{G}_{R_i}$, we have

$$\int_0^T Q(t)\hat{g}_{R_i}(t)dt \le \hat{G}_{R_i} \int_0^T Q(t)dt,\ i = 1, \ldots, N,$$

and therefore

$$\frac{\hat{\phi}_{R_i}}{\hat{G}_{R_i}} \le \int_0^T Q(t)dt,\ i = 1, \ldots, N.$$

Finally, by taking the maximum on both parts of the last inequality over all i we obtain the lower estimate for the total mass m. The theorem is proved.

4. **Estimation of the average intensity of a non-stationary source**. It is easy to estimate the average intensity $\overline{Q}$ of an unknown non-stationary point source over a time interval $(0, T)$ if its location r_0 in the domain D is known. Indeed, summing formula (7.2.7) over j from 1 to M and using the mean value theorem we obtain

$$\sum_{j=1}^{M} \phi_j(\mathbf{R}) = \sum_{j=1}^{M} \phi(\mathbf{R}, t_j) = \sum_{j=1}^{M} \int_0^T Q(t) g_j(r_0, t)dt$$
$$= \int_0^T Q(t) \left\{ \sum_{j=1}^{M} g_j(r_0,\ t) \right\} dt = \overline{Q} \sum_{j=1}^{M} \int_0^T g_j(r_0,\ t)dt$$

where $\overline{Q}$ can be considered as the average value of the emission intensity $Q(t)$ in the interval $(0, T)$. Then

$$\overline{Q} = \left\{ \sum_{j=1}^{M} \phi_j(\mathbf{R}) \right\} \left\{ \sum_{j=1}^{M} \int_0^T g_j(r_0,\ t)dt \right\}^{-1} \tag{7.4.11}$$

It should be noted that all the results obtained in this section (and in Sect. 7.5) can be generalized to the case when the mean value of errors is different from zero: $\frac{1}{M}\sum_{j=1}^{M} \delta\phi_j \neq 0$. Moreover, the relevant proofs are similar.

7.5 Identification of the Location of Permanent and Instantaneous Sources

In Sect. 7.4, we assumed that the position r_0 of a constant point source and the position r_0 and instant of emission t_e of an instantaneous point source are known. This allowed us to determine the unknown intensities Q_c and Q_e of these two sources as well. We now describe a method for identifying the location r_0 of a permanent point source, as well as the location r_0 and moment of emission t_e of an instantaneous point source. We use formula (7.4.1) and condition (7.4.2) again. To do this, we need to define some nonlinear functions, the so-called *localization functions*, whose zeros will determine both r_0 andt_e. Once these two parameters have been determined, formulas (7.4.5) and (7.4.9) can be used to identifying the intensity of both sources.

1. **Permanent source**. Let the point source have a constant intensity: $Q(t) = Q_c = Const,\ 0 < t < T$.

Definition 7.5.1 Let **R** be a useful monitoring site. The parameter $\widehat{\phi}_R$ and function $\widehat{g}_R(r)$ are defined as follows:

$$\hat{\phi}_R = \sum_{j=1}^{M} \phi_j \text{ and } \hat{g}_R(r) = \int_0^T \left\{ \sum_{j=1}^{M} g_j(r,t) \right\} dt$$

Theorem 7.5.1 *Let* $\mathbf{R}, \mathbf{R}_1, \ldots, \mathbf{R}_N$ *be useful monitoring sites. Then the source position* r_0 *is a zero of the localization function*

$$\Phi(r) = \ln \left\{ 1 + \sum_{i=1}^{N} \left[\hat{\phi}_R \hat{g}_{R_i}(r) - \hat{\phi}_{R_i} \hat{g}_R(r) \right]^2 \right\} \tag{7.5.1}$$

Besides, the emission intensity is $Q_c = \widehat{\phi}_R / \widehat{g}_R(r_0)$.

Proof By applying Eq. (7.4.5) for each useful monitoring site we obtain

$$Q_c = \frac{\hat{\phi}_R}{\hat{g}_R(r_0)} \text{ and } Q_c = \frac{\hat{\phi}_{R_i}}{\hat{g}_{R_i}(r_0)},\ i = 1, \ldots, N$$

Then

$$\frac{\hat{\phi}_{R_i}}{\hat{g}_{R_i}(r_0)} = \frac{\hat{\phi}_R}{\hat{g}_R(r_0)}, \text{ or } \left(\hat{\phi}_R \hat{g}_{R_i}(r_0) - \hat{\phi}_{R_i} \hat{g}_R(r_0) \right)^2 = 0,\ i = 1, \ldots N$$

Summing the last equations over i from 1 to N yields

$$\sum_{i=1}^{N} \left(\hat{\phi}_R \hat{g}_{R_i}(r_0) - \hat{\phi}_{R_i} \hat{g}_R(r_0) \right)^2 = 0 \tag{7.5.2}$$

It follows from (7.5.2) that $\Phi(r_0) = 0$. Once the source position r_0 has been determined from $\Phi(r_0) = 0$, the emission rate Q_c is calculated using Eq. (7.4.5), which takes into account the pollutant concentration data and the adjoint functions at the monitoring site **R**. The theorem is proved.

2. **Instantaneous source**. In case of an instantaneous point source (7.2.4), the analysis is similar. However, unlike the previous case, we now must determine the three parameters, namely, the intensity Q_e, the site r_0 and the instant t_e of the emission.

Definition 7.5.2 Let **R** be a useful monitoring site. The parameter $\widehat{\phi}_R$ and the function $\widehat{g}_R(r, t)$ are defined as follows:

$$\hat{\phi}_R = \sum_{j=1}^{M} \phi_j \quad \text{and} \quad \hat{g}_R(r,t) = \sum_{j=1}^{M} g_j(r,t).$$

Theorem 7.5.2 *Let* **R**, **R**$_1$, ..., **R**$_N$ *be useful monitoring sites. The point* (r_0, t_e) *is the zero of the following localization function*

$$\Phi(r,t) = \ln\left\{1 + \sum_{i=1}^{N}\left[\hat{\phi}_R \hat{g}_{R_i}(r,t) - \hat{\phi}_{R_i}\hat{g}_R(r,t)\right]^2\right\} \tag{7.5.3}$$

Besides, the emission intensity is $Q_e = \widehat{\phi}_R / \widehat{g}_R(r_0, t_e)$.

Proof By applying Eq. (7.4.9) for each useful monitoring site we obtain

$$Q_e = \frac{\hat{\phi}_R}{\hat{g}_R(r_0,t_e)} \text{ and } Q_e = \frac{\hat{\phi}_{R_i}}{\hat{g}_{R_i}(r_0,t_e)},\ i = 1, \ldots, N$$

Then

$$\frac{\hat{\phi}_{R_i}}{\hat{g}_{R_i}(r_0,t_e)} = \frac{\hat{\phi}_R}{\hat{g}_R(r_0,t_e)}, \text{ or } \left(\hat{\phi}_R \hat{g}_{R_i}(r_0,t_e) - \hat{\phi}_{R_i}\hat{g}_R(r_0,t_e)\right)^2 = 0,\ i = 1, \ldots N$$

Summing the last equations over i from 1 to N yields

$$\sum_{i=1}^{N}\left(\hat{\phi}_R \hat{g}_{R_i}(r_0,t_e) - \hat{\phi}_{R_i}\hat{g}_R(r_0,t_e)\right)^2 = 0 \tag{7.5.4}$$

It follows from this equation that $\Phi(r_0, t_e) = 0$.

Once the location r_0 and the instant t_e of the emission from $\Phi(r_0, t_e) = 0$, the emission rate Q_e is calculated using Eq. (7.4.9), which takes into account pollutant concentration data and adjoint functions at the monitoring site **R**. The theorem is proved.

Note that in the localization functions (7.5.1) and (7.5.3) a logarithm function is introduced to rescale the values of the sums (7.5.2) and (7.5.4) for arbitrary points r_0. This makes it possible to more accurately identify the zero of the localization function. Note also that the feasible spaces containing zeros of the localization functions are defined by Eqs. (7.4.5) and (7.4.9). Indeed, since the source intensity is positive and the monitoring sites are useful ($\sum_{j=1}^{M} \phi_j(\mathbf{R}) > 0$), the terms $\widehat{g}_R$ must also be positive. Thus, the feasible space for the monitoring sites $\mathbf{R}_1, \ldots, \mathbf{R}_N$ can be defined as

$$F_c = \{r_0 \in D : \prod_{i=1}^{N} \hat{g}_{R_i}(r_0) > 0\} \tag{7.5.5}$$

in case of a source of constant intensity, and as

$$F_e = \left\{(r_0, t_e) \in D \times (0, T) : \prod_{i=1}^{N} \hat{g}_{R_i}(r_0, t_e) > 0\right\} \tag{7.5.6}$$

in case of an instantaneous source. In the general case, knowledge of feasible spaces (7.5.5) and (7.5.6) is useful in practice, since narrowing the search area for zeros of localization functions reduces the computational costs.

7.6 Tikhonov Regularization for Identification of Time-Varying Emissions

We now present a method for recovering the time-varying emissions. Let us assume that the emission rate of a point source varies with time, and $\phi_j(\mathbf{R}) = \phi(\mathbf{R}, t_j) + \delta\phi_j, j = 1, 2, \ldots, M$ at the monitoring site $\mathbf{R}$ (see 7.4.1). We now propose the Tikhonov regularization method for solving the following inverse problem:

$$\text{Minimize } J(Q) = \frac{1}{2} \int_0^T \left(\frac{dQ}{dt}\right)^2 dt \tag{7.6.1}$$

$$\text{Subject to} : Q(t) \geq 0, \quad 0 < t < T, Q(0) = Q(T) = 0 \tag{7.6.2}$$

$$\text{and } -\varepsilon \leq \phi_j(\mathbf{R}) - \int_0^{t_j} Q(t) g_j(r_0, t) dt \leq \varepsilon, \quad j = 1, .., M \tag{7.6.3}$$

Functional (7.6.1) is minimized in order to filter out errors $\delta\phi_j$ and determine the intensity of emissions. In view of the presence of these errors, the difference between the terms in the constraints (7.6.3) is considered to be nonzero, and a positive parameter ε is introduced to expand the feasible space of solutions to the inverse problem. Examples described in Sect. 7.8 show that this parameter affects the smoothness of the solution, and its optimum value coincides with the maximum value of errors $\delta\phi_j$.

Note that both the objective function and the constraints directly depend on the control variable Q, and the discrete version of problem (7.6.1)–(7.6.3) will be an optimization problem for many real variables (quadratic programming problem). The problem will be formulated below, and some particular examples of this problem will be solved in Sect. 7.7 using the *quadprog* procedure in MATLAB [29].

1. **Existence and uniqueness of a regularized solution.** Let us now prove the existence and uniqueness of a regularized solution of problem (7.6.1)–(7.6.3) [30]. Denote by H the Hilbert space of real-valued absolutely continuous functions in $[0, T]$ such that $q(0) = q(T) = 0$ and $q\prime(t) \equiv \partial q/\partial t$ is square integrable [31]. The inner product and norm in H are defined as follows:

$$(p, q)_H = \int_0^T p'(t)q'(t)dt \text{ and } \|q\|_H = \sqrt{(q, q)_H} = \sqrt{\int_0^T \left[q'(t)\right]^2 dt}. \quad (7.6.4)$$

The unknown emission rate $Q(t)$ belongs to space H, since it is considered to be a non-negative continuously differentiable function, i.e., $Q(t) \in C^1[0, T]$ and $Q(0) = Q(T) = 0$.

The feasible space for the variational problem (7.6.1)–(7.6.3) is defined as

$$\begin{aligned} \mathrm{F}_\varepsilon = \{ & q \in H : q(t) \geq 0 \text{ in } [0, T] \text{ and} \\ & -\varepsilon \leq \phi_j(R) - \int_0^{t_j} q(t) g_j(r_0, t)dt \leq \varepsilon,\ j = 1, .., M\} \end{aligned} \quad (7.6.5)$$

where $\varepsilon \geq \varepsilon^* = \max_j|\delta\phi_j|$. Note that, by virtue of (7.2.11), $Q \in \mathrm{F}_\varepsilon$ for all $\varepsilon \geq \varepsilon^*$, and therefore F_ε is a non-empty set from the space H. Let us now consider the main properties of the feasible space (7.6.5) and give a minimum distance. These results will be used later to prove the existence and uniqueness of a regularized solution of the inverse problem.

Theorem 7.6.1 *A non-empty closed convex set in a uniformly convex Banach space has a unique point closest to a given point* [32].

Lemma 7.6.1 *The space H is a uniformly convex Banach space* [32].

Lemma 7.6.2 *The feasible space* F_ε *is a convex set in* H.

Proof Indeed, let $q_1, q_2 \in \mathrm{F}_\varepsilon$ and $\lambda \in (0,1)$. Then $\lambda q_1 + (1 - \lambda)q_2 \in H$ and $\lambda q_1 + (1 - \lambda)q_2 \geq 0$ in $[0, T]$. On the other hand, for each j ($j = 1, .., M$), we have

$$\left|\phi_j(\mathbf{R}) - \int_0^{t_j} \left[\lambda q_1(t) + (1-\lambda) q_2(t)\right] g_j(r_0, t) dt\right|$$
$$\leq \lambda \left|\phi_j(\mathbf{R}) - \int_0^{t_j} q_1(t) g_j(r_0, t) dt\right| + (1-\lambda)\left|\phi_j(\mathbf{R}) - \int_0^{t_j} q_2(t) g_j(r_0, t) dt\right|$$
$$\leq \lambda\varepsilon + (1-\lambda)\varepsilon = \varepsilon.$$

Therefore, $\lambda q_1 + (1-\lambda) q_2 \in \mathrm{F}_\varepsilon$ and F_ε is convex.

Lemma 7.6.3 *Any sequence that converges in the norm of space H converges uniformly.*

Proof Let $q(t)$ be a function of H. Then

$$\begin{aligned} 2|q(t)| &= \left|\int_0^t q'(\tau) d\tau - \int_t^T q'(\tau) d\tau\right| \\ &= \left|\int_0^t \operatorname{sgn}(t-\tau) q'(\tau) d\tau + \int_t^T \operatorname{sgn}(t-\tau) q'(\tau) d\tau\right| \\ &= \left|\int_0^T \operatorname{sgn}(t-\tau) q'(\tau) d\tau\right| \end{aligned}$$

where $t \in [0, T]$, and $\operatorname{sgn}(t-\tau) = 1$ if $\tau < t$ and $\operatorname{sgn}(t-\tau) = -1$ if $\tau > t$. Consequently,

$$2|q(t)| \leq \sqrt{\int_0^T \left[\operatorname{sgn}(t-\tau)\right]^2 d\tau}\, \|q\|_H = \sqrt{T}\|q\|_H \tag{7.6.6}$$

Here we used the Schwarz inequality. Thus, any sequence that converges in H converges uniformly.

Lemma 7.6.4 *The feasible space F_ε is a closed set in H.*

Proof We must show that $\overline{\mathrm{F}}_\varepsilon \subset \mathrm{F}_\varepsilon$. Let $q \in \overline{\mathrm{F}}_\varepsilon$. Then, there exists a sequence $\{q_n\}_{n=1}^{\infty}$ in F_ε such that $\|q_n - q\|_H \to 0$ when $n \to \infty$, and by Lemma 7.6.3, the sequence $\{q_n\}_{n=1}^{\infty}$ converges to q uniformly. Besides,

$$\left|\phi_j(R) - \int_0^{t_j} q(t) g_j(r_0, t) dt\right| = \left|\phi_j(R) - \int_0^{T} q(t) g_j(r_0, t) dt\right|$$
$$\leq \left|\phi_j(R) - \int_0^{T} q_n(t) g_j(r_0, t) dt\right| + \int_0^T |q(t) - q_n(t)| g_j(r_0, t) dt.$$

Let $C = \max_j \int_0^T g_j(r_0, t)dt < \infty$. Using (7.6.3) and (7.6.6) in the last estimate leads to

$$\left|\phi_j(\mathbf{R}) - \int_0^{t_j} q(t)g_j(r_0, t)dt\right| \leq \varepsilon + 0.5\sqrt{T}\|q - q_n\|_H \int_0^T g_j(r_0, t)dt$$
$$\leq \varepsilon + 0.5\sqrt{T}C\|q - q_n\|_H,$$

and $\left|\phi_j(\mathbf{R}) - \int_0^{t_j} q(t)g_j(r_0, t)dt\right| \leq \varepsilon$ in the limit when $n \to \infty$. Thus, all integral constraints in (7.6.5) are satisfied.

To prove that $q(t) \geq 0$, we use the rule of contradiction. Let $q(t_0) < 0$ for some $t_0 \in (0, T)$. Since the sequence $\{q_n\}_{n=1}^{\infty}$ converges uniformly on $[0, T]$, $|q_n(t_0) - q(t_0)| \to 0$ as $n \to \infty$, and there is a number N such that $|q_N(t_0) - q(t_0)| < |q(t_0)|/2$. This implies the inequality $q_N(t_0) < q(t_0)/2 < 0$, which contradicts the fact that $q_N \in \mathrm{F}_\varepsilon$. Thus, $q \in \mathrm{F}_\varepsilon$. The lemma is proved.

Obviously, the zero function $q_c(t) \equiv 0$ is the only constant function in H, and $q_c(t) \notin \mathrm{F}_\varepsilon$ (see the next lemma), and the regularized solution Q_ε of the inverse problem cannot be either zero or constant.

Lemma 7.6.5 *If $\phi_K(\mathbf{R}) > \varepsilon^* = \max_j|\delta\phi_j|$ for some K $(1 \leq K \leq M)$ then $q_c \notin \mathrm{F}_\varepsilon$ for any ε from the interval $\varepsilon^* \leq \varepsilon < \phi_K(\mathbf{R})$.*

Proof Let's use the rule of contradiction again. Let $q_c \in \mathrm{F}_\varepsilon$. Then, the K-th integral constraint in (7.6.5) gives

$$\phi_K(\mathbf{R}) - \int_0^{t_K} q_c g_K(r_0, t)dt \leq \varepsilon,$$

i.e., $\phi_K(\mathbf{R}) \leq \varepsilon$, which contradicts the inequality $\varepsilon < \phi_K(\mathbf{R})$, and hence $q_c \notin \mathrm{F}_\varepsilon$.

Note that condition $\phi_K(\mathbf{R}) > \varepsilon^*$ is quite reasonable and means that the values of $\phi(\mathbf{R}, t_j)$ are more significant than the errors in data.

Theorem 7.6.2 *Let $\phi_K(\mathbf{R}) > \varepsilon^* = \max_j|\delta\phi_j|$ for some K $(1 \leq K \leq M)$. Then the variational problem (7.6.1)–(7.6.3) has a unique non-trivial solution for any $\varepsilon \in [\varepsilon^*, \phi_K(\mathbf{R}))$.*

Proof By Lemma 7.6.1, H is a uniformly convex Banach space, and by virtue of (7.2.11), $Q \in \mathrm{F}_\varepsilon$ for all $\varepsilon \geq \varepsilon^*$, i.e., F_ε is a non-empty set. Moreover, by Lemmas 7.6.2 and 7.6.4, the feasible space F_ε is a closed convex set in H.

On the other hand, $q_c \notin \mathrm{F}_\varepsilon$ since $\varepsilon < \phi_K(\mathbf{R})$. Therefore, by Theorem 7.6.1, there is a unique function Q_ε in the feasible space F_ε that minimizes the distance to the function q_c. By virtue of (7.6.1), Q_ε also minimizes the functional $J(q)$ on the feasible space F_ε. Therefore, Q_ε is a regularized solution of the inverse problem.

7.7 Quadratic Programming Problem

To numerically solve the variational problem (7.6.1)–(7.6.3) and estimate the unknown emission rate $Q(t)$ of a new source, we now introduce the problem of quadratic programming. Let $t_l = l \cdot \Delta t$ be the nodes of the regular grid on the interval $[0, T]$, ($l = 0, 1, \ldots, L, \Delta t \cdot L = T$). As basic functions we take linear splines

$$\gamma_l(t) = \begin{cases} 1 - \left|\frac{t-t_l}{\Delta t}\right| \text{ if } \left|\frac{t-t_l}{\Delta t}\right| < 1 \\ 0, \qquad\qquad \text{otherwise} \end{cases}, \; l = 0, 1, \ldots, L, \tag{7.7.1}$$

besides, the extreme functions γ_0 and γ_L are additionally equal to zero outside the interval $[0, T]$. The functions (7.7.1) have the following useful properties:

$$\gamma_l(t_j) = \delta_{lj} = \begin{cases} 1, \; l = j \\ 0, \; l \neq j \end{cases}, \quad l, j = 0, 1, \ldots, L, \tag{7.7.2}$$

$$\begin{gathered} \int_0^T \left(\frac{d\gamma_0}{dt}\right)^2 dt = \int_0^T \left(\frac{d\gamma_L}{dt}\right)^2 dt = \frac{1}{\Delta t}, \\ \int_0^T \left(\frac{d\gamma_l}{dt}\right)^2 dt = \frac{2}{\Delta t}, \quad l = 1, \ldots, L-1, \end{gathered} \tag{7.7.3}$$

$$\int_0^T \left(\frac{d\gamma_{l-1}}{dt}\right)\left(\frac{d\gamma_l}{dt}\right) dt = -\frac{1}{\Delta t}, \quad l = 1, \ldots, L \tag{7.7.4}$$

and

$$\int_0^T \left(\frac{d\gamma_j}{dt}\right)\left(\frac{d\gamma_l}{dt}\right) dt = 0, \quad |j - l| > 1 \tag{7.7.5}$$

The emission rate $Q(t)$ will be sought as a linear combination

$$Q(t) = \sum_{l=1}^{L-1} Q_l \gamma_l(t) \tag{7.7.6}$$

where $Q_j = Q(t_j), j = 1, \ldots, L-1$ (note that due to (7.7.2), $Q_0 = Q(0) = 0$, $Q_L = Q(T) = 0$). Substituting (7.7.6) into (7.6.1) and taking into account the properties (7.7.3)–(7.7.5), we obtain

$$J(\boldsymbol{Q}) = \frac{1}{2}\boldsymbol{Q}^T \boldsymbol{H} \boldsymbol{Q}$$

where $\boldsymbol{Q} = (Q_1, Q_2, \ldots, Q_{L-1})^T$ is a vector of Euclidean space $\mathbb{R}^{L-1}$, and $\boldsymbol{H}$ is a tridiagonal, symmetric and positive definite matrix of order $L-1$, whose entries $\boldsymbol{H}_{jl} = \int_0^T \gamma'_j \gamma'_l dt$ are determined by inner products, i.e.,

$$\boldsymbol{H} = \frac{1}{\Delta t}\begin{bmatrix} 2 & -1 & 0 & 0 & \cdots & 0 \\ -1 & 2 & -1 & \ddots & & 0 \\ 0 & -1 & 2 & \ddots & 0 & \vdots \\ 0 & \ddots & \ddots & \ddots & -1 & 0 \\ \vdots & & 0 & -1 & 2 & -1 \\ 0 & 0 & \cdots & 0 & -1 & 2 \end{bmatrix}$$

Without loss of generality, we will assume that the sampling times coincide with the grid nodes in the interval $(0, T)$, and hence $M = L$. Thus, substituting (7.7.6) into (7.2.11), we obtain

$$\phi(\mathbf{R}, t_j) = \sum_{l=1}^{j} a_{jl} Q_l, \quad j = 1, .., L \tag{7.7.7}$$

where

$$a_{jl} = \int_{t_{l-1}}^{t_{l+1}} \gamma_l(t) g_j(r_0, t) dt \tag{7.7.8}$$

Then the discrete form of the quadratic programming problem (7.6.1)–(7.6.3) is

$$\text{Minimize } J(\boldsymbol{Q}) = \frac{1}{2}\boldsymbol{Q}^t \boldsymbol{H}\boldsymbol{Q} \tag{7.7.9}$$

$$\text{Subject to : } Q_l \geq 0, \quad l = 1, \ldots, L-1, Q_0 = Q_L = 0 \tag{7.7.10}$$

$$\text{and } -\varepsilon \leq \phi_j - \sum_{l=1}^{j} a_{jl} Q_l, \leq \varepsilon, \quad j = 1, .., L \tag{7.7.11}$$

where the coefficients a_{jl} are determined by formula (7.7.8). Note that to simplify constraints (7.7.11), the coefficients a_{jl} can be calculated using second-order formulas (for example, using the trapezoidal rule):

$$\begin{aligned} a_{jj} &= \tfrac{\Delta t}{2} g_j(r_0, t_j), j = 1, \ldots, L-1, \text{ and} \\ a_{jl} &= \Delta t \cdot g_j(r_0, t_l), l = 1, \ldots, j-1, \; l < j \leq L \end{aligned} \tag{7.7.12}$$

Note also that problem (7.7.9)–(7.7.11) has a solution, since the feasible space defined by relations (7.7.10) and (7.7.11) is a nonempty compact set in the space $\mathbb{R}^{L-1}$ and $J(\boldsymbol{Q})$ is continuous. In addition, since $\boldsymbol{H}$ is a positive definite matrix, $J(\boldsymbol{Q})$ is a strictly convex function, and hence the solution of problem (7.7.9)–(7.7.11) is unique in the convex feasible space (7.7.10) and (7.7.11) [32]. This optimal solution can be computed efficiently using the *lsqlin* or *quadprog* routines from MATLAB [29], provided that all adjoint functions are already defined.

7.8 Examples of Identification of Parameters of Unknown Sources

To demonstrate the effectiveness of the quadratic programming problem (7.7.9)–(7.7.11), this section present some numerical examples. The examples contain data errors (7.4.1). Due to the fact that the inverse problem is ill-posed, determining the intensity $Q(t)$ of a new source is a difficult task, and problems (7.6.1)–(7.6.3) and (7.7.9)–(7.7.11) must be solved. Note that the examples below use synthetic data with uniformly distributed errors.

Example 7.8.1 We start with a simple zero-dimensional dispersion model (box model) for the pollutant concentration [22],

$$\frac{d\phi}{dt} + \left(\sigma + \frac{1}{V}ua^2\right)\phi = \frac{1}{V}Q(t), \quad 0 < t < T \tag{7.8.1}$$

$$\phi(0) = 0 \tag{7.8.2}$$

It is assumed that at any moment the pollutant forms a homogeneous mixture with air in a region $D = [0, a] \times [0, a] \times [0, a]$ that is a cube with edge length a and volume $V = a^3$. The wind speed u is horizontal, and σ denotes the chemical conversion coefficient of the pollutant. Let us try to estimate $Q(t)$ directly from (7.8.1) and (7.8.2). For this purpose, in (7.8.1) two well-known finite difference formulas of the second order of approximation are used to obtain

$$\frac{\phi_{j+1} - \phi_{j-1}}{2\Delta t} + r\frac{\phi_{j+1} + \phi_{j-1}}{2} = \frac{1}{V}Q_j$$

where $r = \sigma + \frac{1}{V}ua^2$, Q_j approximates $Q(t_j)$ at the time points $t_j = j\Delta t$ ($j = 0, 1, \ldots, L$), $\Delta t = T/L$ is the grid step between observational data, and $\{\phi_j\}_{j=0}^{L}$ is the time series of pollutant concentrations. Grouping the terms in the last equation, we get

$$Q_j = \frac{V}{2}\left\{\left(r + \frac{1}{\Delta t}\right)\phi_{j+1} + \left(r - \frac{1}{\Delta t}\right)\phi_{j-1}\right\}, \quad j = 1, \ldots, L-1 \tag{7.8.3}$$

To check Eq. (7.8.3) consider the input signal (emission rate)

$$Q(t) = \begin{cases} 0,\ 0 \le t < 2 \\ q,\ 2 \le t \le 4 \\ 0,\ 4 < t \le 10 \end{cases} \tag{7.8.4}$$

and output signal (pollutant concentration)

$$\phi(t) = \begin{cases} 0,\ \ 0 \le t < 2 \\ \frac{q}{rV}(1 - exp(-r(t-2))),\ \ 2 \le t \le 4 \\ \frac{q}{rV}(exp(-r(t-4)) - exp(-r(t-2))),\ \ 4 < t \le 10 \end{cases} \tag{7.8.5}$$

for the dispersion model (7.8.1) and (7.8.2). Figure 7.5 shows both signals for $a = 1\,\text{km}$, $u = 0.5\,\text{kmh}^{-1}$, $\sigma = 0.0001\,\text{h}^{-1}$, $T = 10\,\text{h}$ and $q = 100\,\text{kgh}^{-1}$. The synthetic time series (7.4.1) is

$$\phi_j = \phi(t_j) + \delta\phi_j,\ \ \Delta t = T/L, t_j = j\Delta t,\ \ j = 0,\ 1,\ \ldots,\ L \tag{7.8.6}$$

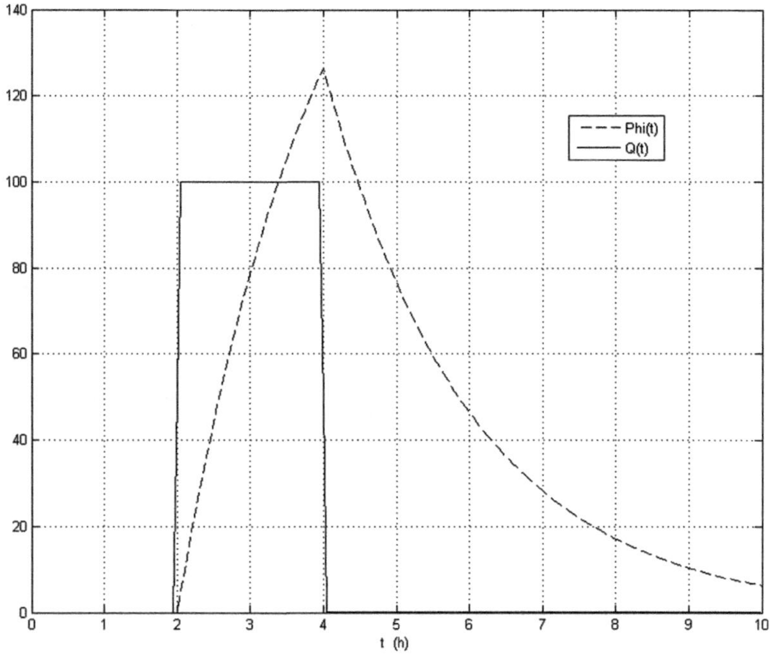

Fig. 7.5 Emission rate $Q(t)$ and pollutant concentration ϕ given by (7.7.4) and (7.7.5)

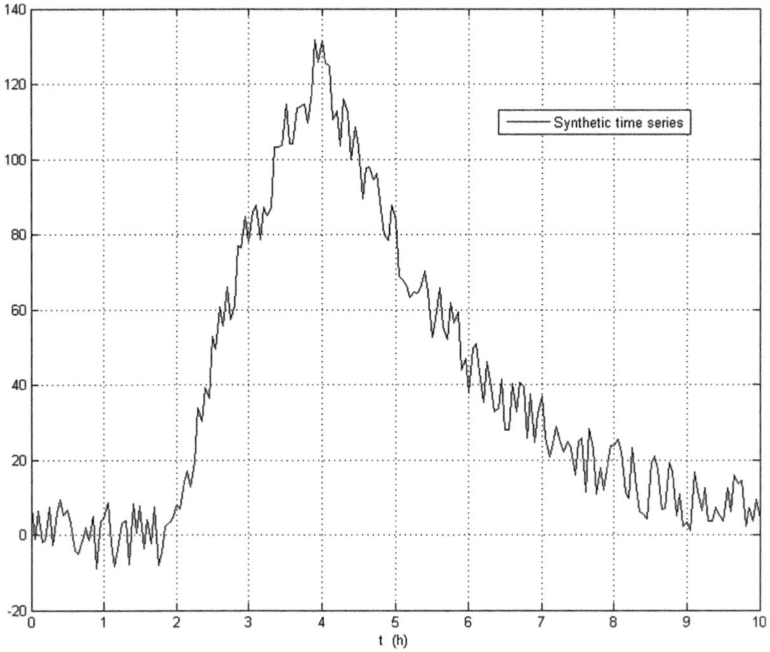

Fig. 7.6 Synthetic time series $\{\phi_j\}_{j=0}^{L}$ for pollutant concentration

where $\phi(t_j)$ are given using (7.8.5), and $\{\delta\phi_j\}_{j=0}^{L}$ are uniformly distributed errors. Figure 7.6 shows the time series (7.8.6) for $\Delta t = 0.05\,\text{h}$ and $\delta\phi_j$ randomly selected from a uniform distribution in the interval $(-0.5, 0.5)$. In this example, the normalized amplitude errors can be up to 15% of the maximum output signal ϕ.

Approximate emission rates obtained using scheme (7.8.3) and such synthetic time series (shown in Fig. 7.6) are presented in Fig. 7.7. It can be seen that the relative errors in the amplitude of emission rate can reach 175% (since the inverse problem is ill-posed), so the true values of emission rate $Q(t)$ are lost.

This result can be explained as follows. Substituting (7.8.6) into Eq. (7.8.3), we obtain

$$|Q(t_j) - Q_j| \leq \frac{V}{2}\delta\phi\left\{\left|r + \frac{1}{\Delta t}| + |r - \frac{1}{\Delta t}\right|\right\} + o\big((\Delta t)^2\big) \leq V\delta\phi\left(r + \frac{1}{\Delta t}\right) + o\big((\Delta t)^2\big)$$

where $\delta\phi = \max_j\{|\delta\phi_j|\}$. Thus, the error in amplitude increases with increasing volume V and sampling frequency $1/\Delta t$. In particular, if the number of observations in the interval $(0, T)$ increases, then $\Delta t \to 0$ and the estimate of $Q(t)$ according to

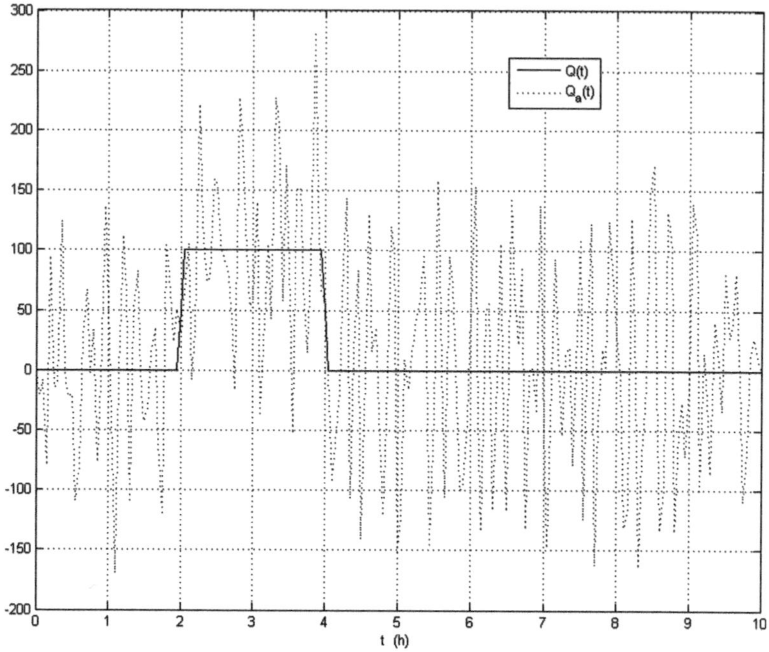

Fig. 7.7 Approximate emission rate $Q_a(t)$ derived from the scheme (7.8.3) and the true emission rate $Q(t)$ determined by (7.8.4)

the scheme (7.8.3) worsens. Other methods described in the literature and based on finite differences show similar undesirable fluctuations [3].

To solve this particular inverse problem, we now consider the quadratic programming problem (7.7.9)–(7.7.11) plus (7.7.12). For the dispersion model (7.8.1) and (7.8.2), an explicit relationship between the unknown emission rate $Q(t)$ and the pollutant concentration ϕ can be established by solving Eq. (7.7.1) using the Laplace transform [33]:

$$\phi(t) = \frac{1}{V} e^{-rt} \int_0^t Q(\xi) e^{r\xi} d\xi, \quad 0 < t < T \tag{7.8.7}$$

where $r = \sigma + \frac{1}{V} u a^2$. Then the corresponding equations (7.2.10) can be written as

$$\phi(t_j) = \int_0^{t_j} Q(t) g_j(t) dt, \quad t_j = j\Delta t, \; j = 1, \ldots, L$$

where $g_j(t) = \frac{1}{V}\exp\{-r(t_j - t)\}$ is the kernel of integral equation, $0 < t < t_j$. In addition, due to the stationary dispersion conditions considered in this example,

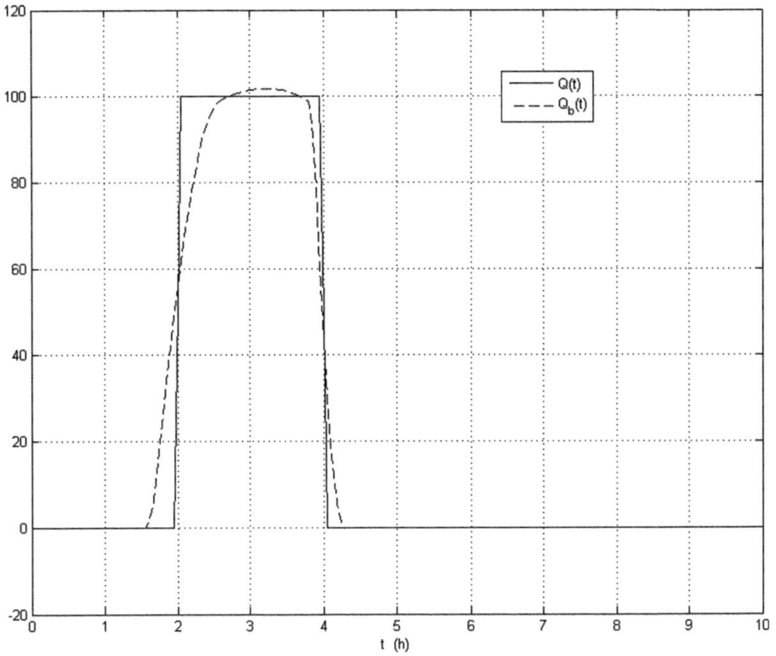

Fig. 7.8 Emission rate $Q_b(t)$ determined by (7.7.9)–(7.7.11) and true emission rate $Q(t)$ determined by (7.8.4)

the basic kernel is $g_\alpha(t) = \frac{1}{V}\exp\{-r(T-t)\}$, $0 < t < T$, which has the property $g_\alpha(T - t_j + t) = g_j(t)$ for $t \leq t_j$ (see Eq. 7.2.12).

Figure 7.8 shows the input signal (7.8.4) and the solution of the quadratic programming problem (7.7.9)–(7.7.11) with (7.7.12). This solution was calculated using the MATLAB *quadprog* routine of [29].

In this example, the time series $\{\phi_j\}_{j=0}^{L}$ shown in Fig. 7.6 was created using (7.8.6). It can be seen that the numerical solution of the regularization method (7.6.1)–(7.6.3) improved the estimation of the emission rate obtained by the simple scheme (7.8.3).

Numerical experiments also show that if the parameter $\varepsilon \to +\infty$ then the solution to problem (7.7.9)–(7.7.11) smooths out and tends to zero, since the feasible space expands so that it contains the global minimum of the objective function J. In the other limiting case when $\varepsilon > \delta\phi$ and $\varepsilon \to \delta\phi^+$, where $\delta\phi = \max_j\{|\delta\phi_j|\} > 0$, the regularization leads to the best approximation of the emission rate. This effect can be observed in Fig. 7.9, which successively shows solutions to problem (7.7.9)–(7.7.11) with synthetic data (presented in Fig. 7.6) for $\varepsilon = e \cdot \delta\phi$ where $e = 1.8$, $e = 1.4$ and $e = 1.0$. The relative errors of the corresponding approximations are equal to 0.32, 0.28 and 0.19.

Finally, when $\varepsilon < \delta\phi$, the solution of the inverse problem becomes unstable (shows fluctuations). Since the uncertainty $\delta\phi$ of real data is unknown, such a sharp

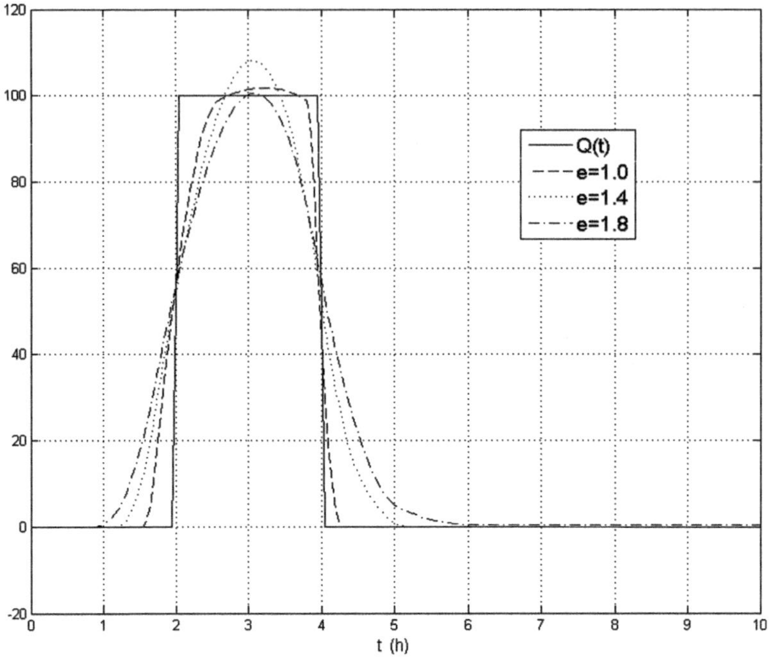

Fig. 7.9 Behavior of the solution of the quadratic programming problem (7.7.9)–(7.7.11) when $\varepsilon \to \delta\phi^+$

change in the behavior of the solution indicates the point at which the optimal value of $\delta\phi$ is reached.

Example 7.8.2 Let us now consider a one-dimensional diffusion–reaction model for the pollutant concentration $\phi = \phi(x, t)$:

$$\frac{\partial \phi}{\partial t} - \mu \frac{\partial^2 \phi}{\partial x^2} + \sigma \phi = Q(t)\delta(x - x_0), 0 < x < 1, 0 < t < T \tag{7.8.8}$$

$$\mu \frac{\partial \phi}{\partial x}(0, t) = 0, t > 0 \tag{7.8.9}$$

$$\mu \frac{\partial \phi}{\partial x}(1, t) = -\zeta \phi(1, t), t > 0 \tag{7.8.10}$$

$$\phi(x, 0) = 0, 0 < x < 1 \tag{7.8.11}$$

where $\mu > 0$ is the diffusion coefficient, $\sigma \geq 0$ is the reaction parameter, and x_0 is the position of the point source in the domain $D = (0,1)$ with emission rate $Q(t)$. Condition (7.8.9) means that there is no impurity flux at $x = 0$ (closed boundary), and condition (7.8.10) means that the impurity flux at the open boundary $x = 1$ is proportional to ϕ with $\zeta > 0$. As a result, the total mass of the pollutant in D increases due to the emission rate Q from the source and decreases due to the outflow at $x = 1$ and exponential decay ($\sigma > 0$) according to the following equation:

$$\frac{\partial}{\partial t}\int_0^1 \phi(x,t)dx = Q(t) - \zeta\phi(1,t) - \int_0^1 \sigma\phi(x,t)dt$$

The diffusion model (7.8.8)–(7.8.11) is well posed because its solution exists, is unique, and continuously depends on the initial condition and forcing. This statement is true because the problem operator A defined as

$$A\phi = -\mu\frac{\partial^2\phi}{\partial x^2} + \sigma\phi$$

is positive semidefinite:

$$(A\phi,\ \phi) = \int_0^1 \mu\left(\frac{\partial\phi}{\partial x}\right)^2 dx + \zeta\phi^2(1,t) + \int_0^1 \sigma\phi^2(x,t)dt \geq 0$$

The nonnegative solution of the Sturm–Liouville problem for the operator of the model (7.8.8)–(7.8.11) is expressed by the Fourier series [34] as:

$$\phi(x,t) = \sum_{k=1}^{\infty}\left(\int_0^t d_k Q(\xi)e^{-(\lambda_k^2\mu+\sigma)(t-\xi)}d\xi\right)\cos(\lambda_k x) \tag{7.8.12}$$

where

$$d_k = 2\cos(\lambda_k x_0)\left[1 + \frac{\mu}{\zeta}\sin^2(\lambda_k)\right]^{-1},\ k = 1,\ 2,\ \ldots,$$

$\{\cos(\lambda_k x)\}_{k=1}^{\infty}$ is an orthogonal system of eigenfunctions, and $\{\lambda_k\}_{k=1}^{\infty}$ is the corresponding set of frequencies (eigenvalues) that are the roots of the equation

$$\zeta\cos(\lambda)\ - \mu\lambda\sin(\lambda) = 0$$

These frequencies can be efficiently calculated by Newton's method [35].

Note that for the numerical experiments of this section we used the Fejér series [36, 37] calculated from the Fourier series (7.8.12), which converges pointwise to $\phi(x,t)$. That is, in practice we take into account in (7.8.12) the Fejér coefficients

$$\hat{d}_k = \left(1 - \frac{(k-1)}{K}\right) d_k, k = 1, 2, \ldots K$$

instead of d_k, where K is the truncation number of the series.

On the other hand, applying the Lagrange identity $(A\phi, g) = (\phi, A^* g)$ to the diffusion model (7.8.8)–(7.8.11), we can formulate the following adjoint model:

$$-\frac{\partial g_j}{\partial t} - \mu \frac{\partial^2 g_j}{\partial x^2} + \sigma g_j = p(x, t), 0 < x < 1, 0 < t < T \quad (7.8.13)$$

$$\mu \frac{\partial g_j}{\partial x}(0, t) = 0, t > 0 \quad (7.8.14)$$

$$\mu \frac{\partial g_j}{\partial x}(1, t) = -\varsigma g_j(1, t), t > 0 \quad (7.8.15)$$

$$g_j(x, T) = 0, 0 < x < 1 \quad (7.8.16)$$

where $p(x, t) = \delta(x - \mathbf{R})\delta(t - t_j)$ is the forcing, and $\mathbf{R}$ is the monitoring point in D. As shown earlier, the use of solutions g_j of the adjoint problem (7.8.13)–(7.8.16) leads to

$$\phi(\mathbf{R}, t_j) = \int_0^{t_j} Q(t) g_j(x_0, t) dt$$

In the last formula, the adjoint function g_j can be expressed as

$$g_j(x_0, t) = \sum_{k=1}^{\infty} c_k e^{-(\lambda_k^2 \mu + \sigma)(t_j - t)} \cos(\lambda_k x_0), t \le t_j \quad (7.8.17)$$

where

$$c_k = 2\cos(\lambda_k \mathbf{R}) \left[1 + \frac{\mu}{\varsigma} \sin^2(\lambda_k)\right]^{-1}$$

is the Fourier coefficient of the series (7.8.17). We note again that in the series (7.8.17) we take the Fejér coefficients

$$\hat{c}_k = \left(1 - \frac{(k-1)}{K}\right) c_k, k = 1, 2, \ldots K$$

instead of c_k, where K is the truncation number of the series.

Further, since in this example the dispersion conditions are stationary, the basic kernel

$$g_\alpha(x_0, t) = \sum_{k=1}^{\infty} c_k e^{-(\lambda_k^2\mu+\sigma)(T-t)}\cos(\lambda_k x_0),\ 0 < t < T \tag{7.8.18}$$

has the property $g_\alpha(x_0, T - t_j + t) = g_j(x_0, t)$ for $t \le t_j$ (see Eq. 7.2.8). Figure 7.10 shows the input signal $Q(t)$ given by (7.8.4) and the corresponding output signal $\phi(\mathbf{R}, t)$ calculated according to (7.8.12) for the following parameters: $x_0 = 0.3\,\text{km}$, $\mathbf{R} = 0.8\,\text{km}$, $\zeta = 1.0\,\text{km}$, $\mu = 0.5\,\text{km}^2$, $\sigma = 0$, $T = 10\,\text{h}$, $q = 100\,\text{kgh}^{-1}$ and $K = 10000$.

The basic kernel $g_\alpha(x_0, t)$ calculated according to (7.8.18) and two examples of adjoint functions $g_j(x_0, t)$ (see Eq. 7.8.17) are shown in Fig. 7.11. In full accordance with Eq. (7.2.8), each adjoint function is a time shift of the basic kernel.

Once the adjoint functions are obtained using the basic kernel, one can consider the quadratic programming problem (7.7.9)–(7.7.11) with (7.7.12) to solve the inverse problem. As in Example 7.8.1, the synthetic pollutant concentration time series $\{\phi_j\}_{j=0}^{L}$ is formed by the formula (7.4.1),

$$\phi_j = \phi\big(\mathbf{R}, t_j\big) + \delta\phi_j,\quad \Delta t = T/L, t_j = j\Delta t,\quad j = 0,\ 1,\ \ldots,\ L,$$

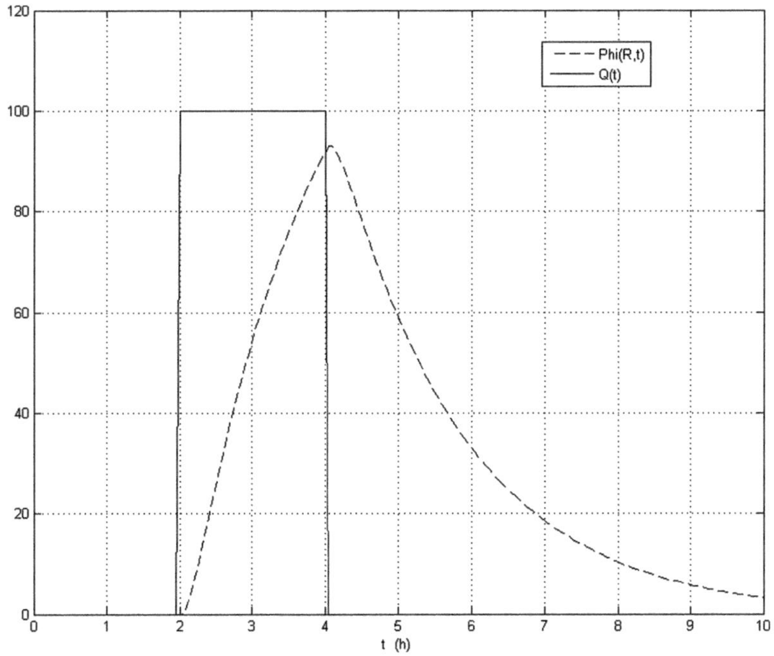

Fig. 7.10 Emission rate $Q(t)$ given by (7.8.4) and the concentration at $x = \mathbf{R}$ calculated according to (7.8.12)

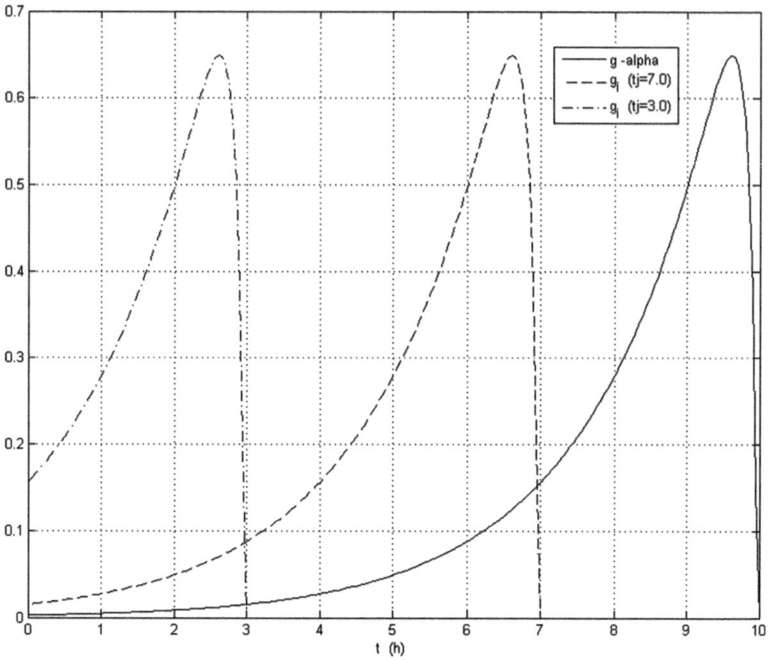

Fig. 7.11 Basic kernel $g_\alpha(x_0, t)$ and adjoint functions $g_j(x_0, t)$ for $t_j = 3.0$ and $t_j = 7.0$

where $\{\delta\phi_j\}_{j=0}^{L}$ are uniformly distributed errors.

Figure 7.12 shows the time series calculated for $\Delta t = 0.05$ h and $\delta\phi_j$ values randomly selected from a uniform distribution in the interval $(-0.5, 0.5)$. Normalized amplitude errors can be up to 15% of the maximum output signal $\phi(\mathbf{R}, t)$.

Figure 7.13 shows the input signal (7.8.4) and the behavior of the solution of the quadratic programming problem (7.7.9)–(7.7.11) plus (7.7.12) when $\varepsilon \to \delta\phi^+$, where $\varepsilon = e \cdot \delta\phi$, $\delta\phi = \max_j\{|\delta\phi_j|\} > 0$ and $e = 8.0$, $e = 4.0$, $e = 2.0$ and $e = 1.0$, successively.

These solutions were calculated using the MATLAB *quadprog* routine [29]. It can be seen that the behavior of the solution depends on ε in the same way as in Example 7.8.1, and for $\varepsilon = \delta\phi$, we obtain the best approximation of the emission rate $Q(t)$ that the method can give.

This example is interesting in that a portion of a pollutant emitted at a point $x_0 = 0.3$ and moment t propagates due to diffusion in both directions (left and right). Besides, one part propagating to the right is registered at the monitoring point $\mathbf{R} = 0.8$ at a moment $t_1 > t$, and the other part propagating to the left is registered at a moment $t_2 > t_1$ as a feedback process (reflection). Reflection is the consequence of the presence of a closed boundary at $x = 0$. The fact is that such feedback can be erroneously interpreted as the presence of another source located to the left of the true source, which can generate an additional impulse outside the time interval (2,4). However, thanks to some information provided by the adjoint solutions, this

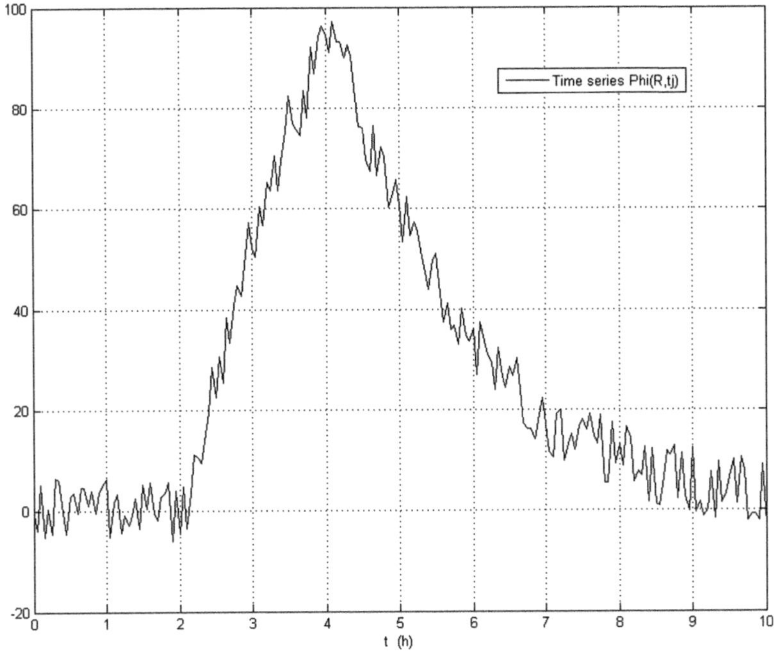

Fig. 7.12 Synthetic time series $\{\phi_j\}_{j=0}^{L}$ for pollutant concentration at $x = \mathbf{R}$

false interpretation is recognized by the regularization process, and finally the only source is correctly determined (see Fig. 7.13).

Example 7.8.3 We now consider the dispersion model (7.8.8)–(7.8.11) in order to estimate the parameters of some constant and instantaneous point sources. For this, the solutions g_j of the adjoint model (7.8.13)–(7.8.16) are calculated using the basic kernel g_α defined by the Fejér-Fourier series (7.8.18). Figure 7.14 shows the basic kernel $g_\alpha(x_0, t)$ and adjoint functions $g_j(x_0, t)$ for the moments $t_j = 3$ and $t_j = 15$. For series (7.8.18), the following parameters were used: $x_0 = 0.5\,\text{km}$, $\mathbf{R} = 0.2\,\text{km}$, $\zeta = 1.0\,\text{km}$, $\mu = 0.1\,\text{km}^2$, $\sigma = 0.01\,\text{h}^{-1}$, $T = 20\,\text{h}$ and $K = 10000$. As in the previous two examples, synthetic data ϕ_j are generated according to Eq. (7.4.1), where the corresponding values $\phi(\mathbf{R}, t_j)$ for each experiment are calculated using the Fejér-Fourier series (7.8.12) and the errors $\delta\phi_j$ are uniformly distributed with zero mean value.

First, consider a constant source located at the point $x_0^* = 0.5\,\text{km}$ with intensity $Q_c = 100\,\text{kgh}^{-1}$. Figures 7.15, 7.16 and 7.17 show the localization function $\Phi(x_0)$ for various monitoring sites $\mathbf{R}$ and $\mathbf{R}_1$. It should be noted that in all cases, the zero of $\Phi(x_0)$ correctly determines the location of the emission source. In addition, Theorem 7.5.1 allows us to correctly calculate the intensity $Q_c = 100$ of the source.

We now consider the case of an instantaneous source that is located at the point $x_0^* = 0.7\,\text{km}$ and has the intensity $Q_e = 120\,\text{kg}$ at the moment $t_e^* = 12\,\text{h}$. Figures 7.18,

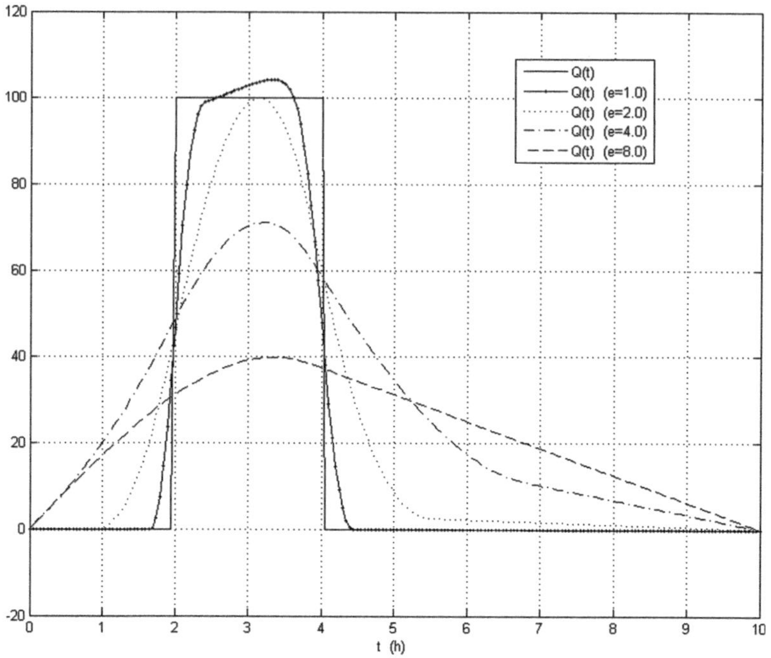

Fig. 7.13 Behavior of the solution of quadratic programming problem (7.7.9)–(7.7.11) when $\varepsilon \to \delta\phi^+$

7.19 and 7.20 show localization functions $\Phi(x_0, t_e)$ for various monitoring sites $\mathbf{R}$ and $\mathbf{R}_1$.

It can be seen that in all cases the zeros of localization function correctly determine the place and moment of the emission. Moreover, Theorem 7.6.2 allows us to correctly calculate the intensity $Q_e = 120$ of the source.

7.9 An Example of Applying the Variational Method for Determining the Intensity of Nonstationary Sources

Let us now apply the variational method (7.6.1)–(7.6.3) to determine the unknown non-stationary emission rate $Q(t)$ of a point source if its position r_0 in the domain D is known. Let us consider a one-dimensional dispersion model and its adjoint in the domain $D = [0,1]$ and time interval $(0, T)$ $(T = 40)$, using the following dimensionless parameters: $U = 0.85e_1$, $v_k^s = 0$, $\mu = 0.5$, $\sigma = 0.1$, $r_0 = 0.3$ and $\mathbf{R} = 0.8$. Also, $Q(t) = 100t^2e^{-0.5t}\sin^2(t)$ if $0 \le t \le 4\pi$, and $Q(t) = 0$ otherwise. Graphs of $Q(t)$ and the solution $\phi(\mathbf{R}, t)$ of the one-dimensional dispersion model for $x = \mathbf{R}$ are shown in Fig. 7.21.

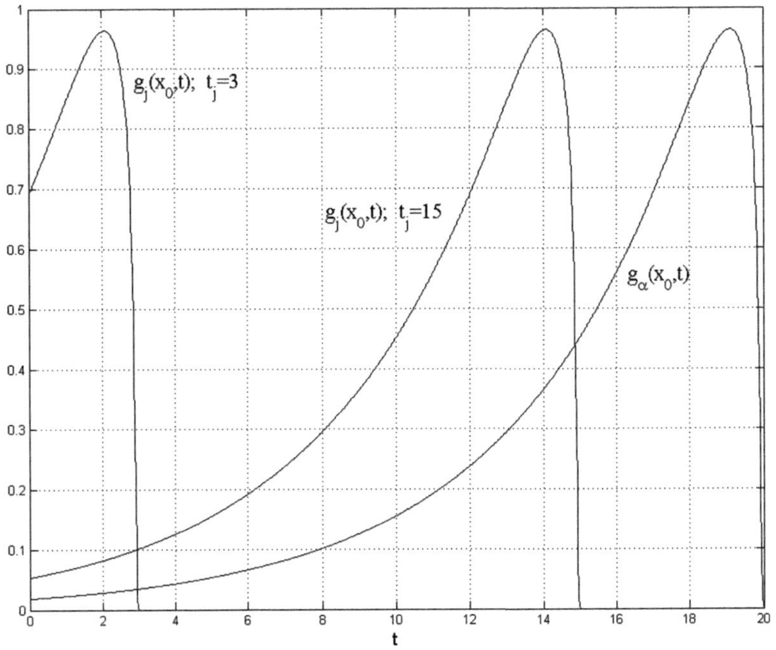

Fig. 7.14 Basic kernel $g_\alpha(x_0, t)$ and adjoint functions $g_j(x_0, t)$ for $t_j = 3$ and $t_j = 15$

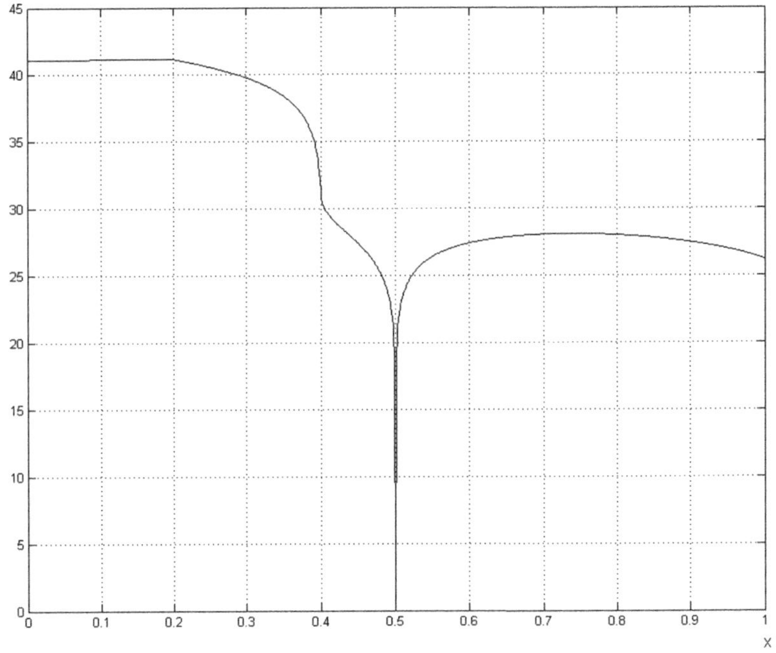

Fig. 7.15 Localization function $\Phi(x_0)$ for the monitoring sites $\mathbf{R} = 0.2$ and $\mathbf{R}_1 = 0.4$

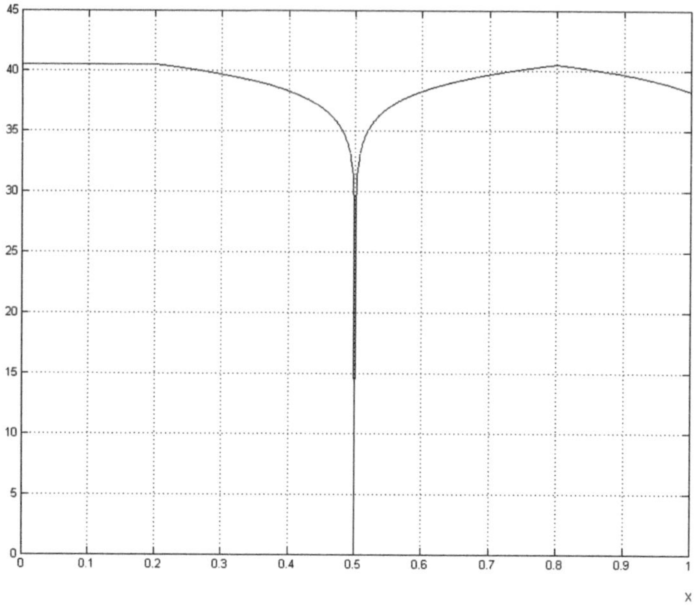

Fig. 7.16 Localization function $\Phi(x_0)$ for the monitoring sites $\mathbf{R} = 0.2$ and $\mathbf{R}_1 = 0.8$

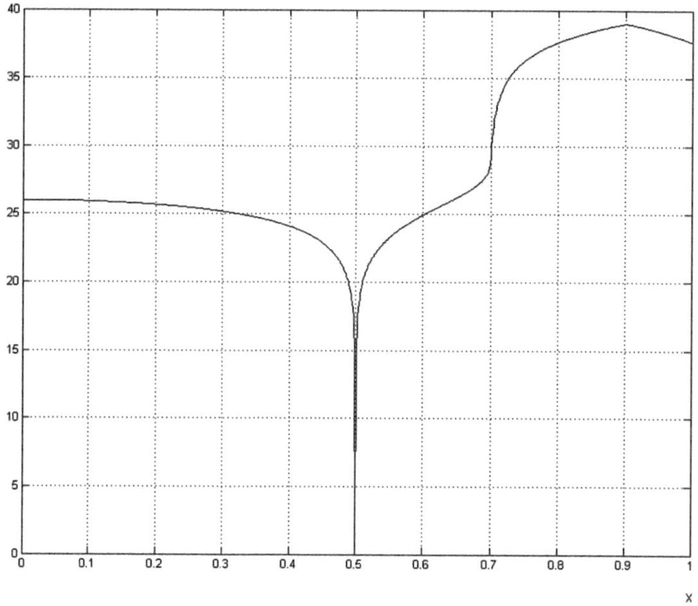

Fig. 7.17 Localization function $\Phi(x_0)$ for the monitoring sites $\mathbf{R} = 0.7$ and $\mathbf{R}_1 = 0.9$

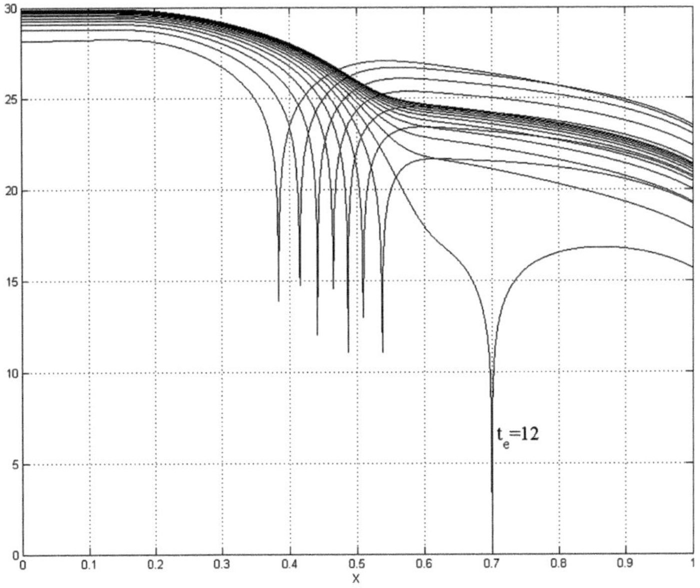

Fig. 7.18 Cross-sections of the localization function $\Phi(x_0, t_e)$ for different t_e. The monitoring sites are $\mathbf{R} = 0.2$ and $\mathbf{R}_1 = 0.5$

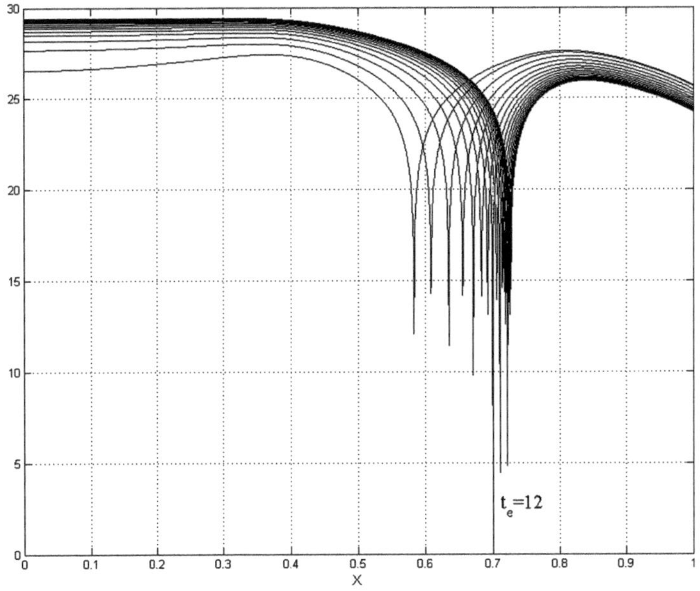

Fig. 7.19 Cross-sections of the localization function $\Phi(x_0, t_e)$ for different t_e. The monitoring sites are $\mathbf{R} = 0.4$ and $\mathbf{R}_1 = 0.8$

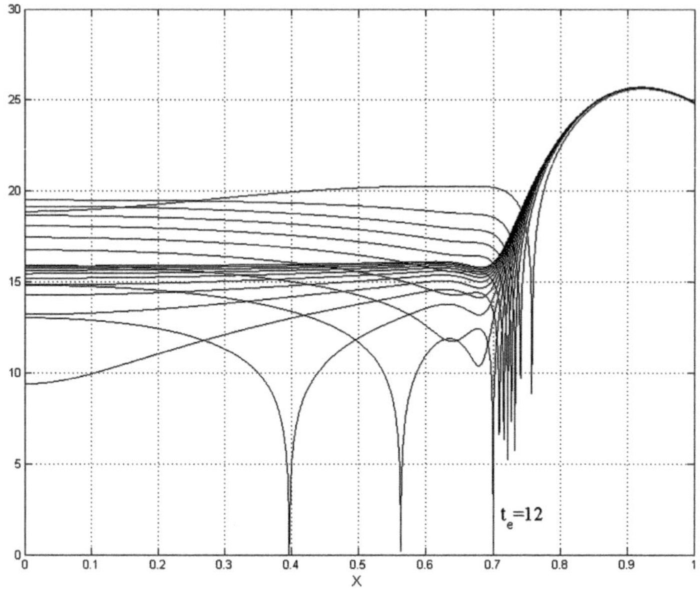

Fig. 7.20 Cross-sections of the localization function $\Phi(x_0, t_e)$ for different t_e. The monitoring sites are $\mathbf{R} = 0.8$ and $\mathbf{R}_1 = 0.9$

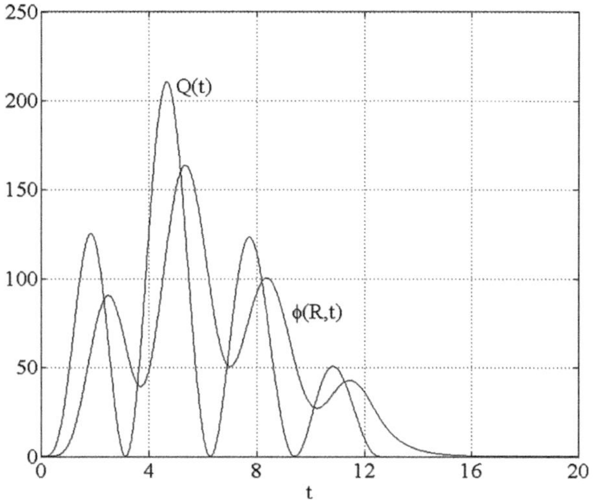

Fig. 7.21 The emission rate $Q(t)$ and the pollutant concentration ϕ at the monitoring site $x = \mathbf{R}$

Since the atmospheric dispersion conditions are stationary, the operators of the direct and adjoint problems do not depend on time. This means that each j-th function $g_j(r_0, t) = g_\alpha(r_0, T - t_j + t)$ in $(0, t_j)$, $(j = 1, 2, \ldots, M)$ is a time shift of the basic kernel $g_\alpha(r, t)$, which is the solution of the adjoint problem under forcing $p(r, t) = \delta(r - \mathbf{R})\delta(t - T)$. The function $g_\alpha(r_0, t)$ and two examples of the adjoint

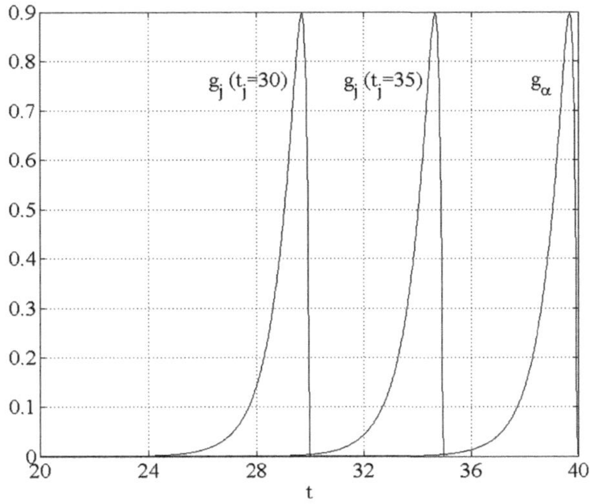

Fig. 7.22 Basic kernel $g_\alpha(r_0, t)$ and adjoint functions $g_j(r_0, t)$ for $t_j = 30$ and $t_j = 35$

functions $g_j(\boldsymbol{r}_0, t)$ are shown in Fig. 7.22. After obtaining the adjoint functions, we can consider the quadratic programming problem (7.7.9)–(7.7.11) for the numerical solution of the inverse problem. The synthetic time series $\{\phi_j\}_{j=1}^{L}$ was generated as $\phi_j(\mathbf{R}) = \phi(\mathbf{R}, t_j) + \delta\phi_j$, where $\delta\phi_j$ were randomly selected from a uniform distribution in the interval $(-0.5, 0.5)$. The error amplitudes were normalized so that they did not exceed 15% of the maximum value of the function $\phi(\mathbf{R}, t)$. Figure 7.23 shows the behavior of the true emission rate $Q(t)$ and three regularized solutions $Q_\varepsilon(t)$ when $\varepsilon = e\varepsilon^*$ approached ε^* from the right, where $\varepsilon^* = \max_j|\delta\phi_j|$, and $e = 2.0$, $e = 1.6$ and $e = 1.002$. Numerical experiments also show that the solution to the quadratic programming problem becomes smoother and tends to zero if $\varepsilon \to +\infty$ (since the feasibility set F_ε expands to such an extent that it contains the global minimum of the objective function). In addition, for $\varepsilon = \varepsilon^*$ we obtain the best approximation to $Q(t)$, which gives the method.

Finally, note that the quadratic programming problem (7.7.9)–(7.7.11) always has a solution since its feasible space is a nonempty compact set in $\mathbb{R}^{L-1}$, and function $J(\boldsymbol{q})$ is continuous. Since $J(\boldsymbol{q})$ is also strictly convex, the solution of problem (7.7.9)–(7.7.11) is unique in the convex feasible space and can be found using the MATLAB *quadprog* routine.

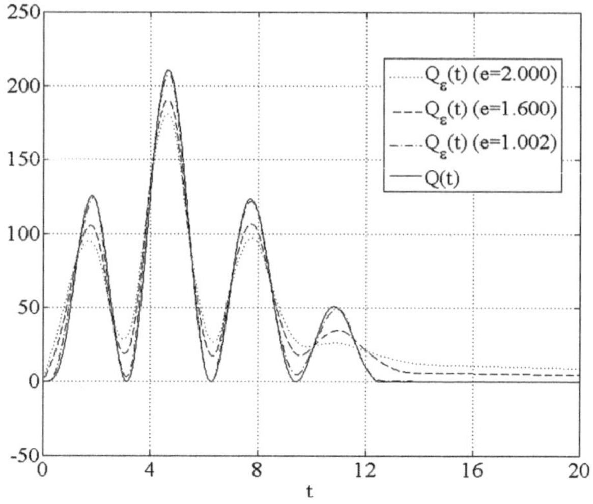

Fig. 7.23 Behavior of regularized solution Q_ε as $\varepsilon \to \varepsilon^*$

References

1. Rao, K.S.: Source estimation methods for atmospheric dispersion. Atmos. Environ. **41**, 6964–6973 (2007)
2. Enting, I.G.: Inverse Problems in Atmospheric Constituent Transport. Cambridge University Press, Cambridge (2002)
3. Groetsch, C.W.: Inverse Problems in the Mathematical Sciences. Vieweg, Braunschweig (1993)
4. Aster, R.C., Borchers, B., Thurber, C.H.: Parameter Estimation and Inverse Problems. Academic Press, Boston (2013)
5. Hadamard, J.: Lectures on Cauchy's Problem in Linear Partial Differential Equations. Yale University Press, New Haven (1923)
6. Tikhonov, A.N., Arsenin, V.Y.: Solution of Ill-Posed Problems. Winston & Sons, Washington (1977)
7. Islam, M.A.: Application of a gaussian plume model to determine the location of an unknown emission source. Water Air Soil Pollut. **112**, 241–245 (1999)
8. Rivera, A.M.: El Cambio Climático. Tercer Milenio (CONACULTA), México (1999)
9. Moreira, D.M., Tirabassi, T., Vilhena, M.T., Carvalho, J.C.: A semi-analytical model for the tritium dispersion simulation in the PBL from the Angra I nuclear power plant. Ecol. Model. **189**, 413–424 (2005)
10. Skiba, Yu.N.: On a method of detecting the industrial plants which violate prescribed emission rates. Ecol. Model. **159**(2–3), 125–132 (2003)
11. Kathirgamanathan, P., McKibbin, R., McLachlan, R.I.: Source release-rate estimation of atmospheric pollution from a non-steady point source-Part1: Source at a known location. Res. Lett. Inf. Math. Sci. **5**, 71–84 (2003)
12. Lushi, E., Stockie, J.M.: An inverse Gaussian plume approach for estimating atmospheric pollutant emissions from multiple sources. Atmos. Environ. **44**, 1097–1107 (2010)
13. Box, G.E.P., Jenkins, G.M., Reinsel, G.C.: Time Series Analysis: Forecasting and Control. Prentice-Hall, Englewood Cliffs (1994)
14. Golestani, A., Gras, R.: Can we predict the unpredictable? Nat. Sci. Reports **4**, 1–6 (2014). Article Number 6834
15. Parra-Guevara, D., Skiba, Yu.N.: Recuperación de la tasa de emisión de una Fuente contaminante: Análisis de la existencia, la unicidad y la estabilidad de las soluciones. Información Tecnológica **27**(5), 251–262 (2016)

16. Fraser, C.G.: The origins of Euler´s variational calculus. Arch. Hist. Exact Sci. **47**(2), 103–141 (1994)
17. Cerdá, E.: Optimización Dinámica. Prentice-Hall, Madrid (2001)
18. Smith, D.R.: Variational Methods in Optimization. Dover Publ., New York (1998)
19. Le, T., Chartrand, R., Asaki, T.: A variational approach to constructing images corrupted by Poisson noise. J. Math. Imaging Vis. **27**, 257–263 (2007)
20. Rudin, L.I., Osher, S., Fatemi, E.: Nonlinear total variation based noise removal algorithms. Physica D **60**, 259–268 (1992)
21. Enting, I.G., Newsam, G.N.: Atmospheric constituent inversion problems: implications for baseline monitoring. J. Atmos. Chem. **11**, 69–87 (1990)
22. Parra-Guevara, D., Skiba, Yu.N.: Elements of the mathematical modelling in the control of pollutants emissions. Ecol. Model. **167**(3), 263–275 (2003)
23. Gasquet, C., Witomski, P.: Fourier Analysis and Applications. Springer-Verlag, NY (1990)
24. Parra-Guevara, D., Skiba, Yu.N.: A strategy for bioremediation of marine shorelines polluted with oil by using several nutrient release points. In: Mathematical Modelling and Numerical Simulation of Oil Pollution Problems. Series "The Reacting Atmosphere", Vol. 2, Ehrhardt, M. (ed.), pp. 25–56. Springer Verlag, Heidelberg (2015)
25. Braun, M.: Differential Equations and Their Applications. Springer-Verlag, NY (1983)
26. Hansen, P.C.: Rank-Deficient and Discrete Ill-Posed Problems: Numerical Aspects of Linear Inversion. SIAM, 3600 University City Science Center Philadelphia (1999)
27. Parra-Guevara, Skiba, Yu.N.: Optimal emission control and identification of an unknown pollution source. J. Phys.: Conf. Series 2090-012142. 17 pp (2021). https://doi.org/10.1088/1742-6596/2090/1/012142.
28. Sharma, L.K., Ghosh, A.K., Nair, R.N., Chopra, M., Sunny, F., Puranik, V.D.: Inverse modeling for aquatic source and transport parameters identification and its application to Fukushima nuclear accident. Environ. Model. Assess. **19**, 193–206 (2014)
29. Venkataraman, P.: Applied Optimization with MATLAB Programming. John Wiley and Sons, New York (2002)
30. Parra-Guevara, D., Skiba, Yu.N., Reyes-Romero, A.: Existence and uniqueness of the regularized solution in the problem of recovery the non-steady emission rate of a point source: application of the adjoint method. In: Rodrigues et al. (eds.) Engineering Optimization IV, pp. 181–186. CRC Press, London (2014)
31. Riesz, F., Sz.-Nagy, B.: Functional Analysis, Dover Publ., New York (1990)
32. Cheney, E.W.: Introduction to Approximation Theory. Chelsea, New York (1966)
33. Godunov, S.K.: Equations of Mathematical Physics. Nauka, Moscow (1971)
34. Broman, A.: Introduction to Partial Differential Equations: From Fourier Series to Boundary-Value Problems. Dover Publ, New York (1989)
35. Kress, R.: Numerical Analysis. Springer-Verlag, New York (1998)
36. Carslaw, H.S.: An Introduction to the Theory of Fourier's series and Integrals. Dover Publ, New York (1952)
37. Fejér, L.: Untersuchungen über Fouriersche reihen. Math. Annalen **58**, 51–69 (1903)

Chapter 8
Numerical Methods for Solving Dispersion Equations

In recent decades, the development of computers and the use of mathematical modeling in the scientific and technical areas caused a revolution in the application of numerical methods. In case of a complicated problem, when it is difficult or impossible to apply analytical, statistical or experimental methods, the use of a numerical method becomes the only way to find a solution. Since a computer can operate only with elements of a finite-dimensional space (numbers, vectors, matrices, etc.), a continuous model should be approximated by a discrete one. For discretizing a continuous problem, one of the finite difference or projection methods can be used. Finally, a system (or systems) of linear algebraic equations is solved using one of the direct or iterative methods.

Obviously, the process of numerical solution introduces various errors. Therefore, a numerical algorithm must be stable with respect to errors. In addition, a numerical method is useful only if the solution of the discrete problem approximates the exact solution of the original continuous problem. In other words, the numerical solution must converge to the exact one when the mesh size tends to zero. For a linear problem, this occurs if the discrete problem properly approximates the original continuous problem (consistency) and the numerical algorithm is stable. This result is known as the Lax equivalence theorem [1]. Therefore, the concepts of approximation, stability and convergence are central ones in the theory of numerical methods.

In this chapter, we will briefly present only some of the basic elements of numerical methods, which can be useful for the numerical solution of various dispersion problems. Section 8.1 explains negative effects of misrepresentation of waves on a grid. Nonlinear instability caused by these effects is discussed in Sect. 8.2. A concept of numerical dispersion is introduced in Sect. 8.3. A relationship between the concepts of approximation, numerical viscosity and stability is considered in Sect. 8.4, while Sect. 8.5 compares the results of using two numerical schemes for the solution of a simple one-dimensional transport problem. The concept of convergence of a numerical solution and the Lax equivalence theorem are given in Sect. 8.6. Section 8.7

Y. N. Skiba and D. Parra Guevara, *Mathematical Methods for the Assessment and Control of Industrial Emissions*, Mathematical Engineering,
https://doi.org/10.1007/978-3-031-58109-0_8

describes the finite difference approximation of the one-dimensional dispersion equation. Section 8.8 provides efficient numerical algorithms for solving complicated multidimensional dispersion models. These algorithms represent a group of powerful and highly efficient methods based on splitting the operator of dispersion model. The main idea is to present the original operator as a sum of operators of more simple structure and split the original complicated problem into a sequence of simpler tasks, called split sub-problems. Besides, each sub-problem is of a type for which simpler and more practical algorithms are available.

Thus, the application of a splitting method allows one to reduce the solution of a complex problem to the successive solution of several simpler problems [1]. A splitting scheme for 2D dispersion equation is given in Sect. 8.9. The efficient Thomas algorithm for the solution of linear systems with a tridiagonal matrix is described in section Sect. 8.10.

8.1 Misrepresentation of Waves on a Grid

We now consider two examples to show that the false representation of waves on a mesh is something typical [2, 3].

Example 8.1.1 Let $i = \sqrt{-1}$, and let x_j be a point of a regular grid of size $h = x_j - x_{j-1}$. We apply the formula $\exp(i\sigma) = \cos\sigma + i\sin\sigma$ and write

$$\exp(ikjh) = \{\exp(i2m\pi)\}^j \exp(ikjh) = \exp\{ijh(k + 2m\pi/h)\},$$

where k and m are any two integers. Thus, it is impossible to distinguish the waves $\exp(ikx)$ and $\exp\{i(k + 2m\pi/h)x\}$ given at grid points jh. Obviously, they have different periods.

When a function is given only by its values at the nodes of a grid, it is very easy to make a wrong conclusion about the behavior of the function. In particular, it is difficult to determine the period of a periodic function. For example, Figs. 8.1 and 8.2 show that the two sine curves $\sin x$ and $\sin 5x$ have the same values at the grid nodes.

Example 8.1.2 Let n be an integer, and let $x_i = ih \quad (i = 0, 1, 2, \ldots, n+1)$ be the points of a regular grid of size h. Let us consider the two sinusoids

$$\sin\frac{2\pi}{(n+1)h}x \quad \text{and} \quad -\sin\frac{2\pi n}{(n+1)h}x$$

whose periods are $(n+1)h$ and $[(n+1)/n]h$, respectively. It is easy to show that at each grid point, these sinusoids have the same values: $f(x_i) = \sin[2\pi\, i/(n+1)]$, $(i = 0, 1, 2, \ldots, n+1)$. Indeed, let $K = \pi/h$ be a wavenumber of the shortest sinusoid

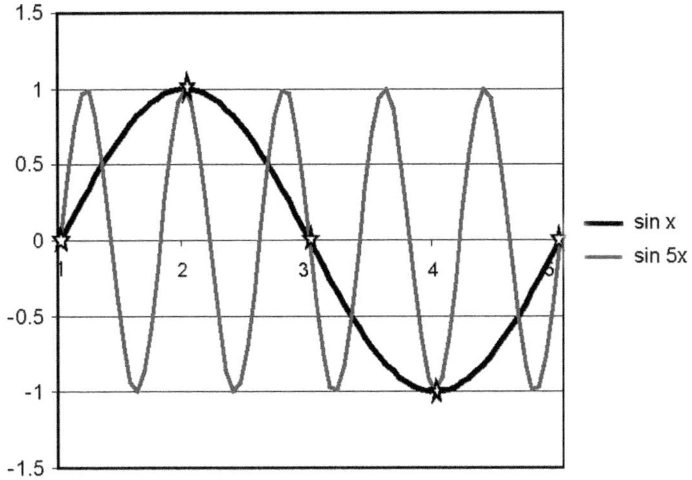

Fig. 8.1 Graphics of functions $\sin x$ and $\sin 5x$ (regular grid)

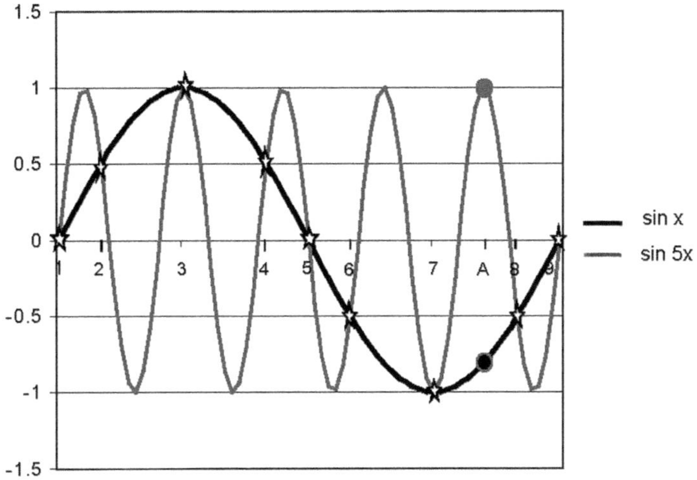

Fig. 8.2 Graphics of functions $\sin x$ and $\sin 5x$ (irregular grid)

that can be represented on the mesh (its period is equal to $2h$). Substitution of K in the identity $\sin kx \equiv \sin[2K - (2K - k)]x$, and the use of formula

$$\sin(\alpha - \beta) = \sin\alpha\cos\beta - \cos\alpha\sin\beta$$

leads to

$$\sin kx = \sin \frac{2\pi}{h} x \cos(\frac{2\pi}{h} - k)x - \cos \frac{2\pi}{h} x \sin(\frac{2\pi}{h} - k)x$$

Since

$$\sin \frac{2\pi}{h} ih = 0 \quad \text{and} \quad \cos \frac{2\pi}{h} ih = 1$$

at $x_i = ih$, we get

$$\text{sin}kih \ = -\ \text{sin}k^* ih \tag{8.1.1}$$

where

$$k^* = 2K - k \tag{8.1.2}$$

It follows from (8.1.2) that if the values of the sinusoids $\sin kx$ and $\sin(2K - k)]x$ are known only at the grid points, then it is impossible to distinguish these sinusoids from each other. There fore, the sinusoid $\sin kx$ can be misinterpreted as the sinusoid $\sin(2K - k)]x$ and vice versa. In particular, if

$$k = \frac{2\pi}{[(n+1)h]} \quad \text{then} \quad k^* = \frac{2\pi}{h} - \frac{2\pi}{(n+1)h} = 2\pi \cdot \left(\frac{n+1}{n}\right)^{-1}, \tag{8.1.3}$$

that is, the sinusoid $\sin kih$ with period $(n+1)h$ can be misinterpreted as the sinusoid $-\sin k^* ih$ with period $[(n+1)/n]h$, and vice versa. For $n = 3$, both sinusoids are shown in Fig. 8.3 [4].

Errors in Fourier series. A smooth odd function on the interval $[-\pi, \pi]$ can be represented by Fourier series:

$$u(x) = \sum_{k=1}^{\infty} a_k \sin kx \tag{8.1.4}$$

If this function is defined only at the grid points x_i, we have

$$u(x_i) = \sum_{k=1}^{\infty} a_k \sin kx_i \tag{8.1.5}$$

In this case, due to the misrepresentation of waves, it is not always possible to correctly determine the amplitudes a_k and, as a result, to correctly identify the wave, whose energy is dominant (whose amplitude is the largest one).

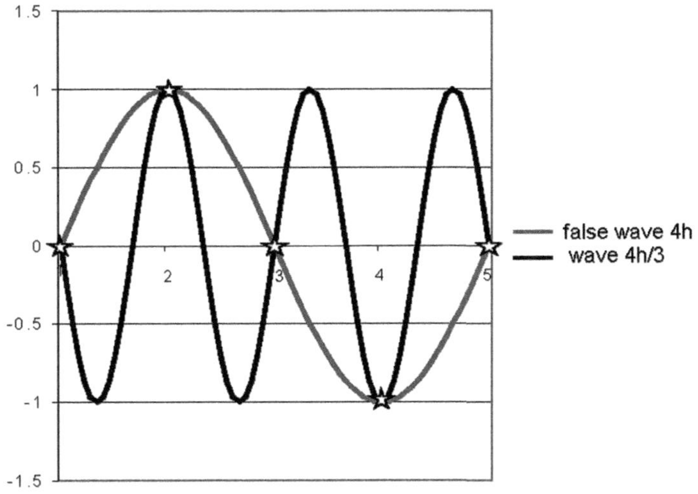

Fig. 8.3 Misrepresentation of wave with period $T = 4h/3$ by the wave with period $T = 4h$

8.2 Nonlinear Instability

Consider now the nonlinear equation

$$u_t + uu_x = 0 \tag{8.2.1}$$

This is a very simple (one-dimensional) case of the Euler equations for an ideal fluid. We now show that the presence of nonlinear term in (8.2.1) causes instability. Indeed, let us search the problem solution in the form of a truncated Fourier series (spectral method):

$$u(t, x) = \sum_{k=1}^{K} a_k(t) \sin kx, \tag{8.2.2}$$

where the sine wave with the maximum wavenumber $K = \pi/h$ is the shortest that can be represented on the mesh. Then the term uu_x can be written as the double sum:

$$uu_x = \sum_{m=1}^{K} \sum_{n=1}^{K} na_m(t)a_n(t) \sin nx \cdot \cos mx \tag{8.2.3}$$

Since

$$\sin nx \cdot \cos mx = \frac{1}{2}[\sin(n + m)x + \sin(n - m)x], \tag{8.2.4}$$

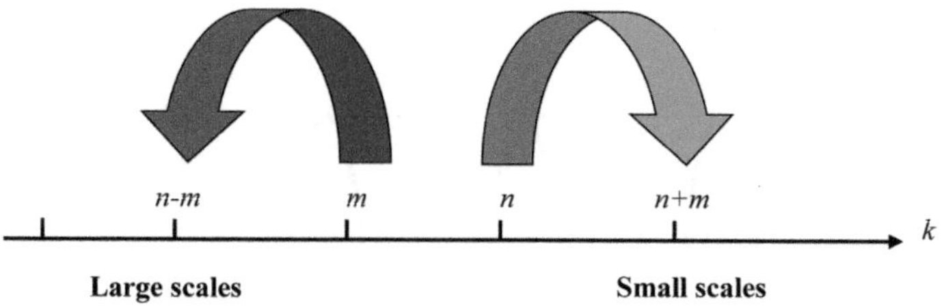

Fig. 8.4 Two energy cascades

the sinusoids $\sin nx$ and $\cos mx$ generate two new sinusoids whose wavenumbers are $n + m$ (greater than n and m) and $n - m$ (less than n, and sometimes less than m). Thus, a perturbation of energy in one part of wavenumbers will generate the cascades of energy in two directions simultaneously: to larger and smaller scales with respect to the scales of initial perturbations (Fig. 8.4).

In spectral methods, when the Fourier series are truncated by an integer $K = \pi/h$, the initial value $u(0, x)$ of solution (8.2.2) contains only sinusoids $\sin kx$ with $k \leq K$. However, due to the nonlinear interaction of waves, sinusoids with wavenumbers $k > K$ will appear, which will be represented, according to (8.1.1) and (8.1.2) as the sinusoids with wavenumbers $k^* = 2K - k < K$. As a result, there appears a false cascade of energy to the spectral interval $(0, K)$ from the spectral interval $k > K$. In particular, when $n = K$ and m is a small number, the formula (8.2.4) generates the waves $\sin(K+m)x$ and $\sin(K-m)x$. The wavenumber $K+m$ of sinusoid $\sin(K+m)x$ is slightly larger than K and such a wave is not taken into account by the formula (8.2.2). However, due to (8.1.1) and (8.1.2), its energy is falsely represented as the energy of the wave $\sin(K-m)x$ that is already included in the series (8.2.2) (Fig. 8.5). Thus, there appears a false cascade of energy from the spectral interval $k > K$ to a spectral interval $[K - \delta, K]$ located near the limit number K. Since there is also a cascade of energy to the interval $[K - \delta, K]$ from the larger scales, these two cascades lead to a false accumulation of energy in the spectral interval $[K - \delta, K]$ (Fig. 8.6). This instability mechanism is known as nonlinear instability.

Fig. 8.5 Cascades of energy nearby the truncation number K due to false representation of waves with wavenumbers $k > K$

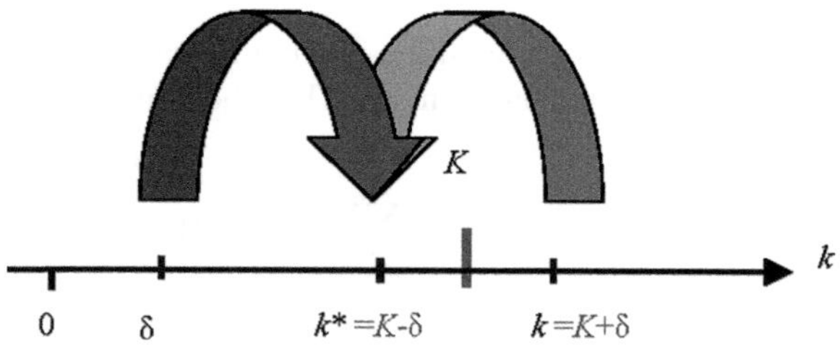

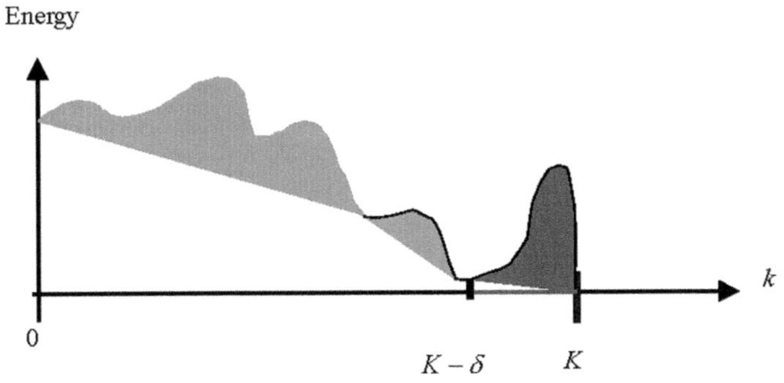

Fig. 8.6 Artificial accumulation of energy in $[K - \delta, K]$ due to false representation of waves with wavenumbers $k > K$

8.3 Numerical Dispersion

We now describe one more negative effect related to misrepresentation of waves on a mesh. In the problem of signal transfer, this effect causes dispersion. Consider the one-dimensional transport equation

$$\frac{\partial \phi}{\partial t} + u\frac{\partial \phi}{\partial x} = 0 \tag{8.3.1}$$

with a constant velocity u in the periodic channel $D = \{0 < x < 1; 0 < t < 1\}$, i.e., $\phi(1, t) = \phi(0, t)$. Suppose that at the initial moment $t = 0$ the solution satisfies the condition

$$\phi(x, 0) = f(x) \tag{8.3.2}$$

where $f(x)$ is a given differentiable function. The exact solution of problem (8.3.1–8.3.2) is

$$\phi(x, t) = f(x - ut) \tag{8.3.3}$$

Thus, along each characteristic line $x - ut = x_0$, the solution $\phi(x, t)$ is invariable:

$$\phi(x, t) = f(x_0) = Const$$

Due to (8.3.3), the initial signal $f(x)$ will not change its original form in the process of propagation along the axis x. In particular, if $f(x)$ is a wave then (8.3.3) describes the propagation of the wave with velocity u.

Since

$$\frac{d}{dx}\mathrm{e}^{\mathbf{i}kx} = \mathbf{i}k\mathrm{e}^{\mathbf{i}kx} \tag{8.3.4}$$

Equation (8.3.1) has a solution in the form of a wave $\phi(x,t) = \mathrm{Re}\{\Phi(t)\exp(\mathbf{i}kx)\}$ where $\Phi(t)$ satisfies the oscillation equation

$$\frac{d\Phi}{dt} + \mathbf{i}ku\Phi = 0$$

with the frequency $\omega(k) = ku$ and invariant phase velocity

$$v(k) = \frac{\omega(k)}{k} = u = Const \tag{8.3.5}$$

Therefore, all the wave solutions $\phi(x,t) = \mathrm{Re}\{\Phi(t)\exp(\mathbf{i}kx)\}$, independently of wavenumber k, propagate at the same velocity. Since the initial solution $\phi(x,0)$ is represented by the Fourier series of sinusoidal waves $\exp(\mathbf{i}kx)$, this explains why the analytical solution $\phi(x,t)$ propagates without any change in its initial shape.

Let us now analyze the propagation of the wave solution $\xi_j(t) = \mathrm{Re}\{\Phi(t)\exp(\mathbf{i}kjh)\}$ of the semi-discrete equation

$$\frac{d\xi_j}{dt} + u\frac{\xi_{j+1} - \xi_{j-1}}{2h} = 0 \tag{8.3.6}$$

which approximates Eq. (8.3.1) at the point x_j of spatial grid $x_i = ih$ with size h ($i = 0, 1, 2, \ldots, K$; $x_0 = 0$ and $x_K = 1$). The derivative $\partial\phi/\partial x$ is approximated in (8.3.6) by the central finite difference [4], and $\xi_j(t)$ denotes the solution value at x_j.

Unlike the continuous case (see (8.3.4)), the discrete derivative of the wave $\exp(\mathbf{i}kx)$ changes its form:

$$\frac{\exp\{\mathbf{i}k(j+1)h - \exp\{\mathbf{i}k(j-1)h}{2h} = \mathbf{i}\frac{\sin(kh)}{h}\exp\{\mathbf{i}kjh\}, \tag{8.3.7}$$

and hence, $\Phi(t)$ satisfies the oscillation equation

$$\frac{d\Phi}{dt} + \mathbf{i}k\left(u\frac{\sin kh}{kh}\right)\Phi = 0 \tag{8.3.8}$$

with the frequency

$$\omega_h(k) = kv_h(k) = k\left(u\frac{\sin kh}{kh}\right)$$

Thus, unlike the continuous case, where the phase velocity $v(k) = u$ does not depend on the wavenumber k, the waves of semi-discrete problem move with the phase velocity

$$v_h(k) = \frac{\omega_h(k)}{k} = u\frac{\sin kh}{kh} \tag{8.3.9}$$

that depends on k. Note that $|v_h(k)| < |u|$ for any k, i.e., in the semi-discrete problem, the velocity $v_h(k)$ of each wave is smaller than the velocity u of the waves in the continuous problem, besides, the value of $v_h(k)$ decreases with the wave length (as k increases). For example, if $kh << 1$ (long waves on a fine grid) then $v_h(k) \approx u[1 - \frac{1}{6}(kh)^2]$, and the error in the phase velocity is of second order with respect to kh. However, this error is rather large for the short waves, besides, the wave with the length of $2h$ (double mesh size) is motionless. Indeed, then $\xi_{j+1} = \xi_{j-1}$ by periodicity, and $d\xi_j/dt = 0$. Note that the error in the group velocity is even larger, since $\frac{\partial \omega_h(k)}{\partial k} = u \cos kh$. This phenomenon is called numerical dispersion of waves of different scales [2, 5].

Example 8.3.1 Let us explain why the scheme

$$\frac{\xi_i^{n+1} - \xi_i^n}{\tau} = -u \frac{\xi_{i+1}^n - \xi_{i-1}^n}{2h} \tag{8.3.10}$$

transmits the signal incorrectly. Suppose that the boundary conditions are periodic: $\xi_N^n = \xi_0^n$. Then, summing (8.3.10) over all points i, we arrive at the equation

$$\sum_{i=1}^{I} \xi_i^{n+1} = \sum_{i=1}^{I} \xi_i^n, \tag{8.3.11}$$

that is, like for the exact solution of Eq. (8.3.1), the mean value of the numerical solution is preserved in time. However, the scheme (8.3.11) significantly alters the local transfer along the x-axis. Indeed, assume that for some n, $\xi_m^n = \delta > 0$, and $\xi_i^n = 0$ for all $i \neq m$. Then for $i = m + 1$, Eq. (8.3.10) accepts the form

$$\frac{\xi_{m+1}^{n+1} - \xi_{m+1}^n}{\tau} = -u \frac{(0 - \delta)}{2h} = \frac{u\delta}{2h}$$

and the zero value $\xi_{\mathrm{m+1}}^n$ at the point (x_{m+1}, t_n) is replaced by the positive value

$$\xi_{\mathrm{m+1}}^{n+1} = \frac{\tau u}{2h} \delta \tag{8.3.12}$$

at the next moment t_{n+1}. It is reasonable and acceptable. However, for $i = m$, we obtain an unreasonable result:

$$\frac{\xi_m^{n+1} - \xi_m^n}{\tau} = -u \frac{0 - 0}{2h} = 0 \quad \text{and} \quad \xi_m^{\mathrm{n+1}} = \xi_m^n = \delta \tag{8.3.13}$$

Thus, the value of solution at the point x_m does not change with time (from t_n to t_{n+1}), as it should be according to (8.3.3), because the solution has the same value along each characteristic line (see Fig. 2.1). And when $i = m - 1$, the result is even worse:

$$\frac{1}{\tau}\left(\xi_{m-1}^{n+1} - \xi_{m-1}^{n}\right) = -\frac{u\delta}{2h},$$

and hence, instead of $\xi_{m-1}^{n+1} = 0$ (expected result by the method of characteristics), the scheme gives the negative value

$$\xi_{m-1}^{n+1} = -\frac{\tau u}{2h}\delta \tag{8.3.14}$$

This is unacceptable. Note that the values $\xi_{m-1}^{n+1} = -\frac{\tau u}{2h}\delta$ and $\xi_{m+1}^{n+1} = \frac{\tau u}{2h}\delta$ decrease with $\tau u/h$. Thus, unlike the exact solution moving to the right ($u > 0$) without changing its form, the initial signal $\xi_m^n = \delta$ is transfered by the scheme (8.3.10) in both directions showing several alterations including false oscillations with negative values.

Suppose that the initial signal is non-negative: $\phi(x, 0) = f(x) \geq 0$. We see that the scheme (8.3.10) is not monotonic, since over time it generates negative values (Fig. 8.7).

In addition, the scheme does not transfer a signal correctly. We showed that both undesirable effects, known as numerical dispersion, are due to the approximation of the convective derivative $u\partial\phi/\partial x$ by means of the central difference formula.

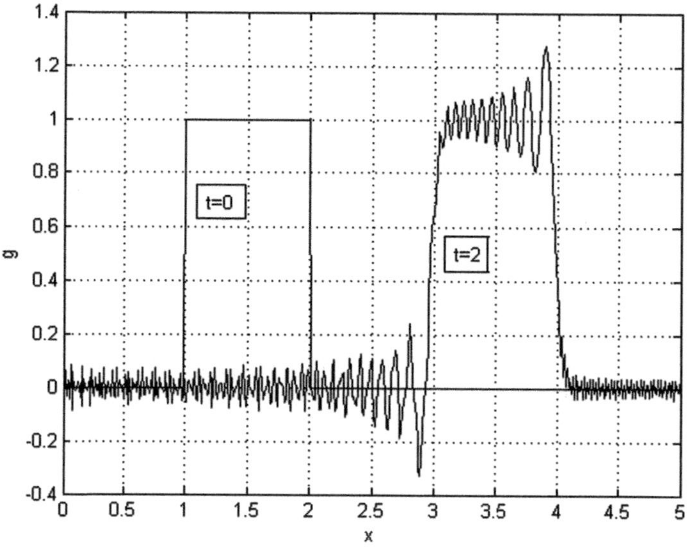

Fig. 8.7 Effect of numerical dispersion

8.4 Approximation, Numerical Viscosity and Stability

In this section, the transport Eq. (8.3.1) is used to show a relashinship that exists between the approximation of the original problem, numerical viscosity and stability of the scheme.

1. **Approximation and numerical viscosity**. We now consider the Godunov scheme [6]

$$\frac{\xi_i^{n+1} - \xi_i^n}{\tau} = -u\frac{\xi_i^n - \xi_{i-1}^n}{h} \tag{8.4.1}$$

that approximates Eq. (8.3.1) with accuracy $O(\tau + h)$. Analyzing this scheme in the same way as the scheme (8.3.6), we obtain

$$\begin{gathered} \xi_{m+1}^n = 0, \quad \xi_{m+1}^{n+1} = \tfrac{\tau u}{h}\delta \\ \xi_m^n = \delta, \quad \xi_m^{n+1} = \left(1 - \tfrac{\tau u}{h}\right)\delta \\ \xi_{m-1}^n = 0, \quad \xi_{m-1}^{n+1} = 0 \end{gathered}$$

The result for ξ_m^{n+1} indicates that the disturbance was moved from its initial location, while the results for ξ_{m+1}^{n+1} and ξ_{m-1}^{n+1} show that the signal is transferred in the correct direction. Thus, the Godunov scheme is monotonic and transfers the signal correctly. The approximation of the transport Eq. (8.3.1) by this scheme is of the first order with respect to τ and h. Indeed, the Taylor series, as applied in a vicinity of point (x_i, t_n) to the terms ξ_i^{n+1} and ξ_{i-1}^n in (8.4.1) leads to

$$\frac{\partial\phi}{\partial t} + u\frac{\partial\phi}{\partial x} = \frac{uh}{2}\frac{\partial^2\phi}{\partial x^2} - \frac{\tau}{2}\frac{\partial^2\phi}{\partial t^2} + O\left(\tau^2 + h^2\right) \tag{8.4.2}$$

Meanwhile, one can observe that due to (8.3.1),

$$\frac{\partial^2\phi}{\partial t^2} = u^2\frac{\partial^2\phi}{\partial x^2}, \tag{8.4.3}$$

and the Godunov scheme approximates with a higher precision, $O\left(\tau^2 + h^2\right)$, the advection–diffusion equation

$$\frac{\partial\phi}{\partial t} + u\frac{\partial\phi}{\partial x} = \mu\frac{\partial^2\phi}{\partial x^2} \tag{8.4.4}$$

where

$$\mu = \frac{u}{2}(h - u\tau) = \frac{uh}{2}\left(1 - \frac{\tau u}{h}\right) \tag{8.4.5}$$

is the numerical viscosity coefficient. Thus, Godunov's scheme (8.4.1) contains artificial (numerical) viscosity.

If $\tau u/h < 1$ then $\mu > 0$, and the Cauchy problem for Eq. (8.4.4) is well posed in the sense of Hadamard, that is, its solution exists, is unique and continuously depends on the initial conditions. The last property manifests the stability of solution. And it is well known that the numerical viscosity is a factor that reduces the amplitude of the numerical solution obtained with the scheme (8.4.1) [7].

If $\tau u/h = 1$ then $\mu = 0$, and the term $O(\tau^2 + h^2)$ in Eq. (8.4.2) is also equal to zero identically. Thus, the scheme (8.4.1) exactly approximates Eq. (8.3.1) if $\tau u/h = 1$, i.e., the numerical solution coincides with the exact solution of Eq. (8.3.1). Apparently, such precision is achieved only when $u = Const$. In the theory of the numerical methods, the condition $\tau u/h \leq 1$, representing a relation between the velocity and mesh sizes h and τ is called Courant condition, or the condition by Courant-Fredrichs-Lewy [5, 7].

Finally, if $\tau u/h > 1$ then scheme (8.4.1) approximates the equation

$$\frac{\partial \phi}{\partial t} + u\frac{\partial \phi}{\partial x} = -|\mu|\frac{\partial^2 \phi}{\partial x^2} \tag{8.4.6}$$

with a negative viscosity coefficient. It is easy to show that the Cauchy problem for Eq. (8.4.6) is ill-posed in the sense of Hadamard, because its solution is unstable with respect to initial errors.

Thus, there is a logical connection between a good approximation of the problem and the Courant condition, and to resolve well the transport Eq. (8.3.1) with the scheme (8.4.1), it is necessary to choose the mesh sizes h and τ such that the Courant condition $\tau u/h \leq 1$ is satisfied. We now show that the same condition guarantees the stability of the Godunov scheme.

2. **Courant Condition and Stability**. Let us investigate the spectral stability of the scheme (8.4.1). The spectral problem

$$u\frac{\omega_k^p - \omega_{k-1}^p}{h} = \lambda_p \omega_k^p \tag{8.4.7}$$

for the operator of the Godunov scheme has the eigenfunctions

$$\omega_k^p = \mathrm{e}^{\mathbf{i}kph} \tag{8.4.8}$$

and eigenvalues

$$\lambda_p = \frac{u}{h}\left(2\sin^2\frac{ph}{2} + \mathbf{i}\cdot\sin(ph)\right) \tag{8.4.9}$$

[1]. We will seek the solution of Eq. (8.3.1) as follows

$$\xi_k^j = \sum_{p=-\infty}^{\infty} \zeta_p^j e^{ikph} \tag{8.4.10}$$

where k and j are indicators of the space–time grid, and the Fourier coefficient ζ_p^j of the solution satisfies the equation

$$\frac{\zeta_p^{j+1} - \zeta_p^j}{\tau} + \lambda_p \zeta_p^j = 0 \tag{8.4.11}$$

Thus,

$$\zeta_p^{j+1} = T_p \zeta_p^j = \left(1 - \tau\lambda_p\right)\zeta_p^j \tag{8.4.12}$$

We now show that the Godunov scheme is conditionally stable, namely, it is stable if the number $r = \tau u/h$ satisfies the Courant condition $r \leq 1$. Indeed, due to (8.4.12), it is necessary to show that

$$\left|1 - \tau\lambda_p\right| \leq 1 \tag{8.4.13}$$

Denoting $\alpha = ph/2$ and using (8.4.9) we obtain

$$\begin{aligned} \left|1 - \tau\lambda_p\right|^2 &= \left(1 - 2r\sin^2\alpha\right)^2 + r^2\sin^2 2\alpha = \\ &= 1 - 4r\left(\sin^2\alpha\right) + r^2\left(4\sin^4\alpha + \sin^2 2\alpha\right) = \\ &= 1 - 4r\sin^2\alpha + 4r^2\sin^2\alpha\left(\sin^2\alpha + \cos^2\alpha\right) = \\ &= 1 - 4r(1 - r)\sin^2\alpha \end{aligned}$$

The term $1-4r(1 - r)\sin^2\alpha$ is always non-negative, since $\underset{0<x<1}{max}\{x(1 - x)\} = 1/4$ and it is less than one if $r \leq 1$. Thus $\left|1 - \tau\lambda_p\right| \leq 1$, and the Godunov scheme is stable if $r \leq 1$. Note that the same condition guarantees the stability in some norms (see Example 8.4.1).

The following scheme illustrates the logical connection between the approximation and stability of a scheme.

3. **Numerical Viscosity and Stability**. We now consider the scheme

$$\frac{\xi_i^{n+1} - \xi_i^n}{\tau} + u\frac{\xi_i^{n+1} - \xi_{i-1}^{n+1}}{h} = 0 \tag{8.4.14}$$

Unlike Godunov's scheme (8.4.1), scheme (8.4.14) is called implicit, since it contains more than one unknown terms (ξ_i^{n+1} and ξ_{i-1}^{n+1}), and hence, in order to find its solution at a moment t_{n+1}, a system of linear equations $A\vec{x} = \vec{b}$ must be solved. It is easy to show that for the transport Eq. (8.3.1), the scheme (8.4.14) is first order accurate in both space and time. At the same time, this scheme is of second-order approximation for the advection–diffusion equation

$$\frac{\partial \phi}{\partial t} + u\frac{\partial \phi}{\partial x} = \mu\frac{\partial^2 \phi}{\partial x^2} \tag{8.4.15}$$

where

$$\mu = \frac{uh + u^2\tau}{2} > 0 \tag{8.4.16}$$

is the viscosity coefficient. The Cauchy problem for Eq. (8.4.15) is well posed in the sense of Hadamard, and scheme (8.4.14) is unconditionally stable. Unlike (8.4.14), the Godunov scheme is explicit and conditionally stable (only for $\tau u/h \le 1$).

Example 8.4.1 We now analyze the stability of the Godunov scheme (8.4.1) using two different norms. We have

$$\xi_i^{n+1} = (1 - r)\xi_i^n + r\xi_{i-1}^n \tag{8.4.17}$$

where $r = \tau u/h$ and $u = Const$ is the velocity. If $r \le 1$ then

$$\left|\xi_i^{n+1}\right| \le (1 - r)\left|\xi_i^n\right| + r\left|\xi_{i-1}^n\right| \tag{8.4.18}$$

and Godunov scheme is stable in the norm $\|\xi^n\|_\infty = \max_i\left|\xi_i^n\right|$:

$$\left\|\xi^{n+1}\right\|_\infty \le \left\|\xi^n\right\|_\infty \tag{8.4.19}$$

It is a direct and simple method to verify stability. However, it has limited application.

We now analyze the stability of the scheme in another norm that allows studying some nonlinear schemes as well. To do this, we square the expression (8.4.18) and sum the result over all grid points. Then we obtain

$$\sum_{i=1}^{I}\left(\xi_i^{n+1}\right)^2 = \sum_{i=1}^{I}\left[(1-r)^2(\xi_i^n)^2 + 2r(1-r)\xi_i^n\xi_{i-1}^n + r^2(\xi_{i-1}^n)^2\right] \tag{8.4.20}$$

Assuming the periodic boundary conditions, we obtain

$$\sum_{i=1}^{I}\left(\xi_{i-1}^n\right)^2 = \sum_{i=1}^{I}\left(\xi_i^n\right)^2 \tag{8.4.21}$$

Using the condition $r \le 1$, formula (8.4.21) and the Schwarz inequality

$$\sum_{i=1}^{I} \xi_{i-1}^{n} \xi_{i}^{n} \leq \left\{ \sum_{i=1}^{I} (\xi_{i}^{n})^{2} \sum_{i=1}^{I} (\xi_{i-1}^{n})^{2} \right\}^{1/2} = \sum_{i=1}^{I} (\xi_{i}^{n})^{2} \tag{8.4.22}$$

we get [5, 8]:

$$\sum_{i=1}^{I} (\xi_{i}^{n+1})^{2} \leq \left[(1-r)^{2} + 2r(1-r) + r^{2} \right] \sum_{i=1}^{I} (\xi_{i}^{n})^{2} \tag{8.4.23}$$

or

$$\left\| \xi^{n+1} \right\|_{2} \leq \left\| \xi^{n} \right\|_{2} \tag{8.4.24}$$

where

$$\left\| u^{n} \right\|_{2} = \left(\sum_{i=1}^{I} (u_{i}^{n})^{2} \right)^{1/2} \tag{8.4.25}$$

is the Euclidean norm of solution. Thus with periodic boundary conditions, the Courant condition $r = \tau u/h \leq 1$ guarantees stability of the scheme (8.4.17) in the norm (8.4.25).

8.5 Example: Comparison of Two Numerical Schemes

Let us consider one-dimensional transport problem

$$\frac{\partial}{\partial t}\phi + u\frac{\partial}{\partial x}\phi = 0, \quad \phi(x, 0) = g(x) \tag{8.5.1}$$

in a periodical channel, where $u > 0$ is a constant velocity.

We now compare the numerical solutions of problem (8.5.1) obtained by the Godunov scheme and Galerkin method to show negative role of the numerical viscosity and Gibbs effect.

1. **Godunov's method**. As an initial condition we take two functions: The gaussian pulse

$$g(x) = \exp\{-(x - 2.5)^{2}/2\} \tag{8.5.2}$$

and the rectangular pulse

$$g(x) = \begin{cases} 1, & \text{if } x \in [0, 5] \\ 0, & \text{otherwise} \end{cases} \tag{8.5.3}$$

First, we solve problem (8.5.1) using the Godunov scheme:

$$\xi_i^{n+1} = (1 - r)\xi_i^n + r\xi_{i-1}^n, \quad \xi_i^0 = g_i, \tag{8.5.4}$$

where $r = \tau u/h \leq 1$. Figures 8.8 and 8.9 show the evolution of Gaussian pulse (8.5.2) and rectangular pulse (8.5.3) at different times. We used a grid with $\tau = 0.001$ and $h = 0.05$, and therefore $r = 0.02$. The drawback of Godunov scheme is a strong damping due to numerical viscosity. As it is seen from Fig. 8.8, the damping reduces the amplitude of the solution by 10% at $t = 5$.

Figure 8.9 shows that the damping in the case of initial rectangular pulse is not as drastic as in the case of Gaussian pulse, remaining less than 10% even up to $t = 40$. Another important feature is that the solution form is reasonably conserved if $t \leq 10$. Also, Fig. 8.10 shows that there is no appreciable difference in the phase velocities of the analytical and numerical solutions.

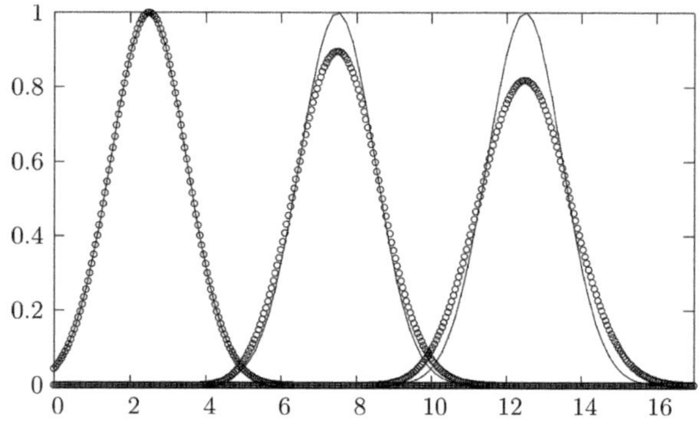

Fig. 8.8 Godunov's scheme ($u = 1$, $\tau = 0.001$, $h = 0.05$, $r = 0.02$). Exact solution (solid line) and numerical solution for $t = 0$, $t = 5$ and $t = 10$;$g(x)$ is the Gaussian pulse (8.5.2)

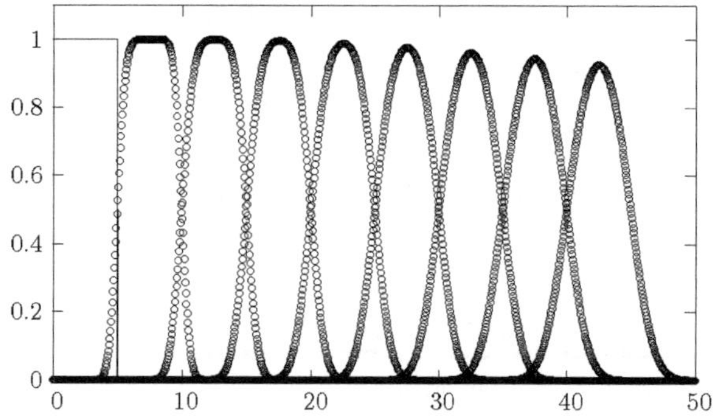

Fig. 8.9 Godunov's scheme ($u = 1$, $\tau = 0.001$, $h = 0.05$, $r = 0.02$). The graph shows the initial condition and the solution obtained at $t = 5n$ where $n = 1, 2, ..., 8$;$g(x)$ is the step function (8.5.3)

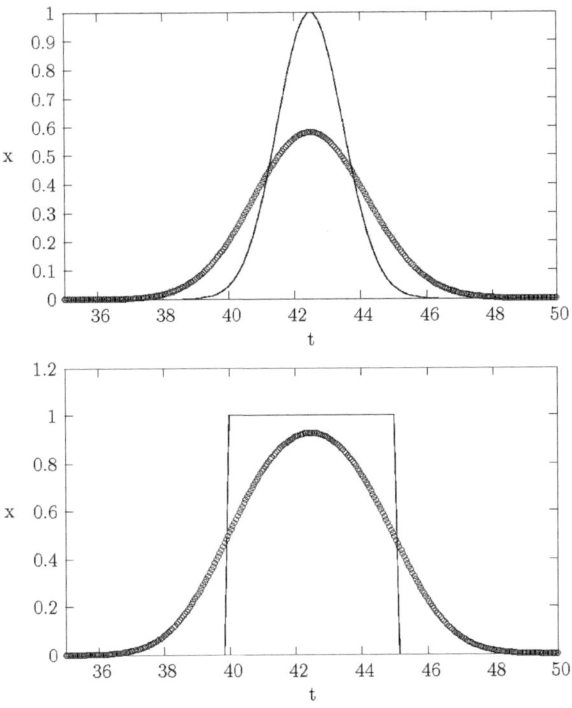

Fig. 8.10 The Gaussian and rectangular pulses approximated by Godunov scheme for $t = 40$. Although in both cases the damping is severe, the phase of the numerical solution still matches with that of the analytical solution

2. **Galerkin Method**. We now solve the problem (8.5.1) by the Galerkin method. A numerical (approximate) solution $\phi(x, t)$ will be sought in the form of a truncated fourier series:

$$\phi(x, t) = \sum_{j=1}^{N} v_j \varphi_j(x - ut), \tag{8.5.5}$$

where $\{\varphi_1(x), \varphi_2(x), ..., \varphi_N(x)\}$ are the basic functions, and v_j are unknown coefficients. This choice satisfies Eq. (8.5.1) identically, and hence, the only problem is to determine the coefficients v_j in order to approximate well the initial condition

$$\phi(x, 0) = \sum_{j=1}^{N} v_j \varphi_j(x) = g(x) \tag{8.5.6}$$

Taking the inner product of both sides of Eq. (8.5.6) with each of the basic functions $\varphi_i(x)$ leads to a linear system

$$A\vec{v} = \vec{f} \tag{8.5.7}$$

where $\vec{v} = \{v_1, v_2, ..., v_N\}^T, \vec{f} = \{f_1, f_2, ..., f_N\}^T, A = \{a_{ij}\}$,

$$f_i = (g(x), \varphi_i(x)), \quad i = 1, 2, ..., N, \tag{8.5.8}$$

$$a_{ij} = (\varphi_i(x), \varphi_j(x)), i, j = 1, 2, ..., N. \tag{8.5.9}$$

Solving the system (8.5.7) we find v_j and therefore the approximate solution (8.5.5).

Gaussian pulse. As the initial condition we take the Gaussian pulse:

$$g(x) = 10 \exp\{-(x - 1.5)^2/2\} \tag{8.5.10}$$

The approximations of the original function (8.5.10) obtained for different truncation numbers N of the Fourier series are shown in Fig. 8.11. The method is highly accurate even with fewer terms in the expansion. Actually, the approximation with only three terms (Fig. 8.11a) gives large fluctuations on both sides of the pulse and damping. The method shows its precisión, since these effects disappear almost completely with only six terms (Fig. 8.11b), and the damping is practically negligible if 9 terms are used (Fig. 8.11c). Obviously, the approximate solution $\phi(x, t)$ at any time $t > 0$ is obtained by formula (8.5.5), that is, the initial graphic is moved with velocity u without changing its shape.

Rectangular pulse. As the initial condition, we now take a rectangular pulse

$$g(x) = \begin{cases} 10, & \text{if } x \in [0, 3] \\ 0, & \text{otherwise} \end{cases} \tag{8.5.11}$$

Again, let us compare several approximations of initial function (8.5.11) obtained with different truncation numbers N of the Fourier series. One can see deformations and a much slower convergence to initial solution $g(x)$ due to Gibbs' effect (Fig. 8.12), which becomes hardly noticeable only at $N \geq 700$.

8.6 Convergence. Lax Equivalence Theorem

We now study the conditions that guarantee the convergence of the numerical solution of the discrete problem to the exact solution of the original continuous problem. Consider in a domain D with boundary S a problem

$$A\phi = f \text{ in } D, \tag{8.6.1}$$

$$L\phi = g \text{ at } S \tag{8.6.2}$$

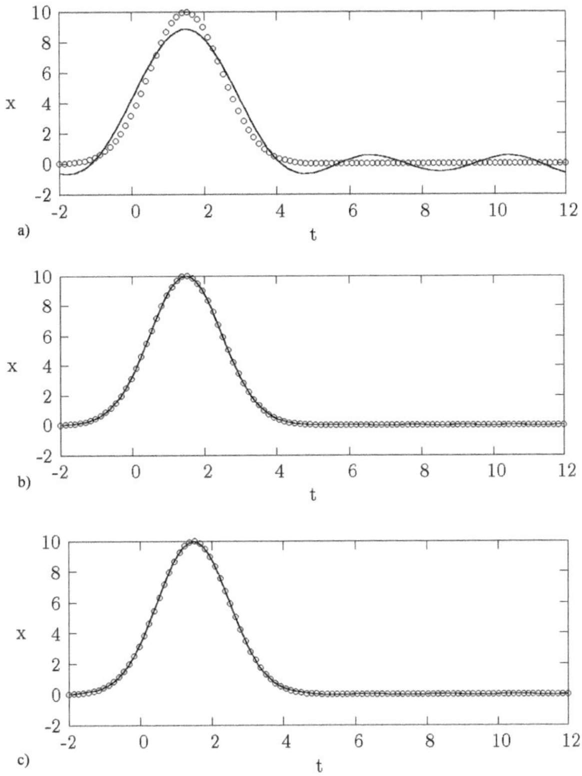

Fig. 8.11 Approximation of initial function $g(x)$ defined by (8.5.10) with a truncated Fourier series; **a** $N = 3$; **b** $N = 6$ and **c** $N = 9$ (solid line). A spatial grid of 100 points was used

with linear differential operators A and L. The solution $\phi(x)$ is defined in a closed domain $D \cup S$, while functions $f(x)$ and $g(x)$ are given functions defined in D and its boundary S, respectively. Equation (8.6.2) represents the boundary condition. Note that (8.6.1), (8.6.2) can also be a non-stationary and multidimensional problem. In this case, $x = (x_1, x_2,\ldots,x_n, t)$. For example, it can be the Dirichlet problem for the two-dimensional Laplace operator:

$$-\left\{\frac{\partial^2}{\partial x^2} + \frac{\partial^2}{\partial y^2}\right\}\phi(x, y) = f(x, y) \text{ if } (x, y) \in D = (0, 1) \times (0, 1) \tag{8.6.3}$$

$$\frac{\partial}{\partial n}\phi(x, y) = 0 \text{ if } (x, y) \in S \tag{8.6.4}$$

where S is the boundary of the domain D. Let us consider the discrete problem

$$A^h\phi^h = f^h \text{ in } D_h \tag{8.6.5}$$

$$L^h\phi^h = g^h \text{ at } S_h \tag{8.6.6}$$

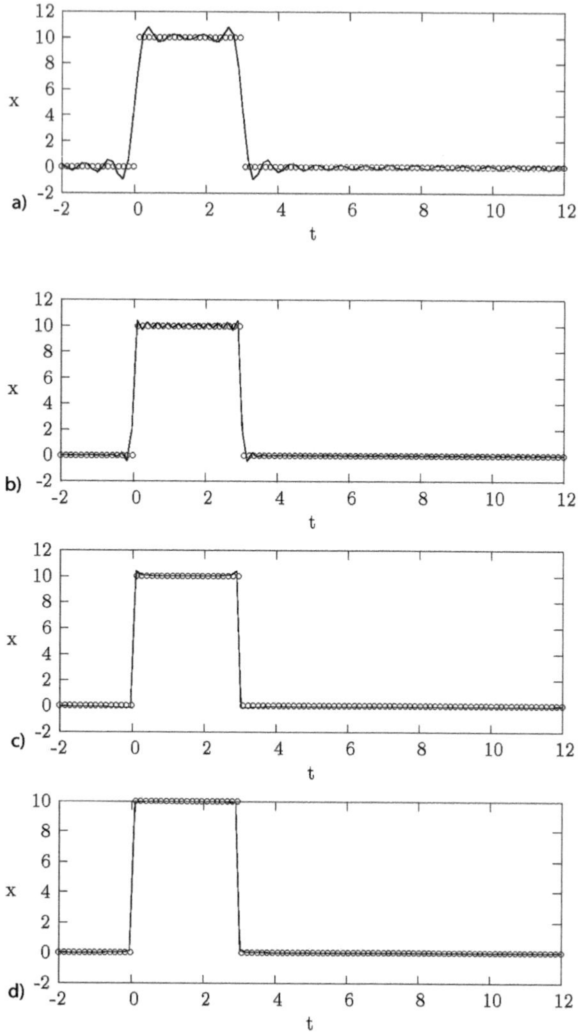

Fig. 8.12 Approximation of initial function $g(x)$ defined by (8.5.11) by a truncated Fourier series; **a** $N = 20$; **b** $N = 50$ y **c** $N = 100$ y **d** $N = 700$ (solid line). A spatial grid of 100 points was used

obtained by approximating the problem (8.6.1), (8.6.2) on grids D_h and S_h of size h. Then A^h and L^h are the matrices that represent the operators A and L in the spaces of grid functions f^h and g^h. It should be noted that grid functions ϕ^h, f^h and g^h belong to the different spaces Φ, F and G, since they are calculated on different grids, namely, on grids $D_h \cup S_h$, D_h and S_h, respectively.

The special interest is the interconnection of the three most important characteristics of each computing process:

(1) the approximation of the differential problem (8.6.1), (8.6.2) by the discrete problem (8.6.5), (8.6.6);
(2) the stability of numerical solution ϕ^h with respect to errors in f^h and g^h;

(3) the convergence of numerical solution ϕ^h to exact solution $\phi(x)$ as the grid step h tends to zero.

Indeed, before applying a numerical method we choose a mesh size h. Furthermore, the discretization error depends on h and it should decrease if a finer mesh is used. Due to the lack of accurate data, the discrete problem usually contains errors in the forcing, and also in the initial and boundary conditions. Moreover, there are always rounding errors in the computations. The stability of a numerical algorithm allows us to calculate solution ϕ^h with good accuracy, as it prevents the uncontrolled increase of all the errors in the calculation process. Finally, in practice we find a numerical solution ϕ^h only for a fixed mesh size h. And it is particularly important to be sure that the numerical solution ϕ^h, found for some mesh size h, converges to the exact solution $\phi(x)$ when h tends to zero.

In this section, we will prove the Lax equivalence theorem [1, 5, 9] which asserts that the solution ϕ^h converges to the solution $\phi(x)$ if the discrete problem is stable and approximates the continuous linear problem (8.6.1), (8.6.2). The importance of the theorem is that while the convergence of a finite difference solution to the solution of the partial differential equation is desirable, this is usually difficult to establish. However, approximation and stability are typically much easier to show than convergence. Therefore, convergence is usually shown using the Lax equivalence theorem.

Definition 8.6.1 It is said that the discrete problem (8.6.5), (8.6.6) approximates the differential problem (8.6.1), (8.6.2) with order $O(h^n)$ with respect to h, if there exist positive constants h_0, M_1 and M_2 independent of h so that

$$\left\|A^h(\phi)_h - f^h\right\|_F \leq M_1 h^n \tag{8.6.7}$$

and

$$\left\|L^h(\phi)_h - g^h\right\|_G \leq M_2 h^n \tag{8.6.8}$$

for any $h < h_0$.

In the formulas (8.6.7) and (8.6.8), $(\phi)_h$ is a projection of the exact solution $\phi(x)$ on the grid points of spaces F and G, respectively. The simplest example of the projection $(\phi)_h$ gives the grid function that accepts the values of continuous solution $\phi(x)$ at the grid points.

Definition 8.6.2 It is said that the discrete problem (8.6.5), (8.6.6) is stable with respect to errors in f^h and g^h if there are positive constants h_0, C_1 and C_2 independent of h such that

$$\left\|\phi^h\right\|_\Phi \leq C_1 \left\|f^h\right\|_F + C_2 \left\|g^h\right\|_G \tag{8.6.9}$$

for any $h < h_0$.

Remark 8.6.1 The independence of the constants C_1 and C_2 of h in (8.6.9) means that the sensitivity of problem with respect to variations in f^h and g^h does not change with the mesh size.

Theorem 8.6.1 (Lax equivalence theorem). *Suppose that the discrete problem (8.6.5), (8.6.6) is stable (according to Definition 8.6.2) and approximate the linear differential problem (8.6.1), (8.6.2) with order $O(h^n)$ with respect to h (according to Definition 8.6.1). Then the numerical solution ϕ^h converges to the exact solution $\phi(x)$ as $h \to 0$ with the rate proportional to h^n:*

$$\left\| (\phi)_h - \phi^h \right\|_{\Phi} \leq (C_1 M_1 + C_2 M_2) h^n \tag{8.6.10}$$

where M_1, M_2, C_1 and *C_2 are the constants from* (8.6.7–8.6.9).

Proof Let h_0 be so small that (8.6.7–8.6.9) are satisfied. Due to the stability, there exists unique solution ϕ^h for any functions f^h and g^h, and any $h < h_0$. Applying A^h to $(\phi)_h - \phi^h$, we obtain

$$A^h\{(\phi)_h - \phi^h\} = A^h(\phi)_h - A^h\phi^h = A^h(\phi)_h - f^h \tag{8.6.11}$$

Similarly,

$$L^h\{(\phi)_h - \phi^h\} = L^h(\phi)_h - g^h \tag{8.6.12}$$

As a result, we get the problem for errors

$$A^h\delta\phi^h = \delta f^h \text{ in } D_h \tag{8.6.13}$$

$$L^h\delta\phi^h = \delta g^h \text{ at } S_h \tag{8.6.14}$$

where

$$\delta\phi^h = (\phi)_h - \phi^h, \quad \delta f^h = A^h(\phi)_h - f^h \quad \text{and} \quad \delta g^h = L^h(\phi)_h - g^h$$

The structures of the problems (8.6.5–8.6.6) and (8.6.13–8.6.14) coincide, and therefore their solutions have the same stability properties. Applying the estimate (8.6.9) to the problem (8.6.13)–(8.6.14), we get

$$\left\|\delta\phi^h_\Phi\right\| \leq C_1 \left\|\delta f^h\right\|_F + C_2 \left\|\delta g^h\right\|_G \leq C_1 \left\|A^h(\phi)_h - f^h\right\|_F + C_2 \left\|L^h(\phi)_h - g^h\right\|_G \tag{8.6.15}$$

Further, by (8.6.7) and (8.6.8), we obtain

$$\left\| (\phi)_h - \phi^h \right\|_\Phi \le C_1 M_1 h^n + C_2 M_2 h^n \le (C_1 M_1 + C_2 M_2) h^n \tag{8.6.16}$$

The theorem is proved.

Remark 8.6.2 The orders of approximation in (8.6.7) and (8.6.8) may be different, for example, n and m. In this case, the convergence rate in (8.6.10) is proportional to $\min\{n, m\}$.

Remark 8.6.3 By Theorem 8.6.1, only the stable schemes make sense, as any unstable scheme does not allow computing the solution due to uncontrolled growth of errors (at least, the rounding errors are always present). In practice it is possible to build several stable schemes that approximate the original continuous problem. Normally the most economical algorithm that allows finding the solution with acceptable accuracy is preferred.

Remark 8.6.4 The Lax theorem is valid for any continuous linear problem. It can be shown that for a nonlinear problem, the conditions (8.6.7)–(8.6.9) are also necessary. However, they are generally not sufficient for the convergence.

Example 8.6.1 [1]. Let us find the solution $\phi(x, t)$ of the Cauchy problem.

$$A\phi \equiv \frac{\partial \phi}{\partial t} + u\frac{\partial \phi}{\partial x} = 0 \tag{8.6.17}$$

$$L\phi \equiv \phi(x, 0) = g(x) \tag{8.6.18}$$

in the unbounded domain $-\infty < x < \infty, \quad 0 \le t \le T$, where $u = Const$, and $g(x)$ is twice continuously differentiable function such that

$$\sup_{-\infty<x<\infty} \left| \frac{\partial^2}{\partial x^2} g(x) \right| = G < \infty \tag{8.6.19}$$

The exact solution of problem (8.6.17), (8.6.18) is

$$\phi(x, t) = g(x - ut), \tag{8.6.20}$$

and hence $\phi(x, t) = Const$ along each characteristic line $x - ut = Const$. Besides, due to (8.6.19), the solution is bounded.

Let us introduce regular grids in space and time with mesh sizes h and τ, respectively: $x_i = ih; t_n = n\tau$; $n = 0, 1, 2, ..., N + 1$; $T = (N + 1)\tau$. Denote as ξ_i^n the value of numerical solution at the grid point (x_i, t_n) and use the explicit Godunov scheme [9]:

$$\left(A^{h,\tau}\xi\right)_i^n \equiv \frac{\xi_i^{n+1} - \xi_i^n}{\tau} + u\frac{\xi_i^n - \xi_{i-1}^n}{h} = 0 \tag{8.6.21}$$

$$(L^h\xi)_i \equiv \xi_i^0 = g_i \tag{8.6.22}$$

Thus, the initial condition (8.6.18) is approximated exactly. However, the scheme (8.6.21) approximates (8.6.17) with the first order (see (8.6.2):

$$\left\|A^{h,\tau}(\phi)_{h,\tau} - A^{h,\tau}\xi^{h,\tau}\right\| \equiv \left\|A^{h,\tau}(\phi)_{h,\tau}\right\| \le M_1(h+\tau) \tag{8.6.23}$$

where $\|f\| = \max\limits_n \sup\limits_i \left|f_i^n\right|$, and grid function $(\phi)_{h,\tau}$ has values $\phi(x_i, t_n)$ at the grid nodes (x_i, t_n).

We now examine the stability of Godunov scheme. Suppose that an error ε_i^n made at the R. H. S. of Eq. (8.6.21) causes an error δ_i^n in the numerical solution ξ_i^n. It leads to the error equation

$$\delta_i^{n+1} = (1-r)\delta_i^n + r\delta_{i-1}^n + \tau\varepsilon_i^n,\ \delta_i^0 = 0 \tag{8.6.24}$$

where $r = \tau u/h$, or

$$\left|\delta_i^{n+1}\right| \le |1-r|\left|\delta_i^n\right| + |r|\left|\delta_{i-1}^n\right| + \tau\left|\varepsilon_i^n\right| \tag{8.6.25}$$

If $|r| \le 1$ (Courant condition) then $|1-r| + |r| = 1$, and

$$\sup_i\left|\delta_i^{n+1}\right| \le \sup_i\left|\delta_i^n\right| + \tau\varepsilon \tag{8.6.26}$$

It follows from (8.6.26) that the Godunov scheme is stable if $|r| \le 1$ (see also example 8.4.1):

$$\|\delta\| \le \tau(N+1)\|\varepsilon\| \le T\|\varepsilon\| \tag{8.6.27}$$

Therefore, by the Lax equivalence theorem,

$$\left\|(\phi)_{h,\tau} - \xi\right\| \le TM_1(h+\tau), \tag{8.6.28}$$

that is, the numerical solution converges to the exact solution of Cauchy problem with the rate proportional to h and τ.

Example 8.6.2 [1]. It is convenient to give the geometric interpretation of the relationship between the Courant condition and the convergence of Godunov scheme. Suppose that $u = 1$ and $T = 1$, i.e., the exact solution $\phi(x,t) = g(x-t)$ is invariable along the characteristic line $x - t = Const$ and, in particular, $\phi(0,1) = \phi(-1,0) = g(-1)$. In this case, the Courant stability condition is $r = \tau/h \le 1$, and the discrete problem (8.6.21), (8.6.22) is reduced to

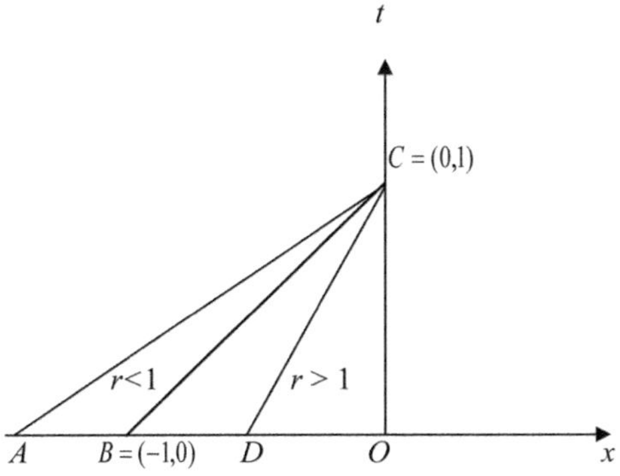

Fig. 8.13 Domains of influence of value ξ_0^{N+1} depending on value of Courant number

$$\xi_i^{n+1} = (1-r)\xi_i^n + r\xi_{i-1}^n, \quad \xi_i^0 = g_i \tag{8.6.29}$$

Since $(N+1)\tau = 1$, ξ_0^{N+1} approximates the value $\phi(0, 1)$ at the point C (Fig. 8.13).

The most favorable situation corresponds to the case $r = 1$ when the scheme (8.6.29) exactly approximates the differential problem, and the characteristic line BC implies that ξ_0^{N+1} depends only on the value $g(-1)$ at the point $B = (-1, 0)$: $\xi_0^{N+1} = g(-1)$ (Fig. 8.13).

If $r < 1$ then according to (8.6.29), the domain of influence of value ξ_0^{N+1} is the entire triangle ACO that includes the point $B = (-1, 0)$ and characteristic line BC (Fig. 8.13). Conversely, when $r > 1$, the domain of influence of the value ξ_0^{N+1} is the triangle DCO, which does not include the characteristic line BC, and therefore the point $B = (-1, 0)$ too (Fig. 8.14).

Consider now the case $r > 1$. Let $\varepsilon > 0$ be a small number such that the interval $(-1-\varepsilon, -1+\varepsilon)$ on the x-axis is outside the interval DO. If we now smoothly modify the function $g(x)$ in interval $(-1-\varepsilon, -1+\varepsilon)$ so that the value of function $g(x)$ at the point $B = (-1, 0)$ is changed, then the value of exact solution $\phi(0, 1) = g(-1)$ will also be changed (due to the method of characteristic lines). Meanwhile, the value

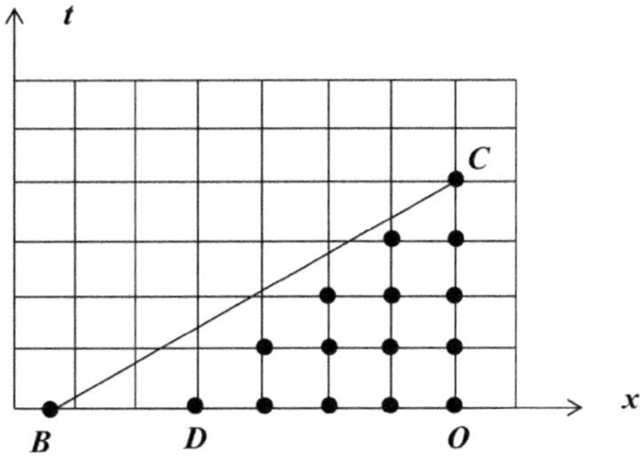

Fig. 8.14 The case $r > 1$. The characteristic line BC and point $B = (-1, 0)$ do not belong to the domain of influence DCO of the value ξ_0^{N+1}

ξ_0^{N+1} of numerical solution will not be changed because the initial solution $g(x)$ is unchangeable in the domain of influence DCO of value ξ_0^{N+1}. Obviously, the same situation is conserved if both h and τ tend to zero maintaining the number r. Thus, in the case $r > 1$ there is no convergence of the numerical solution to the exact one. By the Lax theorem, the lack of convergence is a direct consequence of the instability of scheme (8.6.29) when $r > 1$.

8.7 Finite Difference Approximation of the 1D Dispersion Equation

In the absence of sources, the one-dimensional dispersion equation is Eq. (8.4.4):

$$\frac{\partial \phi}{\partial t} + u\frac{\partial \phi}{\partial x} = \mu\frac{\partial^2 \phi}{\partial x^2} \tag{8.7.1}$$

To discuss the combined effect of advection and diffusion, let's consider the problem of accuracy arising solely from the approximation of spatial derivatives. If the spatial first derivative is approximated by an upstream difference (with $u > 0$) and the second derivative is approximated by a standard three-point stencil, the resulting differential–difference approximation to (8.7.1) will be

$$\frac{d\phi_i}{dt} + u\frac{\phi_i - \phi_{i-1}}{h} = \mu\frac{\phi_{i+1} - 2\phi_i + \phi_{i-1}}{h^2} \tag{8.7.2}$$

where $h = \Delta x$ is the mesh size. It is easy to see that the Eq. (8.7.2) is an $O(h^2)$ -approximation of the modified equation

$$\frac{\partial \phi}{\partial t} + u\frac{\partial \phi}{\partial x} = \mu(1+\nu)\frac{\partial^2 \phi}{\partial x^2} \tag{8.7.3}$$

where $\nu = uh/(2\mu)$ is called the numerical viscosity, it is a measure of the relative influence of advection and diffusion at the smallest spatial scales resolved on the grid. Comparison of the modified Eq. (8.7.3) with the original advection–diffusion Eq. (8.7.1) shows that the differential–difference approximation (8.7.2) gives an unaccurate approximation to the diffusion term unless $\nu \ll 1$.

As the horizontal resolution increases, the number ν decreases. Nevertheless, in many problems involving low-viscosity flow, this grid size can be several orders of magnitude smaller than the physical scales of primary interest, so it is not possible to resolve scales in which molecular viscosity dominates numerical diffusion without exceeding the resources of the most advanced computers.

To accurately represent the diffusion term in low-viscosity flow, it is necessary to use a less diffusive approximation to $u\partial\phi/\partial x$, such as a centered difference

$(u_{i+1/2}\phi_{i+1} - u_{i-1/2}\phi_{i-1})/2h$ if u is variable, or a higher-order one-sided difference. The error for such differences is made up of terms containing higher order derivatives. Spatial derivatives of high even order are more scale-selective in their damping than the second derivative operator $\partial^2\phi/\partial x^2$.

If $f(x, t)$ represents the pollution sources, then the diffusion problem becomes

$$\frac{\partial \phi}{\partial t} + u\frac{\partial \phi}{\partial x} = \mu\frac{\partial^2 \phi}{\partial x^2} + f \quad , \ \varphi(x, 0) = g(x) \tag{8.7.4}$$

After discretization in space, the problem accepts a semi-discrete form:

$$\frac{d\vec{\varphi}}{dt} + A\vec{\varphi} = \vec{f}\ , \qquad \vec{\varphi}(0) = \vec{g} \tag{8.7.5}$$

To discretize problem (8.7.5) in time, we may introduce a regular time grid $\{t_j\}$ of small size τ and approximate problem (8.7.5) in each time interval (t_j, t_{j+1}) in the form

$$\frac{\vec{\phi}^{j+1} - \vec{\phi}^{j}}{\tau} + A\left\{\alpha\vec{\phi}^{j+1} + (1-\alpha)\vec{\phi}^{j}\right\} = \vec{f}^{j}, \quad \vec{\phi}^{0} = \vec{g} \tag{8.7.6}$$

where α is a parameter, $0 \leq \alpha \leq 1$. If $\alpha = 0$, then the numerical scheme

$$\frac{\vec{\phi}^{j+1} - \vec{\phi}^{j}}{\tau} + A\vec{\phi}^{j} = \vec{f}^{j}, \quad \vec{\phi}^{0} = \vec{g}, \tag{8.7.7}$$

is called explicit. Its solution $\vec{\phi}^{j+1} = B\vec{\phi}^{j} + \tau S\vec{f}^{j}\ ,\ \ \vec{\phi}^{0} = \vec{g}$ where $B = E - \tau A,\ \ S = E$ and E is the identity matrix, does not require solving the system of linear algebraic equations. However, this scheme is conditionally stable, usually placing limits on the form of the space–time grid. If $\alpha \neq 0$, then the numerical scheme is called implicit, since its solution already reduces to solving a system of linear algebraic equations.

The best known implicit schemes of the family (8.7.6) are the scheme

$$\frac{\vec{\phi}^{j+1} - \vec{\phi}^{j}}{\tau} + A\vec{\phi}^{j+1} = \vec{f}^{j}, \quad \vec{\phi}^{0} = \vec{g} \tag{8.7.8}$$

for $\alpha = 1$, and the Crank-Nicolson scheme for $\alpha = 0.5$:

$$\frac{\vec{\phi}^{j+1} - \vec{\phi}^{j}}{\tau} + A\left\{\frac{\vec{\phi}^{j+1} + \vec{\phi}^{j}}{2}\right\} = \vec{f}^{j+1/2}, \quad \vec{\phi}^{0} = \vec{g} \tag{8.7.9}$$

Implicit schemes are unconditionally stable if the matrix A is positive semidefinite. Now let's prove this statement for the Crank-Nicolson scheme using two lemmas.

Lemma 8.7.1 (Kellogg's lemma) [1]. *Let A be a positive semi-definite matrix, and let $\sigma \geq 0$. Then*

$$\left\| (E - \sigma A)(E + \sigma A)^{-1} \right\| \leq 1 \tag{8.7.10}$$

in the spectral norm. If A is positive definite and $\sigma > 0$ then $(E - \sigma A)(E + \sigma A)^{-1} < 1$.

Lemma 8.7.2 [1]. *Let A be a positive semi-definite matrix, and let $\sigma \geq 0$. Then*

$$(E + \sigma A)^{-1} \leq 1 \tag{8.7.11}$$

in the spectral norm. If A is positive definite and $\sigma > 0$ then $(E + \sigma A)^{-1} < 1$.

Theorem 8.7.1 *Crank-Nicolson scheme (8.7.9) is stable respect to small perturbations in the initial condition and forcing.*

Proof Indeed, the scheme can be written as

$$\vec{\phi}^{k+1} = B\vec{\phi}^k + \tau S\vec{f}^{k+\frac{1}{2}}$$

where $B = (E + \frac{\tau}{2}A)^{-1}\left(E - \frac{\tau}{2}A\right)$ and $S = (E + \frac{\tau}{2}A)^{-1}$. The use of the Euclidian norm in the last equation leads to

$$\left\| \vec{\phi}^{j+1} \right\| \leq \|B\| \cdot \left\| \vec{\phi}^j \right\| + \tau \|S\| \cdot \left\| \vec{f}^{j+\frac{1}{2}} \right\| \tag{8.7.12}$$

We define $\left\| \vec{f} \right\| = \max\limits_j \left\| \vec{f}^{j+1/2} \right\|$, and take into account that

$$(E + \frac{\tau}{2}A)^{-1}\left(E - \frac{\tau}{2}A\right) = \left(E - \frac{\tau}{2}A\right)(E + \frac{\tau}{2}A)^{-1} \tag{8.7.13}$$

Indeed, applying the matrix $(E + \frac{\tau}{2}A)^{-1}\left(E - \frac{\tau}{2}A\right)$ to the identity

$$(E - \frac{\tau}{2}A)^{-1}\left(E - \frac{\tau}{2}A\right) = \left(E + \frac{\tau}{2}A\right)(E + \frac{\tau}{2}A)^{-1}$$

and using the commutativity of matrices $\left(E - \frac{\tau}{2}A\right)$ and $\left(E + \frac{\tau}{2}A\right)$, we get (8.7.13).

Applying Lemmas 8.7.1 and 8.7.2 to (8.7.12) leads to

$$\left\| \vec{\phi}^{k+1} \right\| \leq \left\| \vec{\phi}^k \right\| + \tau \left\| \vec{f} \right\|$$

It follows from the last inequality that $\left\|\vec{\phi}^k\right\| \le \left\|\vec{\phi}^0\right\| + k\tau\left\|\vec{f}\right\|$. Since $k\tau \le T$ we obtain

$$\left\|\vec{\phi}^k\right\| \le \|\vec{g}\| + T\left\|\vec{f}\right\|$$

The last inequality proves the unconditional stability of the scheme (8.7.9) with respect to small perturbations of the forcing and the initial condition. Theorem is proved.

Among all schemes of the family (8.7.6), only the Crank-Nicolson scheme has a second-order approximation in time. In addition, this is the only scheme that conserves the solution norm provided that the matrix A is skew-symmetric and $f(x,t) = 0$. Indeed, consider the scalar product $(\cdot,\cdot)$ of each term of the equation

$$\frac{\vec{\phi}^{j+1} - \vec{\phi}^j}{\tau} + A\left\{\frac{\vec{\phi}^{j+1} + \vec{\phi}^j}{2}\right\} = 0$$

with vector $\left(\vec{\phi}^{j+1} + \vec{\phi}^j\right)/2$. Then the property $\left(A\vec{\phi}, \vec{\phi}\right) = 0$ leads to $\left\|\vec{\phi}^{j+1}\right\| = \left\|\vec{\phi}^j\right\|$ where $\left\|\vec{\phi}\right\| = (\vec{\phi}, \vec{\phi})^{1/2}$. The scheme is especially usefull when modeling a closed system that conserves energy.

8.8 Splitting Method

We now consider two splitting methods for the the solution of evolution problem

$$\begin{aligned} &\frac{d\vec{\phi}}{dt} + A(t)\vec{\phi} = f \quad \text{in} \quad (0, T) \\ &\vec{\phi}(0) = \vec{g} \end{aligned} \tag{8.8.1}$$

where

$$A(t) = A_1(t) + A_2(t) \tag{8.8.2}$$

and $A(t)$, $A_1(t)$ and $A_2(t)$ are positive semi-definite and time-dependent matrices:
$(A(t)\vec{x}, \vec{x}) \ge 0$ and $(A_i(t)\vec{x}, \vec{x}) \ge 0$ for any $\vec{x} \ne 0$, $i = 1, 2$

1. **Homogeneous problem** ($\vec{f} = 0$). Supposing that the elements of matrices A_i are sufficiently smooth functions, we approximate them in each small-time interval (t_j, t_{j+1}) as follows:

$$\Lambda^j = A\left(t_{j+\frac{1}{2}}\right),\ \Lambda_i^j = A_i\left(t_{j+\frac{1}{2}}\right),\ i = 1, 2 \tag{8.8.4}$$

Yanenko's scheme. In each small-time interval (t_j, t_{j+1}), the componentwise splitting algorithm by Yanenko [10] consists in the consecutive application of Crank-Nicolson schemes for every split operator Λ_i^j:

$$\frac{\vec{\phi}^{j+\frac{1}{2}} - \vec{\phi}^j}{\tau} + \Lambda_1^j \frac{\vec{\phi}^{j+\frac{1}{2}} + \vec{\phi}^j}{2} = 0$$

$$\frac{\vec{\phi}^{j+1} - \vec{\phi}^{j+1/2}}{\tau} + \Lambda_2^j \frac{\vec{\phi}^{j+1} + \vec{\phi}^{j+1/2}}{2} = 0 \tag{8.8.5}$$

After exclusion of the auxiliary vector $\vec{\phi}^{j+1/2}$, the system of equations reduces to

$$\vec{\phi}^{j+1} = T^j \vec{\phi}^j \tag{8.8.6}$$

where

$$T^j = (E + \frac{\tau}{2}\Lambda_2^j)^{-1}\left(E - \frac{\tau}{2}\Lambda_2^j\right)(E + \frac{\tau}{2}\Lambda_1^j)^{-1}\left(E - \frac{\tau}{2}\Lambda_1^j\right) \tag{8.8.7}$$

Approximation. First, we analyze the approximation. Suppose that the time step τ is so small that

$$\frac{\tau}{2}\left\|\Lambda_i^j\right\| < 1 \tag{8.8.8}$$

and expand the operator T^j into the power series:

$$T^j = E - \tau\Lambda^j + \frac{\tau^2}{2}\left[\left(\Lambda_1^j\right)^2 + 2\Lambda_2^j\Lambda_1^j + \left(\Lambda_2^j\right)^2\right] - O(\tau^3) \tag{8.8.9}$$

If operators Λ_i^j commute, i.e., if

$$\Lambda_1^j\Lambda_2^j = \Lambda_2^j\Lambda_1^j \tag{8.8.10}$$

then (8.8.9) can be written as

$$T^j = E - \tau\Lambda^j + \frac{\tau^2}{2}\left(\Lambda^j\right)^2 - O(\tau^3) \tag{8.8.11}$$

Thus, if operators Λ_i^j commute then the first three terms of the transition operators of the Yanenko scheme and the Crank-Nicolson scheme coincide. Since the Crank-Nicolson scheme is of second-order approximation in τ, Yanenko's scheme (8.8.5) also approximates problem (8.8.1) with accuracy $O(\tau^2)$ if the operators Λ_i^j commutve, and only with accuracy $O(\tau)$ if the operators Λ_i^j are non-commutative [1].

Stability. Let $\|A\|$ and $\|\vec{x}\|$ the spectral norm of matrix A and the Euclidean norm of vector $\vec{x}$, respectively. Due to (8.8.7),

$$T^j = T_1^j T_2^j \tag{8.8.12}$$

where

$$T_i^j = (E + \frac{\tau}{2}\Lambda_i^j)^{-1}\left(E - \frac{\tau}{2}\Lambda_i^j\right) \tag{8.8.13}$$

and the stability of scheme (8.8.5) in the Euclidean norm follows from (8.8.6) and inequality

$$\|T^j\| \leq \left\|T_1^j\right\| \left\|T_2^j\right\| \leq 1 \tag{8.8.14}$$

which is valid due to Lemma 8.7.1.

Marchuk's scheme. In general, the matrices Λ_i^j do not commute. We know that in this case the Yanenko scheme is only first-order accurate method in τ, and for reducing the approximation errors it is necessary to use rather small step τ, i.e., to make more arithmetic operations.

We now consider the scheme by Marchuk, which is second-order accurate method in τ, even if Λ_i^j do not commute [1]. This scheme represents a symmetrized version of Yanenko's scheme in each double time step interval $\left(t_{j-1}, t_{j+1}\right)$:

$$\begin{aligned}
\frac{\vec{\phi}^{j-1/2}-\vec{\phi}^{j-1}}{\tau} + \Lambda_1^j \frac{\vec{\phi}^{j-1/2}+\vec{\phi}^{j-1}}{2} &= 0 \\
\frac{\vec{\phi}^{j}-\vec{\phi}^{j-1/2}}{\tau} + \Lambda_2^j \frac{\vec{\phi}^{j}+\vec{\phi}^{j-1/2}}{2} &= 0 \\
\frac{\vec{\phi}^{j+1/2}-\vec{\phi}^{j}}{\tau} + \Lambda_2^j \frac{\vec{\phi}^{j+1/2}+\vec{\phi}^{j}}{2} &= 0 \\
\frac{\vec{\phi}^{j+1}-\vec{\phi}^{j+1/2}}{\tau} + \Lambda_1^j \frac{\vec{\phi}^{j+1}+\vec{\phi}^{j+1/2}}{2} &= 0
\end{aligned} \tag{8.8.15}$$

where

$$\Lambda_i^j = A_i\left(t_j\right) \tag{8.8.16}$$

and $\vec{\phi}^{j-1/2}$, $\vec{\phi}^{j}$ and $\vec{\phi}^{j+1/2}$ are auxiliary vectors. It follows from (8.8.15) that

$$\vec{\phi}^{j+1} = T^j \vec{\phi}^{j-1} \tag{8.8.17}$$

where T^j is the transition matrix,

$$\begin{aligned}
T^j &= (E + \tfrac{\tau}{2}\Lambda_1^j)^{-1}\left(E - \tfrac{\tau}{2}\Lambda_1^j\right)(E + \tfrac{\tau}{2}\Lambda_2^j)^{-1}\left(E - \tfrac{\tau}{2}\Lambda_2^j\right) \\
&\times(E + \tfrac{\tau}{2}\Lambda_2^j)^{-1}\left(E - \tfrac{\tau}{2}\Lambda_2^j\right)(E + \tfrac{\tau}{2}\Lambda_1^j)^{-1}\left(E - \tfrac{\tau}{2}\Lambda_1^j\right) \\
&= E - 2\tau\Lambda^j + \tfrac{(2\tau)^2}{2}(\Lambda^j)^2 - O\left(\tau^3\right)
\end{aligned} \tag{8.8.18}$$

Approximation. The order of approximation of matrix T^j given by (8.8.18) is $O(\tau^2)$. Indeed, the Crank-Nicolson scheme in the same time interval is

$$\frac{\vec{\phi}^{j+1} - \vec{\phi}^{j-1}}{2\tau} + \Lambda^j \frac{\vec{\phi}^{j+1} + \vec{\phi}^{j-1}}{2} = 0 \tag{8.8.19}$$

and therefore

$$\vec{\phi}^{j+1} = K^j \vec{\phi}^{j-1} = \left[E - 2\tau\Lambda^j + \frac{(2\tau)^2}{2} \left(\Lambda^j\right)^2 \right] \vec{\phi}^{j-1} \tag{8.8.20}$$

where $\Lambda^j = A(t_j)$. It is easy to see that matrix T^j coincides up to order $O(\tau^2)$ with the transition matrix K^j of Crank-Nicolson scheme. It should be noted that the scheme (8.8.15) is the second-order accurate scheme only due to its symmetry in the interval (t_{j-1}, t_{j+1}). Thus, Marchuk's scheme increases the order of approximation in the case of non-commutative matrices Λ_i^j [11].

Stability. The transition operator in (8.8.17) is

$$T^j = T_1^j T_2^j T_2^j T_1^j \tag{8.8.21}$$

where

$$T_i^j = (E + \frac{\tau}{2}\Lambda_i^j)^{-1} \left(E - \frac{\tau}{2}\Lambda_i^j\right) \tag{8.8.22}$$

Thus, matrices T_i^j have the same structure as the matrices (8.8.13) with the particularity that in (8.8.22), Λ_i^j are defined by (8.8.16) instead of (8.8.4). Using the Euclidean vector norm and the spectral matrix norm again, we obtain

$$\left\| \vec{\phi}^{j+1} \right\| \leq \left\| T^j \right\| \cdot \left\| \vec{\phi}^{j-1} \right\| \tag{8.8.23}$$

and, due to Lemma 8.7.1,

$$\left\| T^j \right\| \leq \left\| T_1^j \right\| \left\| T_2^j \right\| \left\| T_2^j \right\| \left\| T_1^j \right\| \leq 1 \tag{8.8.24}$$

Thus,

$$\left\| \vec{\phi}^{j+1} \right\| \leq \left\| \vec{\phi}^{j-1} \right\| \tag{8.8.25}$$

and therefore

$$\left\| \vec{\phi}^{j} \right\| \leq \left\| \vec{g} \right\| \tag{8.8.26}$$

Thus, scheme (8.8.15) is stable.

Remark 8.8.1 If A is a skew-symmetric matrix, that is,

$$\left(A\vec{\phi}, \vec{\phi}\right) = 0 \text{ for any } \vec{\phi} \neq 0 \tag{8.8.27}$$

then homogeneous problem (8.8.1) conserves the solution norm:

$$\left\|\vec{\phi}(t)\right\| = \left\|\vec{\phi}(0)\right\| = \|\vec{g}\|, \quad t \in (0, T) \tag{8.8.28}$$

In this case, to keep this law, it is important to represent the original matrix as the sum of skew-symmetric matrices:

$$\left(A_i\vec{\phi}, \vec{\phi}\right) = 0 \text{ for any } \vec{\phi} \neq 0,\ i = 1, 2 \tag{8.8.29}$$

Indeed, if (8.8.27) is satisfied then due to (8.8.4) and (8.8.16),

$$\left(\Lambda_i^j\vec{\phi}, \vec{\phi}\right) = 0 \text{ for any } \vec{\phi} \neq 0 \tag{8.8.30}$$

($i = 1, 2;\ j = 1, 2, \ldots$). Therefore,

$$\left\|\vec{\phi}^{j+1}\right\| = \left\|\vec{\phi}^{j}\right\| = \|\vec{g}\| \tag{8.8.31}$$

for the Yanenko scheme (8.8.5), and

$$\left\|\vec{\phi}^{j+1}\right\| = \left\|\vec{\phi}^{j-1}\right\| = \|\vec{g}\| \tag{8.8.32}$$

for the Marchuk scheme (8.8.15). Note that the conservation laws (8.8.31) and (8.8.32) ensure the stability, and therefore they are especially important when the solution of a problem is searched for a long-time interval $(0, T)$. These laws are a direct consequence of using the Crank-Nicolson schemes in (8.8.5) and (8.8.15) at each splitting step. The use of any schemes different from the Crank-Nicolson schemes will break the laws (8.8.31) and (8.8.32).

2. **Inhomogeneous problem** ($\vec{f} \neq 0$). In each double time step interval (t_{j-1}, t_{j+1}), we consider the scheme

$$\begin{aligned}
(E + \tfrac{\tau}{2}\Lambda_1^j)\vec{\phi}^{j-1/2} &= (E - \tfrac{\tau}{2}\Lambda_1^j)\vec{\phi}^{j-1} \\
(E + \tfrac{\tau}{2}\Lambda_2^j)(\vec{\phi}^j - \tau\vec{f}^j) &= (E - \tfrac{\tau}{2}\Lambda_2^j)\vec{\phi}^{j-1/2} \\
(E + \tfrac{\tau}{2}\Lambda_2^j)\vec{\phi}^{j+1/2} &= (E - \tfrac{\tau}{2}\Lambda_2^j)(\vec{\phi}^j + \tau\vec{f}^j) \\
(E + \tfrac{\tau}{2}\Lambda_1^j)\vec{\phi}^{j+1} &= (E - \tfrac{\tau}{2}\Lambda_1^j)\vec{\phi}^{j+1/2}
\end{aligned} \tag{8.8.33}$$

where Λ_i^j is determined by (8.8.16), and $\vec{f}^j = \vec{f}(t_j)$. It can be shown [1] that it is also a second-order accurate scheme. Note that the scheme (8.8.33) can be written in the different form:

$$
\begin{aligned}
(E + \tfrac{\tau}{2}\Lambda_1^j)\vec{\phi}^{j-2/3} &= (E - \tfrac{\tau}{2}\Lambda_1^j)\vec{\phi}^{j-1} \\
(E + \tfrac{\tau}{2}\Lambda_2^j)\vec{\phi}^{j-1/3} &= (E - \tfrac{\tau}{2}\Lambda_2^j)\vec{\phi}^{j-2/3} \\
\vec{\phi}^{j+1/3} &= \vec{\phi}^{j-1/3} + 2\tau\vec{f}^j \\
(E + \tfrac{\tau}{2}\Lambda_2^j)\vec{\phi}^{j+2/3} &= (E - \tfrac{\tau}{2}\Lambda_2^j)\vec{\phi}^{j+1/3} \\
(E + \tfrac{\tau}{2}\Lambda_1^j)\vec{\phi}^{j+1} &= (E - \tfrac{\tau}{2}\Lambda_1^j)\vec{\phi}^{j+2/3}
\end{aligned}
\tag{8.8.34}
$$

3. **The general case of splitting**. We now consider the general case when the matrix $A(t)$ of problem

$$
\frac{d\vec{\phi}}{dt} + A(t)\vec{\phi} = \vec{f}(t) \text{ in } (0, T)
$$

$$
\vec{\phi}(0) = \vec{g} \tag{8.8.35}
$$

can be presented as the sum

$$
A = A_1 + A_2 + \ldots + A_n \tag{8.8.36}
$$

of n positive semi-definite matrices A_i which can be associated with a physical process or a spatial coordinate. In each double time step interval $\left(t_{j-1}, t_{j+1}\right)$, the method byMarchuk has the form

$$
\begin{aligned}
(E + \tfrac{\tau}{2}\Lambda_1^j)\vec{\phi}^{j-[(n-1)/n]} &= (E - \tfrac{\tau}{2}\Lambda_1^j)\vec{\phi}^{j-1} \\
(E + \tfrac{\tau}{2}\Lambda_2^j)\vec{\phi}^{j-[(n-2)/n]} &= (E - \tfrac{\tau}{2}\Lambda_2^j)\vec{\phi}^{j-[(n-1)/n]} \\
&\cdots\cdots\cdots \\
(E + \tfrac{\tau}{2}\Lambda_n^j)(\vec{\phi}^j - \tau\vec{f}^j) &= (E - \tfrac{\tau}{2}\Lambda_n^j)\vec{\phi}^{j-1/n} \\
(E + \tfrac{\tau}{2}\Lambda_n^j)\vec{\phi}^{j+1/n} &= (E - \tfrac{\tau}{2}\Lambda_n^j)(\vec{\phi}^j + \tau\vec{f}^j) \\
&\cdots\cdots\cdots \\
(E + \tfrac{\tau}{2}\Lambda_2^j)\vec{\phi}^{j+[(n-1)/n]} &= (E - \tfrac{\tau}{2}\Lambda_2^j)\vec{\phi}^{j+[(n-2)/n]} \\
(E + \tfrac{\tau}{2}\Lambda_1^j)\vec{\phi}^{j+1} &= (E - \tfrac{\tau}{2}\Lambda_1^j)\vec{\phi}^{j+[(n-1)/n]}
\end{aligned}
\tag{8.8.37}
$$

where $\Lambda_i^j = A_i\left(t_j\right)$, $i = 1, \ldots, n$. Another algorithm equivalent to scheme (8.8.37) is

$$
\begin{aligned}
(E + \tfrac{\tau}{2}\Lambda_i^j)\vec{\phi}^{j-[(n+1-i)/(n+1)]} &= (E - \tfrac{\tau}{2}\Lambda_i^j)\vec{\phi}^{j-[(n+2-i)/(n+1)]} \\
i &= 1, 2, \ldots, n; \\
\vec{\phi}^{j+[1/(n+1)]} &= \vec{\phi}^{j-[1/(n+1)]} + 2\tau\vec{f}^j; \\
(E + \tfrac{\tau}{2}\Lambda_{n-i+2}^j)\vec{\phi}^{j+[i/(n+1)]} &= (E - \tfrac{\tau}{2}\Lambda_{n-i+2}^j)\vec{\phi}^{j+[(i-1)/(n+1)]} \\
i &= 2, 3, \ldots, n+1
\end{aligned}
\tag{8.8.38}
$$

8.9 Splitting Scheme for the 2D Dispersion Problem

Unconditionally stable second-order symmetrized two-cycle splitting schemes for the solution of dispersion problem (2.4.1–2.4.6) and its adjoint problem (3.4.9–3.4.12) were used in [12] (see also [13]). The direct and adjoint discrete problems are compatible in the sense that their operators satisfy the Lagrange identity, and therefore the direct and adjoint pollution estimates remain valid in discrete form.

Using the continuity equation $\nabla \cdot U = 0$, operator A of problem (2.4.1)–(2.4.6) (3.4.9)–(3.4.12) can be written as $A = A_1 + A_2$ where

$$A_1\phi = \frac{1}{2}\frac{\partial}{\partial x}(u\phi) + \frac{1}{2}u\frac{\partial \phi}{\partial x} + \frac{1}{2}\sigma\phi - \frac{\partial}{\partial x}\mu\frac{\partial \phi}{\partial x} \tag{8.9.1}$$

$$A_2\phi = \frac{1}{2}\frac{\partial}{\partial y}(v\phi) + \frac{1}{2}v\frac{\partial \phi}{\partial y} + \frac{1}{2}\sigma\phi - \frac{\partial}{\partial y}\mu\frac{\partial \phi}{\partial y} \tag{8.9.2}$$

Each 1D split operator A_i $(i = 1, 2)$ is positive definite [13]. The splitting of the adjoint model is performed in the similar way. The functions are discretized using different grids:

$$\phi_{ij} = \phi(x_i, y_j),\ u_{i\pm 1/2,j} = u(x_{i\pm 1/2}, y_j),\ v_{i,j\pm 1/2} = v(x_i, y_{j\pm 1/2})$$
$$\mu_{i\pm 1/2,j} = \mu(x_{i\pm 1/2}, y_j),\ \mu_{i,j\pm 1/2} = \mu(x_i, y_{j\pm 1/2})$$

The second-order approximations of operators A_i and the continuity Eq. (2.4.3) have the following forms:

$$\begin{aligned}\left(A_1^h\phi\right)_{ij} &= \frac{1}{2\Delta x}\left[u_{i+1/2,j}\phi_{i+1,j} - u_{i-1/2,j}\phi_{i-1,j}\right] + \frac{1}{2}\sigma\phi_{ij}\\ &- \frac{1}{(\Delta x)^2}\left[\mu_{i+1/2,j}(\phi_{i+1,j} - \phi_{ij}) - \mu_{i-1/2,j}(\phi_{ij} - \phi_{i-1,j})\right]\end{aligned} \tag{8.9.3}$$

$$\begin{aligned}\left(A_2^h\phi\right)_{ij} &= \frac{1}{2\Delta y}\left[v_{i,j+1/2}\phi_{i,j+1} - v_{i,j-1/2}\phi_{i,j-1}\right] + \frac{1}{2}\sigma\phi_{ij}\\ &- \frac{1}{(\Delta y)^2}\left[\mu_{i,j+1/2}(\phi_{i,j+1} - \phi_{ij}) - \mu_{i,j-1/2}(\phi_{ij} - \phi_{i,j-1})\right]\end{aligned} \tag{8.9.4}$$

$$\frac{1}{\Delta x}\left[u_{i+1/2,j} - u_{i-1/2,j}\right] + \frac{1}{\Delta y}\left[v_{i,j+1/2} - v_{i,j-1/2}\right] = 0 \tag{8.9.5}$$

The adjoint matrices $(A_i^h)^*$ are obtained if the functions $u,\ v,\ \phi$ in (8.9.3) and (8.9.4) are replaced by functions $-u,\ -v,\ g$, respectively. Let us now briefly show the discretization of the boundary condition only at one boundary point (see details in [13]).

Let $M = (x_{1/2}, y_j)$ be the left boundary point of line y_j for a number j, and let $u_{1/2,j} > 0$ be a positive value of the velocity component u at the point M. Since the boundary is parallel to the component v of velocity vector $U = (u, v)$ at the point M, the outward unit normal at this point is $n = (-1, 0)$, and $U_n = U \cdot n = -u_{1/2,j} < 0$.

Thus, M is the point of the input boundary S^-, and therefore conditions (2.4.5) and (3.4.11) are approximated as

$$\mu_{1/2,j}\frac{\phi_{0,j}-\phi_{1,j}}{\Delta x}+u_{1/2,j}\frac{\phi_{1,j}+\phi_{0,j}}{2}=0\,,\quad g_{1,j}=g_{0,j} \tag{8.9.6}$$

Note that all the matrices A_i^h and $(A_i^h)^*$ are positive definite, besides, $(A_i^h)^* = (A_i^*)^h, (i = 1, 2)$.

In each double time step interval (t_{n-1}, t_{n+1}), the numerical scheme for the transport Eqs. (2.4.1)–(2.4.6) consists of the five equations that must be solved one by one:

$$\begin{aligned}
\Phi\big[n-\tfrac{3-i}{3}\big]-\Phi\big[n-\tfrac{4-i}{3}\big]&=-\tfrac{\tau}{2}A_i^h\big(\Phi\big[n-\tfrac{3-i}{3}\big]+\Phi\big[n-\tfrac{4-i}{3}\big]\big),\ (i=1,2)\\
\Phi\big[n+\tfrac{1}{3}\big]-\Phi\big[n-\tfrac{1}{3}\big]&=2\tau F[n]\\
\Phi\big[n+\tfrac{4-i}{3}\big]-\Phi\big[n+\tfrac{3-i}{3}\big]&=-\tfrac{\tau}{2}A_i^h\big(\Phi\big[n+\tfrac{4-i}{3}\big]+\Phi\big[n+\tfrac{3-i}{3}\big]\big),\ (i=2,1)
\end{aligned} \tag{8.9.7}$$

The scheme for the adjoint problem (3.4.9)–(3.4.12) in (t_{n-1}, t_{n+1}) is

$$\begin{aligned}
G\big[n+\tfrac{3-i}{3}\big]-G\big[n+\tfrac{4-i}{3}\big]&=-\tfrac{\tau}{2}(A_i^h)^*\big(G\big[n+\tfrac{3-i}{3}\big]+G\big[n+\tfrac{4-i}{3}\big]\big),\ (i=1,2)\\
G\big[n-\tfrac{1}{3}\big]-G\big[n+\tfrac{1}{3}\big]&=2\tau P[n]\\
G\big[n-\tfrac{4-i}{3}\big]-G\big[n-\tfrac{3-i}{3}\big]&=-\tfrac{\tau}{2}(A_i^h)^*\big(G\big[n-\tfrac{4-i}{3}\big]+G\big[n-\tfrac{3-i}{3}\big]\big),\ (i=2,1)
\end{aligned} \tag{8.9.8}$$

where Φ and G are vectors which approximate the exact solutions ϕ and g at the grid points. It should be noted that the adjoint scheme (8.9.8) is realized in the opposite time direction. Also, vectors $\Phi[n+1]$, $\Phi[n-1]$, $G[n+1]$ and $G[n-1]$ represent the values of numerical solutions, while the others (with $n \pm i/3$; $i = 0, 1, 2$) are auxiliary vectors of the splitting algorithm. Vector $F[n]$ (or $P[n]$) is the approximation at the moment t_n of the forcing f (adjoint forcing p) at the spatial grid points. In each interval (t_{n-1}, t_{n+1}), both numerical algorithms are unconditionally stable:

$$\|\Phi[n+1]\| \le \|\Phi[n-1]\| + 2\tau\|F[n]\| \tag{8.9.9}$$

$$\|G[n-1]\| \le \|G[n+1]\| + 2\tau\|P[n]\| \tag{8.9.10}$$

where $\|.\|$ is the Euclidean vector norm. In the particular case that $\sigma = 0$, $\mu = 0$, $F[n] = 0$, $P[n] = 0$, and $U_n = 0$ at the boundary S, all the matrices A_i^h and $(A_i^h)^*$ are skew-symmetric, and each of the schemes (8.9.7) and (8.9.8) has two conservation laws [14]:

$$\|\Phi[n+1]\| = \|\Phi[n-1]\|,\quad \|G[n-1]\| = \|G[n+1]\| \tag{8.9.11}$$

and

$$\sum_i \|\Phi_i[n+1]\| V_i = \sum_i \|\Phi_i[n-1]\| V_i \tag{8.9.12}$$

$$\sum_i \|G_i[n-1]\| V_i = \sum_i \|G_i[n+1]\| V_i \tag{8.9.13}$$

where $\Phi_i[n]$ and $G_i[n]$ are the components of the vectors $\Phi[n]$ and $G[n]$, respectively, and V_i is the elementary volume of the i-th grid point. Each of the one-dimensional equations of schemes (8.9.7) and (8.9.8) is resolved by Thomas's factorization method for the tridiagonal matrices (see the next section).

8.10 Thomas's Factorization Algorithm

We now consider a famous method by Thomas [15, 16] to resolve a linear system with a tridiagonal matrix. Such systems typically occur after discretization of boundary value problems for the one-dimensional second-order differential equations. The general form can be written as

$$a_i\phi_{i-1} - b_i\phi_i + c_i\phi_{i+1} = f_i, \ \ i = 1, 2, 3..., N-1 \tag{8.10.1}$$

with the boundary conditions

$$\phi_0 = \alpha_0\phi_1 + \beta_0$$

$$\phi_N = \alpha_N\phi_{N-1} + \beta_N \tag{8.10.2}$$

where $\phi_0, \phi_1, \ldots, \phi_N$ are unknown, while a_i, b_i, c_i, f_i and $\alpha_0, \beta_0, \alpha_N, \beta_N$ are known parameters. Furthermore, assume that the system parameters satisfy the following conditions:

$$|b_i| > |a_i| + |c_i|, i = 1, 2, ..., N-1, \tag{8.10.3}$$

$$|\alpha_0| < 1, |\alpha_N| \leq 1 \tag{8.10.4}$$

It is shown later that conditions (8.10.3) and (8.10.4) guarantee the existence of unique solution of the problem (8.10.1), (8.10.2), and allow us to find the solution by Thomas direct method that is based on the LU factorization of matrix, and is stable and economical. The system (8.10.1), (8.10.2) can also be written in vector form:

$$A\vec{\phi} = \vec{f} \tag{8.10.5}$$

where $\vec{\phi} = (\phi_0, \phi_1, \ldots, \phi_N)^T$ is an unknown vector, $\vec{f} = (\beta_0, f_1, f_2, \ldots, f_{N-1}, \beta_N)^T$ is the given vector, and

$$A = \begin{bmatrix} 1 & -\alpha_0 & 0 & 0 & \ldots & 0 & 0 & 0 \\ a_1 & -b_1 & c_1 & 0 & \ldots & 0 & 0 & 0 \\ 0 & a_2 & -b_2 & c_2 & \ldots & 0 & 0 & 0 \\ \ldots & \ldots & \ldots & \ldots & \ldots & \ldots & \ldots & \ldots \\ 0 & 0 & 0 & 0 & \ldots & a_{N-1} & -b_{N-1} & c_{N-1} \\ 0 & 0 & 0 & 0 & \ldots & 0 & -\alpha_N & 1 \end{bmatrix} \tag{8.10.6}$$

is the tridiagonal matrix of order $n + 1$.

Substituting the first condition (8.10.2) into the first equation of system (8.10.1), we obtain

$$a_1(\alpha_0\phi_1 + \beta_0) - b_1\phi_1 + c_1\phi_2 = f_1$$

or

$$\phi_1 = \alpha_1\phi_2 + \beta_1 \tag{8.10.7}$$

where

$$\alpha_1 = \frac{c_1}{b_1 - a_1\alpha_0}, \ \beta_1 = \frac{a_1\beta_0 - f_1}{b_1 - a_1\alpha_0} \tag{8.10.8}$$

By introducing the expression (8.10.7) in the second equation of system (8.10.1), we obtain the relation between ϕ_2 and ϕ_3. Therefore, let us search the solution in the form

$$\phi_{k-1} = \alpha_{k-1}\phi_k + \beta_{k-1} \tag{8.10.9}$$

By introducing the expression (8.10.9) into the k-th equation of system (8.10.1) we get

$$a_k(\alpha_{k-1}\phi_k + \beta_{k-1}) - b_k\phi_k + c_k\phi_{k+1} = f_k$$

Solving this equation with respect to ϕ_k we obtain

$$\phi_k = \alpha_k\phi_{k+1} + \beta_k \tag{8.10.10}$$

where

$$\alpha_k = \frac{c_k}{b_k - a_k\alpha_{k-1}}, \ \beta_k = \frac{a_k\beta_{k-1} - f_k}{b_k - a_k\alpha_{k-1}} \tag{8.10.11}$$

Therefore, the coefficients of Eqs. (8.10.10), that link adjacent values ϕ_k and ϕ_{k+1} $(k = 1, 2, ..., N-1)$ can be determined by means of recurrent relations (8.10.11), since α_0 and β_0 are given by (8.10.2).

Substituting the expression $\phi_{N-1} = \alpha_{N-1}\phi_N + \beta_{N-1}$, given by (8.10.10) for $k = N-1$, in the second boundary condition (8.10.2) we obtain

$$\phi_N = \alpha_N(\alpha_{N-1}\phi_N + \beta_{N-1}) + \beta_N \tag{8.10.12}$$

where α_N and β_N are known coefficients from formula (8.10.2), while α_{N-1} and β_{N-1} are calculated by formulas (8.10.11). It follows from (8.10.12) that

$$\phi_N = \frac{\beta_N + \alpha_N\beta_{N-1}}{1 - \alpha_N\alpha_{N-1}} \tag{8.10.13}$$

Then, the remaining unknowns $\phi_{N-1}, \phi_{N-2}, ..., \phi_0$ are calculated by using the back substitution algorithm (the formula (8.10.10)).

The conditions (8.10.3) and (8.10.4) guarantee that the calculations with formulas (8.10.11) and (8.10.13) are correct, since their denominators are not reduced to zero.

Indeed, suppose that for some k $(0 \le k < N-1)$ the inequality $|\alpha_{k-1}| < 1$ is valid. For example, $|\alpha_0| < 1$. Since $|b_k| - |a_k| > 0$ due to (8.10.3), we get

$$|b_k - a_k\alpha_{k-1}| \ge |b_k| - |a_k| \cdot |\alpha_{k-1}| > |b_k| - |a_k| > 0, \tag{8.10.14}$$

and hence,

$$|\alpha_k| = \frac{|c_k|}{|b_k - a_k\alpha_{k-1}|} \le \frac{|b_k| - |a_k|}{|b_k| - |a_k| \cdot |\alpha_{k-1}|} < 1$$

Thus, by mathematical induction,

$$|\alpha_k| < 1 \tag{8.10.15}$$

for every k $(k = 0, 1, 2, ..., N-1)$, and hence,

$$|b_k - a_k\alpha_{k-1}| > 0 \text{ for every } k, \tag{8.10.16}$$

and (8.10.3) leads to

$$1 - \alpha_N\alpha_{N-1} > 0, \tag{8.10.17}$$

i.e., the denominators of the expressions (8.10.11) and (8.10.13) never become zero during the calculation process.

References

1. Marchuk, G.I.: Methods of numerical mathematics. Springer-Verlag, New York (1975)
2. Skiba, Yu.N.: Métodos y esquemas numéricos: Un análisis computacional. Dirección General de Publicaciones y Fomento Editorial, UNAM, México (2005)
3. Skiba, Yu.N., Parra-Guevara, D.: Efectos negativos de la aproximación y representación falsa de ondas sobre una malla. Miscelanea Matemática **43**, 133–151 (2006)
4. Mesinger, F., Arakawa, A.: Numerical methods used in atmospheric models. GARP publica1087 tions series, no. 17, vol.1. WMO (1976)
5. Morton, K., Mayers, D.: Numerical solution of partial differential equations. Cambridge University Press, Cambridge (1994)
6. Kress, R.: Numerical analysis. Springer-Verlag, New York (1998)
7. Durran, D.L.: Numerical methods for wave equations in geophysical fluid dynamics. Springer1092 Verlag, New-York (1999)
8. Liu, M., Carroll, J.J.: A high-resolution air pollution model suitable for dispersion studies In complex terrain. Mon. Wea. Rev. **124**, 2396–2409 (1996)
9. Godunov, S.K., Ryabenkii, V.S.: Difference schemes. Elsevier, Amsterdam (1987)
10. Yanenko, N.N.: The method of fractional steps. Springer-Verlag, Berlin (1971)
11. Strang, G.: On the construction and comparison of difference schemes. SIAM J. Numer. Anal. **1098**(5), 506–517 (1968)
12. Marchuk, G.I., Skiba, Yu.N., Protsenko, I.G.: Method of calculating the evolution of Random hydrodynamic fields on the basis of adjoint equations. Izv. Atmos. Ocean. Phys. **21**(2), 87–92 (1985)
13. Skiba, Yu.N.: Balanced and absolutely stable implicit schemes for the main and the adjoint 1103 pollutant transport equations in the limited area. Rev. Int. Contam. Ambient. **9**(2), 39–51 (1993)
14. Skiba, Yu.N.: Air pollution estimates. World Resource Review **9**(4), 542–556 (1997)
15. Ortega, J.M., Poole, W.G., Jr.: An Introduction to numerical methods for differential equations. Pitman Publishing Inc., Marksfield, MA (1981)
16. Thomas, L.H.: Elliptic problems in linear differential equations over a network, Watson Sci. Comput. Lab Report, Columbia University, New York (1949)

Bibliography

1. Álvarez-Vázquez, L.J., García-Chan, N., Martínez, A., Vázquez-Méndez, M.E.: Optimal control of urban air pollution related to traffic flow in road networks. Math. Control Related Fields **8**(1), 177–193 (2018)
2. Beilke, S., Elshout, A.J.: Acid deposition. In: Proceedings of the CEC workshop, D. Reidel Publishing Company (1983)
3. Brandt, J., Christensen, J.H., Frohn, L.M., Zlatev, Z.: Operational air pollution forecast modelling using the THOR system. Phys. Chem. Earth (B) **26**(2), 117–122 (2001)
4. Bravo, H.A., Sosa, R.E., Torres, R.J.: Ozono y lluvia ácida en la Ciudad de México. Ciencias **22**, 33–40 (1991)
5. Caselli, M.: La Contaminación Atmosférica. Siglo Veintiuno, México (1996)
6. COESE.: Factores de Emisión para Vehículos Automotores. Comisión Estatal de Ecología, Gobierno del Estado de Jalisco (1995)
7. Cooper, W.W., Hemphill, H., Huang, Z., Li, S., Lelas, V., Sullivan, D.W.: Survey of mathematical programming models in air pollution management. Eur. J. Oper. Res. **96**(1), 1–35 (1997)
8. Dang, Q.A., Ehrhardt, M., Tran, G.L., Le, D.: Mathematical modelling and numerical algorithms for simulation of oil pollution. Environ. Model. Assess. **17**(3), 275–288 (2012)
9. Elliott, A.J.: Shear diffusion and the spread of oil in the surface layers of the North Sea. Deutsche Hydrograph. Zeitung **39**, 113–137 (1986)
10. Elliott, A.J., Dale, A.C., Proctor, R.: Modelling the movement of pollutants in the UK shelf seas. Mar. Pollut. Bull. **24**, 614–619 (1992)
11. EPA-420-R-92-009: Procedures for Emission Inventory Preparation, Volume IV: Mobile Sources. United States Environmental Protection Agency (1992)
12. Feng, L., Fei, H., Jiang, Z.: Adjoint method for the optimum planning of industrial pollutant sources. Sci. China Ser. D: Earth Sci. **48**(8), 1270–1279 (2005)
13. Freitas, F., Assis, J.J., Almeida, T., Oliveira, I., Reis, E.: Simulation optimization for analysis of sustainable logistics systems. Pesqu. Oper. **37**(1), 145–171 (2017)
14. Fu, L., Hao, J., He, D., He, K., Li, P.: Assessment of vehicular pollution in China. J. Air Waste Manag. Assoc. **51**, 658–668 (2001)
15. Gavrilov, V.P., Gormatyuk, Yu.K.: Pollutant dispersion from stationary sources in the surface layer of the atmosphere. Sov. Meteorol. Hydrol. **2**, 28–37 (1989)
16. Gerdes, R.: A primitive equation ocean circulation model using a general vertical coordinate transformation. 1. Description and testing of the model. J. Geophys. Res.: Oceans 98(C8), 14683–14701 (1993)
17. Goddard, H.C.: Promoting urban sustainability: the case for a tradable supplementary licence system for vehicle use. Urban Stud. **36**, 2317–2331 (1999)

Y. N. Skiba and D. Parra Guevara, *Mathematical Methods for the Assessment and Control of Industrial Emissions*, Mathematical Engineering,
https://doi.org/10.1007/978-3-031-58109-0

18. Godish, T.: Air Quality. Lewis Publishers Inc., Chelsea, Michigan (1991)
19. Goldfarb, D., Idnani, A.: A numerically stable dual method for solving strictly convex quadratic programs. Math. Program. **27**, 1–33 (1983)
20. Golub, H., Ortega, J.M.: Scientific Computing and Differential Equations. Introduction to Numerical Methods. Academic Press, Boston (1992)
21. Greenberg, H.J.: Mathematical programming models for environmental quality control. Oper. Res. **43**(4), 578–623 (1995)
22. Hassan, A.A., Crowther, J.M.: A simple model of pollutant concentrations in a street canyon. Environ. Monitor. Assess. **52**, 269–280 (1998)
23. Hill, M.K.: Understanding of Environmental Pollution. Cambridge University Press, Cambridge (1997)
24. Hobbs, P.V.: Introduction to Atmospheric Chemistry. Cambridge University Press, Cambridge (2000)
25. Jhih-Shyang, S.: An Optimization model for photochemical air pollution control. Eur. J. Oper. Res. **106**, 1–14 (1998)
26. Kantha, L.H., Clayson, C.A.: Numerical Models of Oceans and Oceanic Processes. Academic Press, San Diego (2000)
27. Kawada, J.: Health effects of air pollutants and their management. Atmos. Environ. **18**, 613–620 (1984)
28. Kennicutt II, M.C., McDonald, S.J., Sweet, S.T.: The Bahia Paraiso spill in Arthur Harbor, Anvers Island. Antarct. J. **27**, 331–340 (1992)
29. Kohn, R.E.: A Linear Programming Model for Air Pollution Control. MIT Press Inc., Cambridge, Mass (1978)
30. Ku, J.-Y., Rao, S.T., Rao, K.S.: Numerical simulation of air pollution in urban areas: model development. Atmos. Environ. 21(1), 201–212 (1987)
31. Lawson, Ch.L., Hanson, R.J.: Solving Least Squares Problem. Prentice-Hall, Englewood Cliffs, New Jersey (1974)
32. Lewis, J., Derber, J.C.: The use of adjoint equations to solve a variational adjustment problem with advective constraints. Tellus **37A**, 309–322 (1985)
33. Liao, K.J., Hou, X.: Optimization of multipollutant air quality management strategies: a case study for five cities in the United States. J. Air Waste Manag. Assoc. (Taylor & Francis) **65**(6), 732–742 (2015)
34. Lupu, C.: Development of optimal filters obtained through convolution methods, used for fingerprint image enhancement and restoration. USV Ann. Econ. Publ. Adm. **2**(20), 156–167 (2014)
35. Lynn, D.: Air Pollution, Threat and Response, Addison-Wesley, Reading (1976)
36. Marchuk, G.I.: Mathematical issues of industrial effluent optimization. J. Meter. Soc. Jpn. **60**, 481–485 (1982)
37. Murphy-Klimova, E., Fisher, B.E.A., Sokhi, R.: Treatment of urban areas within a regional transport model of sulphur and nitrogen oxides. Environ. Monitor. Assess. **52**, 213–224 (1998)
38. Nikolaev, Y.A., Plakunov, V.K.: Microbial biofilms: potential of using in waste water treatment. Water: Chem. Ecol. **2**, 11–13 (2008)
39. Nocedal, J. Wright, S.J.: Numerical Optimization. Springer Series in Operations Research and Financial Engineering. Springer Science + Business Media, New York (2006)
40. Panos, G., Seinfeld, J.H.: Mathematical modeling of turbulent reacting plumes I: general theory and model formulation. Atmos. Environ. **20**, 1791–1807 (1986)
41. Parra-Guevara, D., Skiba, Yu.N.: Adjoint approach to estimate the non-steady emission rate of a point source. Int. J. Eng. Res. Appl. **3**(6), 763–776 (2013)
42. Parra-Guevara, D., Skiba, Yu.N., Peña-Maciel, D.: Controlling the forcing of the linear transport equation to meet air quality norms at every point. Int. J. Appl. Math. **30**(6), 527–545 (2017)
43. Parra-Guevara, D., Skiba, Yu.N., Pérez-Sesma, A.: A linear programming model for controlling air pollution. Int. J. Appl. Math. **23**(3), 549–569 (2010)
44. Parrish, D.D., Trainer, M., Hereid, D., Williams, E.J., Olszyna, K.J., Harley, R.A., Meagher, J.F., Fehsenfeld, F.C.: Decadal change in carbon monoxide to nitrogen oxide ratio in US vehicular emissions. J. Geophys. Res. Atmosp. **107**(D12), art. no. 4140 (2002)

45. Phillips, N.A.: A coordinate system having some special advantages for numerical forecasting. J. Meteorol. **14**, 184–185 (1957)
46. Proctor, R., Elliott, A.J., Flather, R.A.: Forecast and hindcast simulations of the Braer oil spill. Mar. Pollut. Bull. **28**(4), 219–229 (1994)
47. Rodhe, H., Crutzen, P., Vanderpol, A.: Formation of sulfuric and nitric acid in the atmosphere during long-range transport. Tellus **33**, 132–141 (1981)
48. Rodrigues, M.S., Kaviski, E.: Daily scheduling of small hydro power plants dispatch with modified particles swarm optimization. Pesqui. Oper. **35**(1), 25–37 (2015)
49. Rossin, A.D., Roberts, J.J.: Episode control criteria and strategy for carbone monoxide. J. Air Pollut. Control Assoc. **22**, 254–259 (1972)
50. Rusco, F.W., Walls, W.D.: Vehicular emissions and control polices in Hong-Kong. Contemp. Econ. Policy **13**, 50–61 (1995)
51. Schneider, D.: Oil spill has long-term effects. Water Envir. Technol. **5**, 30–32 (1993)
52. Skiba, Yu.N.: The derivation and applications of the adjoint solutions of a simple thermodynamic limited area model of the atmosphere-ocean-soil system. World Resour. Rev. **8**(1), 98–113 (1996)
53. Skiba, Yu.N., Adem, J.: A balanced and absolutely stable numerical thermodynamic model for closed and open oceanic basins. Geofís. Internac. **34**(4), 385–393 (1995)
54. Teller, A.: The use of linear programming to estimate the cost of some alternative air pollution abatement policies. Number 20. In: Proceedings of the IBM Scientific Computing Symposium on Water and Air Research Management, pp. 345–353. IBM, New York (1968)
55. Thunis, P., Degraeuwe, B., Pisoni, E., Meleux, F., Clappier, A.: Analyzing the efficiency of short-term air quality plans in European cities, using the CHIMERA air quality model. Air Qual. Atmos. Health **10**(2), 235–248 (2017)
56. Venn, A.J., Lewis, S.A., Cooper, M., Hubbard, R., Britton, J.: Living near a main road and the risk of wheezing illness in children. Am. J. Respir. Crit. Care Med. **164**(12), 2177–2180 (2001)
57. Wolff, G.A., Preston, M.R., Harriman, G., Rowland, S.J.: Some preliminary observations after the wreck of the oil tanker Braer in Shetland. Mar. Pollut. Bull. **26**, 567–571 (1993)
58. Yu, L.E., Hildemann, L.M., Ott, W.R.: A mathematical model for predicting trends in carbon monoxide emissions and exposures on urban arterial highways. J. Air Waste Manag. Assoc. **46**, 430–440 (1996)
59. Yu, S., Li, P., Wang, L., Wu, Y., Wang, S., Liu, K., Zhu, T., Zhang, Y., Hu, M., Zeng, L., Zhang, X., Cao, J., Alapaty, K., Wong, D.C., Pleim, J., Mathur, R., Rosenfeld, D., Seinfeld, J.H.: Mitigation of severe urban haze pollution by a precision air pollution control approach. Sci. Rep. Nat. **8**, 8151 (2018)
60. Zakarin, E.A., Mirkarimova, B.M.: GIS-based mathematical modeling of urban air pollution. Izv. Atmos. Ocean. Phys. **36**, 334–342 (2000)
61. Zanneti, P.: Air Pollution Modelling. Theories, Computational Methods and Available Software, Computational Mechanics Publication, Van Nostrand Reinhold, New York (1990)
62. ZoBell, C.E.: Microbial degradation of oil: present status, problems, and perspectives. In: Ahearn, Meyers (eds.) The Microbial Degradation of Oil Pollutants, Publication No. LSU-SG-73–01, Proceedings Center for Wetlands Resource, pp. 3–16. Louisiana State University, Baton Rouge, LA (1973)

Index

Y. N. Skiba and D. Parra Guevara, *Mathematical Methods for the Assessment and Control of Industrial Emissions*, Mathematical Engineering,
https://doi.org/10.1007/978-3-031-58109-0